AF790275

BIOMONITORING:
GENERAL AND APPLIED ASPECTS ON REGIONAL AND GLOBAL SCALES

Tasks for vegetation science 35

SERIES EDITORS

A. Kratochwil, *University of Osnabrück, Germany*
H. Lieth, *University of Osnabrück, Germany*

The titles published in this series are listed at the end of this volume.

Biomonitoring: General and Applied Aspects on Regional and Global Scales

Edited by

C.A. BURGA

University of Zürich,
Department of Geography,
Zürich, Switzerland

and

A. KRATOCHWIL

University of Osnabrück,
Department of Biology/Ecology,
Osnabrück, Germany

KLUWER ACADEMIC PUBLISHERS

DORDRECHT / BOSTON / LONDON

Library of Congress Cataloging-in-Publication Data.

ISBN 978-90-481-5621-4

Published by Kluwer Academic Publishers,
P.O. Box 17, 3300 AA Dordrecht, The Netherlands.

Sold and distributed in North, Central and South America
by Kluwer Academic Publishers,
101 Philip Drive, Norwell, MA 02061, U.S.A.

In all other countries, sold and distributed
by Kluwer Academic Publishers,
P.O. Box 322, 3300 AH Dordrecht, The Netherlands.

Printed on acid-free paper

Printed in the Netherlands

Contents

Preface

This volume contains a selection of 14 articles dealing with different aspects of biomonitoring and their relation to questions of global change. During the last 10 - 15 years, vegetation changes due to various causes have been more intensively studied in biological and environmental sciences. Especially aspects of global warming lead to a great variety of tasks for vegetation science (see e.g. the articles by Grabherr, Gottfried & Pauli; Carraro, Gianoni, Mossi, Klötzli & Walther; Walther; Defila; Stampfli & Zeiter; Röthlisberger; Burga & Perret and Möller, Wüthrich & Thannheiser).

The different aspects of applied biomonitoring related to (possible) environmental changes concern various ecosystems, e.g. Central European beechwoods, Insubrian evergreen broad-leaved forests, thermophilous lowland deciduous forests, dry grasslands of the lower montane belt of the Ticino Alps, alpine mountain peaks of Switzerland and Austria, Swiss alpine timberline ecotones, and high arctic tundra vegetation.

The volume is divided into three parts: A. General aspects of biomonitoring (contributions by Klötzli; Wildi and Labasch & Otte), B. Examples of applied biomonitoring in Germany and Switzerland (articles by Hakes; Herpin, Siewers, Kreimes & Markert; Defila; Stampfli & Zeiter; Röthlisberger and Ruoss, Burga & Eschmann), and C. Aspects of global change in the Alps and in the high arctic tundra (Grabherr, Gottfried & Pauli; Burga & Perret; Carraro, Gianoni, Mossi, Klötzli & Walther; Walther and Möller, Wüthrich & Thannheiser).

The first article by Klötzli concerns general aspects of biomonitoring. Indeed, biomonitoring in the form of vegetation mapping and observation of permanent plots is a traditional field of vegetation science. Different ways of biomonitoring are presented, its aims and tasks, its methods and its possibilities of statistical evaluation of data sets.

The second contribution by Wildi deals with statistical design and analysis in long-term vegetation monitoring, i.e. succession theories and related methods for time series analysis. Succession, understood as any directional change of vegetation, can be distinguished at three different levels of perception: pattern, process and mechanism. The author describes different methods and gives examples.

Labasch's and Otte's contribution deals with administration levels and tasks of nature conservation efficiency control. Different methods and application possibilities are discussed, the methodology of nature conservation efficiency control on various administrative levels is outlined.

Six contributions give examples of applied biomonitoring in Germany and Switzerland: Recent changes in the vegetation structure and site conditions of nutrient-rich beech forests in central Germany were studied by Hakes. Within a nine-year observation period, remarkable temporal variation in vegetation structure reflecting significant environmental changes could be ascertained.

In the framework of two national monitoring programmes in Germany Herpin, Siewers, Kreimes & Markert investigated changes in heavy metal concentrations in moss (lead was used as an example). This contribution describes mainly methods available to evaluate bioindication data.

Defila's paper deals with phytophenological series as a possible contribution to vegetation monitoring. Based on a remarkable tradition of phenological observations in

Burga & Kratochwil (eds.), BIOMONITORING, 1-3

Switzerland, climate-related trends for the regions Ticino and Engadine were calculated with the national data from 1951 – 1998.

Aspects of species responses to climatic variation and land-use change in grasslands of southern Switzerland have been investigated by Stampfli & Zeiter. For 10 years they monitored the species composition in dry grasslands of high species diversity at experimental sites in the southern Alps. Responses of the abundant herb species to stochastic factors, abandonment and mowing after abandonment of man's influence were examined. Only slow shifts of species composition could be recognized.

Röthlisberger's phytophenological case study in central Switzerland (Canton of Zug) of the time of flowering during the last mild winters shows a clear shifting from the early spring to the winter season. These shifting effects could be recognized on the levels of species diversity and distribution (i.e. colonization of new ecological niches).

The contribution by Ruoss, Burga & Eschmann contains a case study with a technical approach of vegetation monitoring on a restoration site of Pilatus Mountain in central Switzerland. Plant recolonization, induced in two steps by planting first seedlings of herbs and afterwards of alpine grass species, has been monitored for 8 years. Considerable fluctuations of both plant species and cover abundance led to average plant cover values between 60 and 95 %.

The third part of this volume deals with climate-related monitoring studies of arctic-alpine and temperate regions of the northern hemisphere. Grabherr, Gottfried & Pauli summarize a long-term monitoring project of mountain peaks in the Alps. Based on historical and recent records of nival summit floras, the authors detected a general trend towards increased species diversity and abundance, i.e. an upward migration of alpine biota during the last decades.

The contribution by Burga & Perret discusses general features of the upper limit of the Swiss Alpine forest and tree limits, with special emphasis on historical, present and future vegetation dynamics within the timberline ecotone. Complex interactions of climatic, physico-geographical, geological, biological and anthropozoogenic factors are influencing the present timberline ecotone. Six sites were studied for a long-term monitoring project, focusing on different models of climatically induced changes within the timberline ecotone.

Carraro, Gianoni, Mossi, Klötzli & Walther detected changes in forest vegetation in the northern and southern Swiss Alps by comparing 300 vegetation relevés of the last decades. One of the crucial factors of the changes in species composition and vegetation structure are mild winters during the last 30 years. These changes concern also formerly introduced exotic evergreen trees and shrubs, which show at present marked semi-natural spread into the indigenous broad-leaved forests of the southern Alps.

This phenomenon of the so-called "laurophyllisation" has been intensively studied by Walther, whose contribution to this volume summarizes the main features. Although evergreen broad-leaved plants have been cultivated in Switzerland for more than 200 years, only in the last decades a dozen of these species succeeded in escaping from the gardens and in spreading into the forests where they have become naturalized. As temperature can be considered as one of the major climatic determinants of plant distribution, the possible link between the observed vegetation shifts and climatic change is discussed.

Möller, Wüthrich & Thannheiser investigated plant community patterns, phytomass and carbon balance in a high arctic tundra ecosystem and the changes under a climate of increasing cloudiness during the last 4 – 5 decades. The measurements and calculations showed that an alteration of the vegetation pattern is much less effective in changing the carbon balance of the tundra vegetation than a change of light or length of the snow-free period.

Most of these contributions are a first attempt at the high complexity of global change, especially global warming. The articles show different assessments and ways for future research in this field, which will be of increasing importance for environmental aspects, especially the biosphere, including administration levels.

Large part of the editorial work was done by Ms. Martina Lemme (Department of Ecology, University of Osnabrück). She linguistically revised all contributions, in close collaboration with the authors. We are much indebted to her for her efforts.

Dr. Mareike Weinert (Department of Ecology, University of Osnabrück) kindly converted the articles, figures and tables into a camera-ready format. The editors would like to thank her for her excellent assistance.

We are grateful to all authors for their valuable contributions and to Kluwer Academic Publishers for their support in printing this volume.

Prof. Dr. Conradin A. Burga
University of Zurich
Department of Geography
Winterthurerstrasse 190
CH-8057 Zürich
Switzerland

Prof. Dr. Anselm Kratochwil
University of Osnabrück
Department of Biology/Ecology
Barbarastrasse 11
D-49069 Osnabrück
Germany

CHAPTER A

GENERAL ASPECTS OF BIOMONITORING

BIOMONITORING – TASKS AND LIMITS

FRANK KLÖTZLI

Geobotanical Institute ETH Zürich, Zürichbergstrasse 28, CH-8044 Zürich, Switzerland

Keywords: Chaotic development, fuzzy ordination, vegetation mapping, permanent plot, succession, time series

Abstract

Biomonitoring in the form of vegetation mapping and observation of permanent plots is an old tradition in Alpine areas. There are time series as old as 80 years and quite a number of 30-year series. In the lowlands especially wetlands have been controlled. The ways of biomonitoring in Central Europe are presented, especially its aims and tasks, its methods and measures, criteria for choice and establishment, and the evaluation of data sets (by autocorrelation, fuzzy ordination, principal coordinate analysis). The appearance of special species and the margin between two vegetation units mostly fluctuate in a non-stochastic way. Types of changes in monitored and/or mapped areas are discussed.

Summary

1) A brief history of biomonitoring and the evaluation of time series are given.
2) Aims and tasks in biomonitoring include:
   - an inventory, including site conditions
   - (yearly) differences in abundances and dynamics of species
   - (yearly) differences in the quality of the surroundings
   - changes in size and pattern of single habitats
   - control of stability and sensitivity of site and vegetation, including damages and disturbances
   - control of conservation measures and definition of such measures.

Burga & Kratochwil (eds.), BIOMONITORING, 5-16

3) Methods and measures touch:
   - the choice of permanent plots (p.p.), sensitivity and buffer zones and ecotonal areas
   - the comparability of sets of p.p. using standardized methods
   - accessibility and markings (GPS) using non-destructive control.
4) Evaluations may start in the field but are normally performed using statistical methods, e.g. a set including autocorrelation, fuzzy ordination and principal coordinate analysis, allowing to assess any difference in the yearly states (similarity). The presentation of vegetation maps is discussed regarding international conventions on signatures, symbols and colours. Moreover block diagrams are used to depict time series.
5) General conclusions are drawn from the behaviour of species in ecotonal and "heart" zones. "Behaviour groups" (synonyms in brackets) include persistent, fluctuating, welling up (rising), diving (decaying) and undecided (intermittent) species, and they are compared with van der Maarel's and van der Valk's suggestions. In yearly drawn vegetation maps, borderlines normally fluctuate around stable heart zones. Such shifts may be important to assess effects of global change. Possibilities to draw "derived maps" for practical purposes are discussed.

Zusammenfassung

1) Ein kurzer Rückblick auf die Entwicklung der Kontrolle und Überwachung lebender Systeme, des "Biomonitoring", sowie ein Überblick über einige wichtigere Zeitreihen werden gegeben.
2) Ziele und Aufgaben des "Biomonitoring" umfassen:
   - die Inventarisierung einschliesslich der Kontrolle der Standortbedingungen
   - die Ermittlung von jährlichen Unterschieden in der Menge sowie der allgemeinen Dynamik von Organismen
   - die Bestimmung von Unterschieden in der Beschaffenheit der Umgebung
   - das Erkennen von Veränderungen in Flächengrösse und Muster einzelner Habitate
   - die Kontrolle von Empfindlichkeit und Stabilität von Standort und Vegetation inkl. Schäden und Störungen
   - die Überwachung von Schutzvorkehrungen und die (Neu-)Definition solcher Mittel.
3) Methoden und (Schutz-)Mittel berühren:
   - die Auswahl von Dauerflächen und deren Ökotone sowie von Empfindlichkeitszonen
   - die Vergleichbarkeit innerhalb von Dauerflächen – Gruppen mit standardisierten Mitteln
   - die Zugänglichkeit, Betretbarkeit und Markierung von Dauerflächen mit standortverträglichen Mitteln.

4) Eine Auswertung der Daten kann bereits mit Feldmethoden vorgenommen werden, sicher aber mit statistischen Methoden, wie z.B. mit der Kombination von Autokorrelation, "fuzzy ordination" und Hauptkoordinatenanalyse; dieses Vorgehen erlaubt von Jahr zu Jahr die Prüfung der Ähnlichkeitsverhältnisse. Die Ausfertigung von Vegetationskarten wird diskutiert und auf die Wahl von Signaturen, Symbolen und Farben (nach internationaler Übereinkunft) verwiesen, ausserdem auf die Verwendung von Blockdiagrammen für die Darstellung von Zeitreihen.

5) Allgemeine Schlussfolgerungen ergeben sich aus dem Verhalten von Arten in ökotonalen und in Kernzonen. "Verhaltensgruppen" umfassen persistente, fluktuierende, aufsteigende und absteigende sowie intermittierend vorkommende Arten; diese Gruppen werden mit den Vorschlägen von van der Maarel und van der Valk verglichen. In jährlich erstellten Vegetationskarten fluktuieren Grenzlinien und -zonen um stabile(re) Kernzonen. Verschiebungen könnten auf Veränderungen der Umweltbedingungen hinweisen, allenfalls auch auf Erwärmungsvorgänge. Möglichkeiten zur Ableitung von Karten für angewandte Zwecke werden vorgestellt.

Keywords:

behaviour group	Verhaltensgruppe
biomonitoring	Testgebietsüberwachung
chaotic development	chaotische Entwicklung
conservation measures	Schutzmassnahmen
derived map	abgeleitete Karte
disturbance	Störung
dynamics of species	Artschwankungen
ecotone	Ökoton
"heart" zone	Kernzone
inventorization	Inventarisierung
non-destructive control	standortverträgliche Kontrolle
permanent plot	Dauerfläche
sensitivity	Empfindlichkeit
shifts	Verschiebungen
similarity	Ähnlichkeit
time series	Zeitreihen
vegetation map	Vegetationskarte

1. Introduction

Biomonitoring in Alpine areas follows an old tradition although at the time of its first attempts it was handled rather punctually. Braun(-Blanquet) (1913) started setting up some permanent plots in the Rhaetian Alps, Lüdi (see e.g. 1940, 1948) initiated a fertilization experiment on "Schynige Platte" (a mountain near Interlaken/BE) in the "Berner Oberland" ("Bernese highland"), later on continued by Hegg (e.g. 1984 a,b) and Hegg et al. (1992). For that purpose he set up a scheme of permanent plots which were regularly visited. Permanent plots were also established and controlled in the Swiss National Park (e.g. Braun-Blanquet 1931; Lüdi 1966). But the largest series has been set up and controlled year by year by Stüssi (1970). He gave a first account of dynamics in the Alpine meadows at the end of his scientific life, comparing them also with the states when he started his studies. Some of these permanent plots are still revisited and the whole series is currently being re-evaluated by a research group of the University of Zürich and the Federal Research Station of Forest, Snow and Landscape (Krüsi et al. 1995, 1996). Also Trepp's plots in a burned area in mountain pine woodland were set up in 1952 and are still under control.

In a regular way and as a regular tool biomonitoring and simultaneous vegetation mapping have been handled since Ellenberg started his work in Zürich in 1958. He urged his scholars to carefully revisit chosen plots, well knowing that changes in vegetation may happen quite rapidly and strongly. Plots were established in forests, grasslands and wetlands, but many of them were no longer controlled when he left for Göttingen (Germany). However, some areas, especially wetland complexes, have been yearly revisited since the early sixties, and several hundred plots in forested areas were reinventorized in 1994 (see below). Various grassland complexes were controlled for didactic purposes, familiarizing students with these methods of observation and inventorization, and some served as a tool to observe succession dynamics. A good deal of these plots are still marked or so well triangulated that they still belong to a series of long-term monitoring projects (e.g. plots on wetland transplants, Klötzli 1987, 1997).

Also our neighbouring countries have set up permanent plots and remapped many conservation areas to assess trends in vegetation development, especially Austria (e.g. H. Wagner from ca. 1950 onwards, cf. Grabherr 1996), Germany (e.g. Schmidt 1974; Schreiber 1997; and many papers by Runge), the Netherlands (e.g. De Smidt 1977, in *Calluna* heathland since 1960; Roozen & Westhoff 1985; and van Dorp et al. 1985, in dunes for more than 50 years; Beeftink 1987, in salt marshes for ca. 25 or more years; Bakker 1989, since 1972 and a number of shorter time series), as well as other European countries and the United States (e.g. Buell, succession studies since 1958, see Buell et al. 1971; Pickett 1982; Duke Forest, in N-Carolina, since 1934, Peet & Christensen 1987). For further references see van der Maarel (1996 a,b).

2. Aims and tasks

The chief aims of biomonitoring may be listed as follows:

1) to assess
    - the inventory of plants and other organisms
    - the site conditions (e.g. groundwater characteristics, nutrients, browsing and trampling damage)
    - the changes and trends in organismic composition from control to control, including the patterns of important organisms
    - the changing in the surroundings of the plot and in lateral influences (e.g. nutrient fluxes)
    - the surface of each habitat and plant community, or the abundancy of an organism (by mapping);
2) to judge the stability and sensitivity of site and vegetation;

3) to control any ongoing damage or any noxious factors, including effects of visitors;
4) to evaluate any influence, positive or negative, due to the necessary conservation measures;
5) to define specific conservation measures.

(For more details see e.g. Ellenberg et al. 1986; Brown & Rowell 1997.)

Table 1. Evaluated permanent plots.

Examples	Observation	- permanent plots (p.p.) - (periodical) vegetation maps (pvm, vm)
Lüneburger Heide (Klötzli 1993) Wetlands of different kind in the heath	(1975 -) 1983 - 97	30 p.p. + vm
"Stilli Rüss" (Klötzli & Zielinska 1995) River Reuss, old river bed	1973 + 1978 - 96	pvm (each year)
Boppelser Weid (Klötzli 1997) Spring fen complex	1961 + 1963 - 97	pvm (each year) + ca. 30 p.p.
Transect N-S, Switzerland (Klötzli 1995; Klötzli et al. 1996) Nature-near forests (lowland to lower montane)	1940 - 70	~300 p.p.
Mkwaja Ranch/Tanzania coast (Klötzli et al. 1995) 100 km S of Tanga, savanna grasslands	1975 - 80 1992 + 94 + 97	recapitulations of ca. 20 p.p. (of 300 p.p.), partly pvm

3. Methods and measures

Normally, the following methods and measures are used whenever permanent plots (or line transects) are established.

3.1. FOR THE PERMANENT PLOT

To work with permanent plots (p.p.) requires some preliminary decisions, especially regarding the homogeneity of a given site. Should the main goal be to assess its stability, then the plot has to be in the centre of a homogeneous surface, i.e. a place with no critical changes in the plant community. Should, however, the main goal be to assess trends or fluctuations in the composition of one or more given plant communities, then an ecotonal area between two homogeneous surfaces with different plant communities is advisable. Changes in water or nutrient regime etc. are better recognizable at the border zones between two different plant communities or sites.

Generally, these data are part of time series (see below). Thus the same methods should be applied to permanent plots, if possible by the same persons.

3.2. CRITERIA FOR A SET OF PERMANENT PLOTS

The plots of a set and, if possible, the sets in a given country, should be fully comparable among each other: there must be a high accordance how to establish and run permanent plots. It is clear that methods may depend on the biome or the plant community (complex). However, such methods should be standardized on an international level. A first attempt was made during the IBP (International Biological Program).

As mentioned above not every type of grassland or forest can be approached by exactly the same methods. Certain vegetation types or, more precisely, plant communities are quite homogeneous on large surfaces while other communities are mostly encountered in a mosaic with neighbouring communities or totally mixed, e.g. in patches.

Therefore publications with standardized methods are indispensable and should be approved by those working in the same scientific field. Furthermore the timing of the control should be generally defined by phenological criteria (for special concept for Bavaria see Pfadenhauer et al. 1986).

3.3. TECHNICAL ASPECTS

In former days plots had to be marked, if possible directly, or from three different angles outside the individual plot. Marking had to be discrete, too, in remote areas or behind fences. To facilitate a revisit, even in our days of GPS, markings are still a valuable way to redetect plots "far from the beaten track". Normally, iron rods may be hidden in the soil, wooden posts may be discretely placed in the corners of a plot, giving these posts some colour or other marks. Large posts should be avoided since they may attract visitors, destruction, trampling or littering being the results.

In any case, artefacts on permanent plots should be prevented, i.e.

- no trampling (or on given lines only), approach on defined, if possible hidden lines
- no sampling on the plot, except when specially marked or defined on the edge of the plots.

Priority gradients - e.g. nutrients, water table or movement, shadow casts etc. - have to be assessed on near-homogeneous neighbouring surfaces, keeping up a high discipline as to measuring time and period (for details see e.g. Ellenberg et al. 1986).

3.4. EVALUATION

A pre-evaluation may be undertaken by a simple comparison of the species groupings or by assessing the average indicator numbers (sensu Ellenberg). This could be directly done in the field, even in comparing more than one relevé.

Also vegetation maps may be compared straight <u>after</u> the map has been completed.

But, especially in the case of extended series, i.e. for our latitudes more than 10 years (Klötzli et al. 1996), more sophisticated statistical methods should be used. According to Wildi & Orloci (1996) and the findings of Langenauer and Marti (pers. comm.) one of the more appropriate ways to analyse long-term series and the long-term behaviour of the vegetation on a given plot is the simultaneous use of three statistical methods, viz.

- autocorrelation
- fuzzy ordination
- principal coordinate analysis.

(Evaluation of data of Klötzli 1993.)

- *Autocorrelation considering "time"*

This analysis determines spatial, temporal and ecological heterogeneity of a sample. (In our case: temporal heterogeneity of the relevés of a permanent plot.)

- *Correlation using "fuzzy ordination"*

This analysis is based on one site factor. On the x-axis we find the relevés ranked according to floristic composition and one site factor. On the y-axis the site factor is placed, scaled from 0 to 1. With the aid of a correlation coefficient the dependence of a vegetation gradient on an observed site factor is measured (cf. also Roberts 1986). (In our case the site factor is time.)

- *Ordination by principal coordinate analysis or similarity*

First a similarity matrix is calculated. On that basis coordinates for every relevé are generated.

A corresponding trend in vegetation development is given by the analysis of one or two data sets (Table 2). Should there be a trend, the values are high in both analyses. If there are simple meteorological fluctuations, the values are low.

Table 2. Limiting values for (in some cases) directed vegetational changes (values on the basis of experience).

		Combination of possible statistical tests		
vegetation change in/of	cases	autocorrelation	"fuzzy ordination"	similarity
directed (in some cases directed) for all 3 analyses	1	≥ 7 (≥ 0.65)	≥ 0.9 (≥ 0.88)	≤ 0.6 (≤ 0.65)
	2	≥ 7 (≥ 0.65)	≥ 0.88 (≥ 0.86)	≤ 0.45 (≤ 0.5)
	3	≥ 8 (≥ 0.75)	≥ 0.92 (≥ 0.9)	-

Then analysis 3 highlights the similarity of relevés and detects possible trends:

- Similarity is small if an alteration of the vegetational composition is coupled with a considerable change in the species spectrum.
- Similarity may be relatively great in a directed vegetational change if this change is based on a directed increase or decrease of the cover values of certain species.

To distinguish directed from undirected changes, the values from the analyses 1 and 2 must be very high.
An indicator analysis of each plot is made to substantiate such interpretations of all three analyses.
Clear directional changes are only given if all three analyses are within the limiting values. The bordering area is given if all three analyses are within the limiting values for a possible directed change (Langenauer and Marti, pers. comm., on the basis of Wildi 1994). With this statistical approach it is possible to clearly distinguish pure successions (definitions in Glenn-Lewin et al. 1992; van der Maarel 1996b; Agnew et al. 1993) and the intensity of a trend versus pure fluctuations. (A model for the evaluation of chaotic fluctuations has been published in Gassmann et al. 2000.)

4. Graphical evaluation

In the graphical evaluation of a vegetation map first exactitude and scale should be defined and/or determined, especially when fine-scale mapping is used, considering areas of 2-5 mm on a given scale, e.g. in highly diverse grasslands.
Furthermore the signatures and colours are determined, colours mostly in the sense of Gaussen (1928, 1936, 1958, 1965; Seibert 1997), the blue parts of the rainbow colours being reserved for cooler and/or wetter sites, the red parts for warm/dry sites. Nutrient levels are best depicted with signatures, or by the use of non-rainbow colours, such as brown or grey tones.
With pure black and white maps (e.g. for print in a publication) strong signatures should be reserved for the wetter sites, weaker ones for the drier sites, using also hatched lines, crossed lines, or pure black or white, and applying symbols where necessary. Nutrient levels may be depicted with additional signatures or symbols which cannot be mistaken for e.g. humidity signatures. "Zebras", mixed units, should be avoided wherever possible.

5. Specific evaluation

Most of our extended time series were evaluated during the early nineties and published in different journals. Often such papers are written in German for political reasons, research being carried out on behalf of a government organisation and reports, including papers, being therefore in the official language.
Table 2 gives an overview of the sites of these time series for Swiss stations, one German and a Tanzanian station. References (see also Introduction) on other time series contain values not exceeding 10 years (non-chrono-sequences, for details see van der Maarel 1996a).

6. General conclusions from the evaluations

According to these evaluations the following main results may be presented:

1) The methods mentioned above are useful, even when time series are rather irregular.

2) Not many species occur steadily, i.e. without any larger fluctuations ("fluctuation" sensu van der Maarel 1996a, non-directional quantitative change as a response to short-term stochastic environmental variables). They may be called persistent species. (I avoid the term "constant" sensu van der Maarel 1996a, because the behaviour of such species is quantitatively not quite "constant", but without a break in their occurrence and only minor differences from a mean value, e.g. from abundancy + to 1).

3) Most species, without considering their behaviour in a succession, do not appear steadily, they undergo unforeseeable (non-stochastic) fluctuations without any obvious or detectable cause that might be attributed to any of those changes.

   These species may be divided into five groups (sensu Klötzli 1995):
   1. persistent species
   2. fluctuating species (fluctuating regularly and major differences from a mean value, e.g. from abundancy 1 to 3)
   3. welling up species
   4. diving species
   5. "undecided" species.

They are partly comparable to groups proposed by van der Maarel (1996 a,b), e.g. No. 1 to "constant" species (see above), No. 3-5 to "occasional" and "pulsating" species, furthermore he divided them into "local" and "circulating" species. However, the point of view on permanent plots is not quite the same, i.e. his "carousel model" is consistent with our plots as well, but our expressions have been chosen for larger areas and data sets. A paper on "mobility" under these circumstances is in preparation (see also van der Maarel & Sykes 1993).

In his summary van der Valk (1992) presents some "catchy" statements which involve our "General Conclusions":

- "Each species becomes established along a certain portion of a coenocline because of a unique combination of factors ..."
  i.e. fluctuating environmental conditions keep these species fluctuating on a given gradient.

- "Because seeds ... will not germinate under a plant canopy, these species are recruited into the vegetation in openings or gaps created by various kinds of disturbances" and
  "Species with vegetative propagation normally colonise new sites by seed. The abundance ... is largely dependent on vegetative propagation"
  i.e. consumers (insects on a strong gradation or mammals, such as wild boar) may be responsible for small gaps which are then repaired by "undecided" or fluctuating species.

- "The absolute abundance of ramets and the age-state spectrum (i.e. number of seedlings, juveniles, mature specimens) of a clonal species can vary from year to year because of differences in environmental conditions"
  i.e. such species are mostly susceptible to slight modifications and are often "fluctuating" or "undecided", partly also "welling up" or "diving", according to the

extent of such a change. Investigating our new findings with this new model, which explains chaotic fluctuations, even abundant species may "dive" in a short time and decennia later come up again after a couple of favourable years (Gassmann et al. 2000).

A prediction of changes is made by the Markov model, assuming that "the present state depends on the previous state" (van Hulst 1979; review in Usher 1992, and in this issue).

4) Borderlines between two plant communities (e.g. associations) are rarely stable, i.e. they change according to the effects of changes in the water regime, including meteorological events (Klötzli & Zielinska 1995; Klötzli 1997).

5) A certain "heart zone" in the more central parts of rather homogenous plant communities is mostly stable. Thus a map from last year may look quite different although some heart zones are still untouched by changes or disturbances. (For spatial patterns see e.g. Palmer 1988.)

6) Even two to three data sets only - not considering each year in an observation period - may be compared with each other if shifts are suspected in the appearance of certain indicator groups. Such changes may be due to changes in one or several site factors. Rising winter temperatures, for example, have allowed the establishment of more thermophilous species, to the detriment of more montane species which have moved to cooler areas. Two large data sets from the sixties and seventies compared with a new set drawn up in the nineties (average difference 30 years) have revealed such differences for Swiss forests which might be attributed to global change (see Klötzli et al. 1996, or also Carraro et al. 1999 for more details).
   In some other specific cases, totally different evaluations are not required when two or more data sets are compared. Thus a similar approach was used to evaluate changes in

   - management maps
   - maps of changes on lake shores
   - maps of changes in riverine complexes
   - maps of changes in Alpine areas, e.g. due to ski-runs or global warming (but for true secondary succession with distinct successive groups of plants ["Verbrachung"] on abandoned fields see also Schreiber 1997).

In the following articles many specific problems of inventorized or mapped areas are presented, thereby discussing the whole framework of these methods of surveying and mapping or, more specifically, of restoration control.

References

Agnew, A.D.Q., Collins, S.L. & van der Maarel, E. 1993. Mechanisms and processes in vegetation dynamics. J. Veg. Sci. 4: 145-278.

Bakker, J.P. 1989. Nature Management by Grazing and Cutting. Geobotany 14. Kluwer Acad. Publ., Dordrecht.

Beeftink, W.G. 1987. Vegetation responses to changes in tidal inundation of salt marshes. pp. 97-117. In: van Andel, J., Bakker, J.P. & Snaydon, R.W. (eds), Disturbance in Grasslands. Causes, Effects and Processes. Geobotany 10. Junk, Dordrecht.

Braun, J. 1913. Die Vegetationsverhältnisse in der Schneestufe der Rätisch-Lepontischen Alpen. Neue Denkschr. Schweiz. Naturforsch. Ges. 48: 156-307.

Braun-Blanquet, J. 1931. Vegetationsentwicklung im Schweizer Nationalpark. Dok. Erforschg. Schweiz. Nat. P.: 1-82.

Brown, A. & Rowell, T.A. 1997. Integrating monitoring with management planning for nature conservation: some principles. Natur u. Landsch. 72: 502-506.

Buell, M.F., Buell, H.F., Small, J.A. & Siccama, T.G. 1971. Invasion of trees in secondary succession on the New Jersey Piedmont. Bull. Torrey Bot. Club 98: 67-74.

Carraro G., Gianoni, P. & Mossi, R. 1999. Climatic influence on vegetation changes: a verification on regional scale of the lamophyllisation. pp. 31-51. In: Klötzli, F. & Walther, G.-R. (eds), Proceedings to the conference on "Recent shifts in vegetation boundaries of deciduous forests, especially due to general global warming". Birkhäuser, Basel.

De Smidt, J.T. 1977. Interaction of *Calluna vulgaris* and the heather beetle (*Lochmaea suturalis*). pp. 179-186. In: Tüxen R. (ed), Vegetation und Fauna. Cramer, Vaduz.

Ellenberg, H., Mayer, R. & Schauermann, J. 1986. Ökosystemforschung - Ergebnisse des Sollingprojektes 1966-1986. Ulmer, Stuttgart.

Gassmann, F., Klötzli, F. & Walther, G.-R. 2000. Simulation of observed types of dynamics of plants and plant communities. J. Veg. Sci. 11: 31-40.

Gaussen, H. 1928. Signes employés dans la construction des cartes des productions végétales. Bull. Soc. Hist. Nat. Toulouse 57: 443-450.

Gaussen, H. 1936. Le choix des couleurs dans les cartes botaniques. Bull. Soc. Bot. France 1936: 474-480.

Gaussen, H. 1958. L'emploi des couleurs en cartographie. Bull. Serv. Carte Phytogéogr., Ser. A, Carte de la Végét. t. III (1): 5-10.

Gaussen, H. 1965. Classification et cartographie de la végétation. Unesco (AVS) No. 169, 20.07.65: Paris.

Glenn-Lewin, D.C. & van der Maarel, E. 1992. Patterns and processes of vegetation dynamics. pp. 11-59. In: Glenn-Lewin, D.C., Peet, R.K. & Veblen, T.T. (eds), Plant Succession - Theory and Prediction, Chapman and Hall, London.

Grabherr, G. 1997. Vegetations- und Landschaftsgeschichte als Grundlage für Natur- und Landschaftsschutz. Ber. d. Reinh.-Tüxen-Ges. 9: 37-48.

Hegg, O. 1984a. Langfristige Auswirkungen der Düngung auf einige Arten des Nardetums auf der Schynigen Platte ob Interlaken. Angew. Bot. 53: 141-146.

Hegg, O. 1984b. 50jähriger Wiederbesiedlungsversuch in gestörten Nardetum-Flächen auf der Schynigen Platte ob Interlaken. Diss. Bot. 72 (Festschrift Welten): 459-479.

Hegg, O., Feller, U., Dähler, W. & Scherrer, C. 1992. Long term influence of fertilization in a Nardetum. Vegetatio 103: 151-158.

Hobbs, H.S. 1983. Markov models in the study of post-fire succession in heathland communities. Vegetatio 56: 17-30.

Klötzli, F. 1987. Disturbance in transplanted grasslands and wetlands. pp. 79-96. In: van Andel, J., Bakker, J.P. & Snaydon, R.W. (eds), Disturbance in Grasslands. Causes, Effects and Processes. Junk, Dordrecht.

Klötzli, F. 1993. Grundsätze ökologischen Handelns. pp. 9-24. In: DVGW-LAWA Kolloqu. Ökologische Wassergewinnung. DVGW Schr. R. Wasser 73.

Klötzli, F. 1995. Projected and chaotic changes in forest and grassland plant communities. Preliminary notes and theses. Ann. Bot. 53: 225-231.

Klötzli, F. 1997. Zur Dynamik von Naturschutzgebieten in der Schweiz. pp. 191-225. In: Erdmann, K.-H. (ed), Internationaler Naturschutz. Springer, Berlin, Heidelberg, New York.

Klötzli, F., Lupi, C., Meyer, M. & Zysset, S. 1995. Veränderungen in Küstensavannen Tansanias. Ein Vergleich der Zustände 1975, 1979 und 1992. Verh. Ges. Ökol. 24: 55-65.

Klötzli, F., Walther, G.-R., Carraro, G. & Grundmann, A. 1996. Anlaufender Biomwandel in Insubrien. Verh. Ges. Ökol. 26: 537-550.

Klötzli, F. & Zielinska, J. 1995. Zur inneren und äusseren Dynamik eines Feuchtwiesenkomplexes am Beispiel der "Stillen Rüss" im Kanton Aargau. Schr. R. Veg.kde., Sukopp-Festschrift 27: 267-278.

Krüsi, B.O., Schütz, M., Grämiger, H. & Achermann, G. 1996. Was bedeuten Huftiere für den Lebensraum Nationalpark? Cratschla/Mitt. aus d. Schweizerischen Nationalpark 4(2): 51-64.

Krüsi, B.O., Schütz, M., Wildi, O. & Grämiger, H. 1995. Huftiere, Vegetationsdynamik und botanische Vielfalt im Nationalpark. Cratschla/Mitt. aus d. Schweizerischen Nationalpark 3(2): 14-25.

Lüdi, W. 1940. Die Veränderungen von Dauerflächen in der Vegetation des Alpengartens Schynige Platte innerhalb des Jahrzehnts 1928/29 - 1938/39. Ber. Geobot. Forsch. Inst. Rübel Zürich, 1939, Zürich 1940: 93-148.

Lüdi, W. 1948. Die Pflanzengesellschaften der Schynige Platte bei Interlaken und ihre Beziehungen zur Umwelt. Veröff. Geobot. Inst. Rübel, Zürich 23.

Lüdi, W. 1966. Lokalklimatische Untersuchungen am Fuornbach (Ova del Fuorn) und am Spöl im Schweizerischen Nationalpark. Erg. wiss. Unters. SNP 10, Nr. 56: 275-337.

Palmer, M.W. 1988. Fractal geometry: a tool for describing spatial patterns of plant communities. Vegetatio 75: 91-102.

Peet, R.K. & Christensen, N.L. 1987. Competition and tree death. Bio-Science 37: 586-594.

Pfadenhauer, J., Poschlod, P. & Buchwald, R. 1986. Überlegungen zu einem Konzept geobotanischer Dauerbeobachtungsflächen für Bayern. Teil I. Ber. ANL 10: 41-60.

Pickett, S.T.A. 1982. Population patterns through twenty years of old field succession. Vegetatio 49: 45-59.

Pickett, S.T.A. & White, P.S. (eds) 1985. The Ecology of Natural Disturbance and Patch Dynamics. Academic Press, Orlando FL.

Roberts, D.W. 1986. Ordination on the basis of fuzzy set theory. Vegetatio 66: 123-131.

Roozen, A.J.M. & Westhoff, V. 1985. A study on long-term salt marsh succession using permanent plots. Vegetatio 61: 23-32.

Schmidt, W. 1974. Die vegetationskundlichen Untersuchungen von Dauerprobeflächen. Mitt. flor.-soz. Arb.gem. NF 17: 103-106.

Seibert, P. 1997. Zur Farbenwahl von Vegetationskarten. Tuexenia 17: 53-58.

Stüssi, B. 1970. Vegetationsdynamik und Dauerbeobachtung. - Naturbedingte Entwicklung subalpiner Weiderasen auf Alp La Schera im Schweizer Nationalpark während der Reservatsperiode 1939-1965. Erg. wiss. Unters. SNP 13, Nr. 61.

Usher, M.B. 1992. Statistical models of succession. pp. 215-248. In: Glenn-Lewin, D.C., Peet, R.K. & Veblen, T.T. (eds), Plant Succession - Theory and Prediction. Chapman and Hall, London.

van der Maarel, E. 1996a. Vegetation dynamics and dynamic vegetation science. Acta Bot. Neerl. 45(4): 421-442.

van der Maarel, E. 1996b. Pattern and process in the plant community: 50 years after A.S. Watt. J. Veg. Sci. 7: 19-28.

van der Maarel, E. & Sykes, M.T. 1993. Small-scale plant species turnover in a limestone grassland: the carousel model and some comments on the niche concept. J. Veg. Sci. 4: 179-188.

van Dorp, D., Boot, R. & van der Maarel, E. 1985. Vegetation succession on the dunes near Oostvoorne, the Netherlands, since 1934, interpreted from air photographs and vegetation maps. Vegetatio 58: 123-136.

van Hulst, R. 1979. On the dynamics of vegetation: Markov chains as models of succession. Vegetatio 40: 3-14.

Wildi, O. & Orloci, L. 1996. Numerical exploration of community patterns. A guide to the use of MULVA-5. 2nd ed. SPB Acad. Publ., Amsterdam.

STATISTICAL DESIGN AND ANALYSIS IN LONG-TERM VEGETATION MONITORING

OTTO WILDI

Swiss Federal Institute for Forest, Snow and Landscape Research, Zürcherstrasse 111, CH-8903 Birmensdorf, Switzerland

Keywords: Autocorrelation, Fuzzy set, Markov chain, mechanism, ordination, pattern, process, sampling design, succession

Abstract

Succession theories and related methods for time series analysis are presented and illustrated by examples using published data. Starting from a rather general definition of succession which is understood as any directional change of vegetation in time, the phenomena can be distinguished at three different levels of perception: pattern, process and mechanism. Pattern is the phenomenon perceived and reflected in the sample data. Process is a grammar generating this pattern (Dale 1980). Both can be accessed by appropriate statistical analysis. Mechanisms are the causes of the changes. Whereas in some cases they can be inferred from patterns and processes, they cannot be observed or measured directly.

The methods and examples described include the statistical design of long-term investigations, the analysis of community dynamics using Markov chains, pattern analysis by ordination, the analysis of temporal autocorrelation and fuzzy ordination for fitting time series data to a time vector. In most methods, the results can be tested for the existence of trends using Monte Carlo simulation. Experience suggests that directional changes can only be interpreted if the data cover a reasonable number of time steps.

Kurzfassung

Sukzessionstheorien und dazugehörige Methoden für die Analyse von Zeitreihendaten werden vorgestellt und an Beispielen von publizierten Daten illustriert. Ausgehend von einer sehr weit gefassten Definition von Sukzession, die als gerichtete Veränderung der Vegetation verstanden wird, können Phänomene auf drei verschiedenen Perzeptionsstufen unterschieden werden: jener der Muster, der Prozesse und der Mechanismen. Das Muster ist das Erscheinungsbild, wie es in den Daten auftritt. Der Prozess ist eine Grammatik, die das Muster zu generieren erlaubt (Dale 1980). Beide können mit geeigneten statistischen Methoden untersucht werden. Mechanismen sind die Ursachen der Veränderungen. Manchmal kann von den Mustern und Prozessen auf Mechanismen geschlossen werden, doch können sie in der Regel nicht direkt beobachtet oder gemessen werden.

Burga & Kratochwil (eds.), BIOMONITORING, 17-39

Die vorgestellten Methoden und Beispiele umfassen das statistische Erhebungskonzept von Langzeituntersuchungen, die Analyse von Gesellschaftsdynamik mittels Markovketten, Mustererkennung mittels Ordination, die Analyse zeitlicher Autokorrelation und eine Fuzzy-Ordination, um Zeitreihendaten einem Zeitvektor anzupassen. In den meisten Methoden können die Resultate auf die Existenz eines Zeittrends hin getestet werden, wobei Monte Carlo-Simulation zum Einsatz kommt. Die Erfahrungen zeigen, dass gerichtete Veränderungen in der Regel nur interpretiert werden können, wenn genügend Zeitschritte dokumentiert sind.

1. Introduction

In this chapter some statistical aspects related to the study of permanent quadrats are presented. Whereas anyone will agree that processing and analysis of data are needed in practice, the suggestions offered in the following can be used in different ways. First of all, they may be used as a technical manual, a help in the application of statistical methods. Secondly, if succession is interpreted as a distinct mechanism, data analysis involves the selection or construction of a model. The model is then tested using field data. The results of these tests aim to accept or reject previously generated hypotheses. In this context, design and analysis involve the most important decisions in long-term investigations. For users who tend to share this point of view, I decided to start with an outline of some theoretical considerations of succession. These are then related to the presented methods and eventually discussed in the phase of interpretation.

Whether statistical methods are used or not, an interpretation of the results is always required. The present contribution focuses on the multivariate view of data, which is rather complex. There may be good reasons to look for simpler solutions. The interpretation of the dynamics of individual species is one example (Austin et al. 1994). In this case, an ecological interpretation of the observed change of the entire system can be achieved if the resource requirements of the species are known by experience or from other sources. However, it involves the risk of misinterpretation since that species interacts with others and the change may be due to another species and not directly to the environment. A further popular path is the use of indicator values. I decided to forego both, the analysis of individual species response curves as well as the use of indicators. The methods finally selected here all consider species interaction. For the use of indicators we refer to original publications (Landolt 1977; Ellenberg et al. 1991) in which the limitations of application are clearly explained.

2. The perception of vegetation dynamics

2.1. THEORETICAL BACKGROUND

Any relevé of an assemblage of plant species only reflects a state in time, which differs from the previous composition and will further change in future. Gleason (1939, p. 108) mentions that "on every spot of ground, the environment varies in time, and consequently vegetation varies in time." As a logical consequence, the same holds for space: "At any given time, the environment varies in space, and consequently vegetation varies in space."

In our context, we are interested in partitioning variation into two components. One component is random fluctuation with or without any further explanation of cause whereas the other is directed and thus open to description, analysis and interpretation. We refer to this second component as succession. Begon et al. (1996, p. 693) define this as "*the non-seasonal, directional and continuous pattern of colonization and extinction on a site by species populations*". They mention that this definition of course exceeds the range of successional sequences described in many textbooks. In our present context, we would even declare seasonal (reversible) changes as being a kind of succession as they might be key events for permanent, long-term changes.

In many investigations, only the species composition is monitored over time. But of course, the environment too changes with time. Quite often, environmental change is caused by the vegetation. Crocker & Major (1955) (cited in Begon et al. 1996) give an example of soil formation and nitrogen increase after glacial retreat in Alaska. In this case, we have an explanation for succession, but how far such an explanation describes reality is an open question. Following the suggestion of Anand (1997) we may distinguish three levels of perception:

(i) pattern. This is the mere description of what can be seen, measured or extracted from the data. The method to analyse this is pattern analysis (Wildi & Orlóci 1996). Two well-known tools in vegetation science are ordered vegetation tables and ordinations. According to Anand, patterns answer the question "what" has been found. Pattern is the "primitive" in the description of a successional event.

(ii) process. A process is the "grammar" (Dale 1980) generating the pattern in a model environment (Anand 1997). This may be closely related to biological processes, such as competition, or it may be merely statistical like regression. Many authors just distinguish pattern and process (e.g. Watt 1947). De Wit & Goudriaan (1974) use patterns and processes for modelling and it remains open if processes exist that cannot be described in mathematical terms.

(iii) mechanism. This answers the question "why" succession occurs. According to Anand (1997) this has the nature of a "message". An example are the three main driving mechanisms of Connell & Slatyer (1977): facilitation, tolerance and inhibition. Unlike processes that can often be expressed as rules, mechanisms are not easily accessible to measurements and subsequent mathematical description, therefore our motivation to distinguish them from processes.

The analysis of vegetation dynamics is almost inevitably linked to the concept of climax. Strongly promoted by Clements (1916), this concept expresses that after any change in species composition a state of equilibrium is reached in which dying individuals are replaced by descendants of the same species. The monoclimax of Clements suggests that under any given climate vegetation will develop in one direction only and end up in one specific state. The concept of Watt (1947) extends the previous one, suggesting that vegetation finally reaches a cycle of permanently alternating states. Whittaker (1953) went in a different direction with his ideas. He introduced the continuity of climax types which vary with environmental conditions. Although this complicates the situation in many ways, it remains a deterministic approach to the recognition of succession.

There are many ways in which succession can be classified. One is primary versus secondary. Primary succession occurs when a new habitat is created, after the retreat of a glacier, the creation of an island or after a volcanic outburst (e.g. Moral 1993; Moral & Bliss 1993). The term secondary succession is used when a system in equilibrium is faced with an abruptly changing factor initiating the development towards a different equilibrium state. This distinction is of much practical significance, e.g. in experimental research where the initial conditions are controlled. In the context of the analysis, the problems remain the same in primary and secondary succession.
Another possibility is the distinction between autogenic and allogenic succession. A process is autogenic if the changes are caused by the plants themselves, competition, facilitation (Connell & Slatyer 1977) or dispersal. It is allogenic if they are due to continuous change of the environment. Because changes in the environment are often plant-induced, these types may function simultaneously. Yet, the distinction is of great interest in the analysis of processes. If only allogenic succession was in progress, a perfect fit between vegetation and site could be observed. In this case, an investigation would be technically the same as gradient analysis (Whittaker 1967). Autogenic succession, on the other hand, may progress with insignificant or even without any change of the environment. Of course, then the problem of analysis is different. Mechanisms and processes must be found in the vegetation itself. Many of the examples shown below refer to this type.
From the point of view of systems theory we may expect many more types of changes in plant communities. In Clements' (1916) view they tend to approach an equilibrium state. This is, in the simplest case, one state. Such a system is said to have a point attractor. In Watt's (1947) view, the system approaches a cyclic attractor. Whittaker (1967) assumes that the final state depends on the initial state. However, from many observations we have good reason to suspect that the outcome of the final state could be a consequence of the initial state and much less of the ecological conditions, as often believed. This brings up the question if the final state of the system is impossible to predict. A well-known example of such a system is the atmosphere. It is considered chaotic because its state cannot be predicted in the long term. Presumably it is governed by the butterfly effect: a flap of the wings of a butterfly changes the initial conditions, and, by a chain of interactions, changes the weather conditions of a continent. We do not yet know if such processes operate in vegetation systems, and therefore they cannot be excluded.

2.2. AN EXAMPLE

What has been outlined so far relies on theoretical frameworks of succession. The elements distinguished have to be identified in any real system that is analysed. An example is the easiest way to explain this. Orlóci et al. (1993) present a time series documenting recovery of a heathland after fire. The data (Figure 1) have originated from Lippe et al. (1985). They include the cover percentages of all important species and the open soil as variables. For analysis and interpretation, the following features distinguished in succession theory can be considered:

1. The data document a secondary succession. At the beginning, immediately after the fire, there are already species present. The term succession is justified as there is an obvious trend in time. This trend is strong at the beginning and slows down towards the end of the time period.

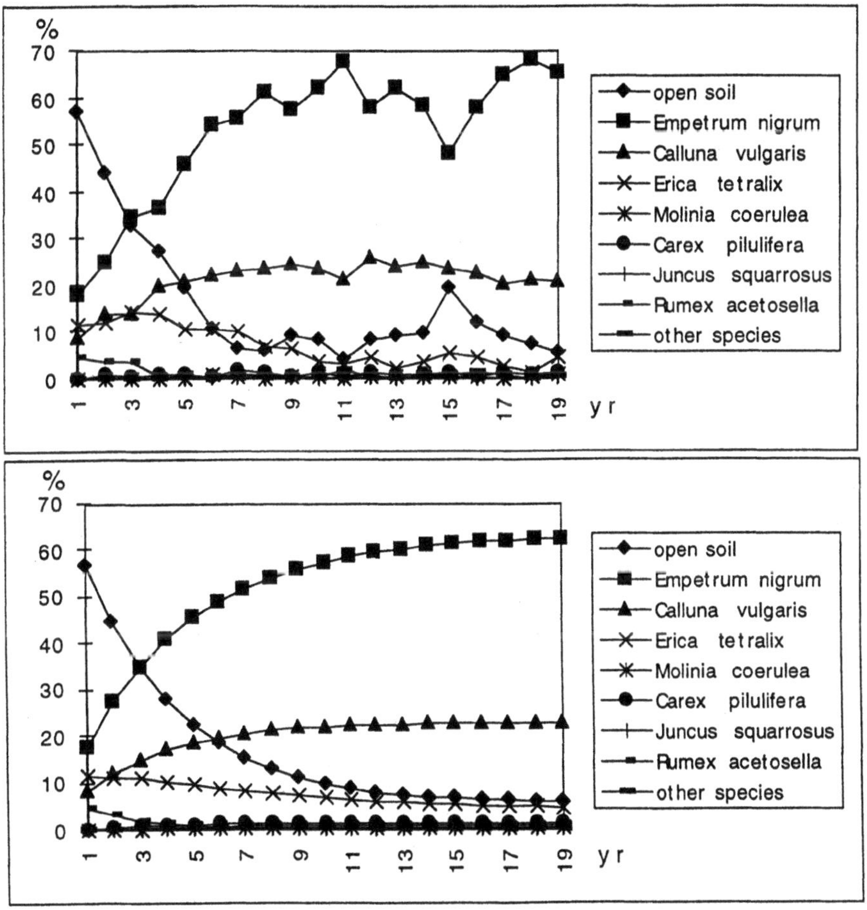

Figure 1. Change of the cover percentage of species in the data of Lippe et al. (1985). Upper graph: field data. Lower graph: simulation.

2. The system seems to reach an equilibrium state after about 8 years. The equilibrium is not a steady state in a strict sense, but the system still fluctuates within a specific range. The data fit a model - to be explained later - whose output is shown in the lower graph of Figure 1. This model represents a real equilibrium state in which also minor variations are outbalanced. As it well fits the field data, the existence of a succession leading towards a stable state of the vegetation can be assumed.

3. The field data and the model are examples for the decomposition of variation into a deterministic component (the model) and unexplained variation (the difference

between the model and the field data). Achieving this distinction is one of the main objectives in the analysis of time series.

4. The example demonstrates that the dynamics is partly unpredictable. In the year 15 a sudden change occurred which might have been the onset of a new development. Most likely, it was caused by a disturbance. The event turns out to be reversible.

5. The change can also be observed at the level of individual species (Figure 1). Species show their typical response curves. Some have similar shape (i.e. increase, decrease). Similar shape also is an indication that there are redundant variables in the data. In many cases, the species also help in the interpretation of the process: if their resource requirements are known, then the change can be interpreted in ecological terms.

In the sections below, it will be shown how some of the typical features found in this and other examples can be revealed and interpreted through statistical analysis.

3. Sampling design and optimisation

The general rules to be observed in the design of long-term observations are the same as for any other ecological investigation (cf. Green 1979). One is that the objective has to be given. Let us assume that the aim of the investigation is to determine (i) if there is a change in species composition over time and (ii) if this change exceeds the natural annual variation that can be expected in stationary systems. Next, the sampling frame has to be defined (Wildi & Orlóci 1996). This requires a definition of the most basic elements. Proceeding from top to bottom, these are:

The sampling space. This is the area that has to be investigated. It is important to note that the final results will only hold for this space and will not be valid outside.
The sampling design used (systematic, random etc.). Like in any other investigation, only a probabilistic sample will yield an unbiased picture of the entire sampling space. In the present context, the temporal scale is of specific interest. The number of time steps available for analysis, the temporal component of sample size, is most influential in statistical testing of time series data (Orlóci et al. 1993).
The biological population. This expresses the way in which vegetation is perceived. Examples are species populations, guilds, species assemblages, life forms, etc. Abstractions of this kind are always involved in investigations. They focus the attention on specific properties of the real system.
The biological unit. This is the smallest element considered in the investigation. It may be an individual, a part of an individual, a cover percentage or something else.
In the recognition of vegetation systems, representativity in space and time should be achieved through the sampling procedure (De Patta Pillar & Orlóci 1993). It is well known that the effort to achieve this goal can be great even if only space is considered (Podani 1984). When the time dimension is added, then the expenditure is multiplied by the number of time steps (Harcombe et al. 1997). There exist ideas of how to reduce this effort. They all point in the same direction: it is believed that variation in space needs not always to be known at any time step, but only after the occurrence of drastic

changes. These can be detected in a reduced set of time series observations (Scott Urquhart et al. 1993).

A very simple scheme is shown in Figure 2. There, the spatial variation is determined at the beginning of the investigation. The plots, s_1 to s_n, represent a complete statistical sample in space. Then, a small subset of relevés which is most representative for the entire sample has to be selected. An optimisation provides the RANK algorithm proposed by Orlóci (1978). In a first step, it selects the relevé that shares a maximum portion of variation with all the other relevés. Then, this variance is removed from the data and a second relevé is retrieved in the same way. The procedure continues until a sufficiently large portion of the total variation is explained by the relevé set chosen. After some time, it may be necessary to reexamine the spatial variation (time step t_5 in Figure 2). This can either be done at regular intervals (e.g. every 5 years) or at a moment when in one of the permanent plots a considerable change is observed. If this happens, it is likely that the spatial variation has also changed and therefore needs to be determined anew. For details of the method we refer to Wildi (1990) and Wildi & Orlóci (1996).

time → / space ↓	time steps								
plots	t_1	t_2	t_3	t_4	t_5	t_6	t_7	...	t_m
s_1	x				x				x
s_2	x	x	x	x	x	x	x	...	x
s_3	x				x				x
s_4	x				x				x
s_5	x				x				x
s_6	x	x	x	x	x	x	x	...	x
...	...				...				...
s_n	x				x				x

Figure 2. Sampling design with reduced sampling effort. Spatial variation is determined at time steps 1, 5, and m only, temporal variation in plots 2 and 6. See explanation in the text for the method to select the plots.

As an example, the selection is applied to a set of 63 relevés (Wildi 1977). The systematic arrangement in the field is shown in Figure 3. Three relevés are selected. The result is typical of small-scale investigations: the first relevé explains as much as 23 percent of the variation. With two relevés, the portion rises to 34 and with 3 to about 39 percent. On closer inspection, not shown here, it can be seen that the three relevés represent rather different vegetation types. Therefore, it can be suggested to use them for permanent annual investigation just as shown for the plots s_2 and s_6 in Figure 2.

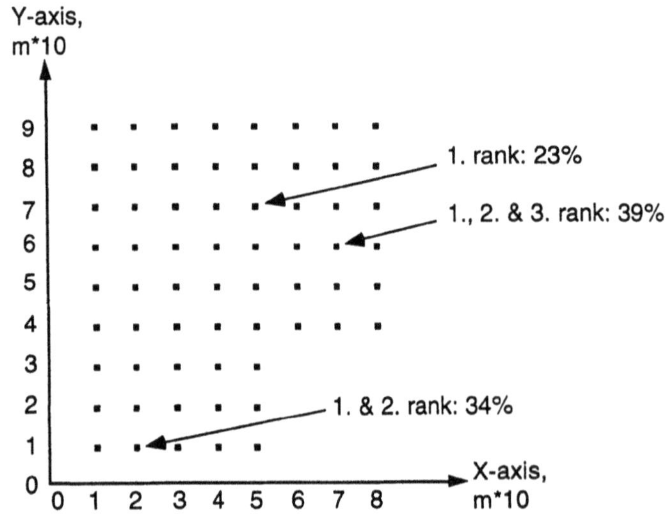

Figure 3. Optimised selection of relevés in a systematic sample of 63 relevés. The 3 selected relevés represent 39 percent of the total spatial variation and were therefore chosen for observation of the temporal variation.

4. Community dynamics and Markov process

There is no process known so far that explains changes within species assemblages in general terms. All mechanisms mentioned earlier play an important role. Invasion and disappearance of the species are usually explained by their different life strategies. Such strategies have been described by Tilman (1988) or Grime (1979). They are essential in understanding the dynamics of entire systems. In many cases, one species is replaced by another. If such changes are not too drastic, like e.g. in autogenic succession, then Markov chains promise to be an excellent approach for modelling.

Changes in permanent plots can be interpreted as replacement processes: several plant populations occupy the same resource, i.e. the physical space. It is a tradition in plant ecology to use cover percentage for quantification of species success. Because "space" can be defined as a fundamental resource exploited by the relevé, cover is a logical (even though two-dimensional only) surrogate to be used for it. If gains and losses of species follow a rule so that any state of the system can be derived from the preceding state, then a Markov process is present (Usher 1981). Win and loss of every species are placed into a transition matrix. This allows to derive the state of the relevé X at time t from the previous time step:

$$X_{t+1} = X_t P \tag{1}$$

The observation vectors X are normal vegetation relevés. Unfortunately, the observation of a conventional permanent plot does not allow to measure the elements of the transition matrix: if a species wins space, then we do not know which other species will lose it. If a species loses ground, we do not know which one will profit from it. A

Markov process, if present, remains undetected. Two plausible assumptions help in this situation (Orlóci et al. 1993):

1. If a species loses part of the main resource, then the other dominating species will most likely profit (i.e. proportionally to their cover);
2. If a species increases its part of the resource, then the remaining dominant species will lose proportionally to their cover.

Both these assertions could be questioned. In succession, a change in the abundance of a species may reflect colonisation of the space by a new species. Similarly, a colonising species may expand its cover at the expense of the species that are least competitive, and which therefore may not be the dominants. This is one more indication that Markov chains cannot be applied when invasion or extinction occur.

The resource "space" is not always entirely occupied by the vegetation. For this reason it may be important to add one more variable to the species list quantifying the open soil. Formally this works like any ordinary species. It is important that the sum of all cover values exactly reaches 100%. This is achieved by the appropriate transformation:

$$x'_{i,t} = \left(x_{i,t} / \sum_{i=1}^{n} x_{i,t} \right) \cdot 100 \qquad (2)$$

In this equation, vector x contains the cover values of species i, t is the present time and n the number of species. In the following artificial example, the sum of cover values is already 100 percent:

	$x_{t=1}$	$x_{t=2}$	$x_{t=3}$
species 1	60	40	10
species 2	25	35	55
species 3	15	25	35
	100	100	100

First, the transition matrix for time step 1 to time step 2 is calculated (Orlóci et al. 1993). For each population i (i.e. species i) a difference results expressing change in time:

$$\text{Diff}(i) = x_{i,1} - x_{i,2} \qquad (3)$$

Positive values of Diff(i) signify a gain, negative ones a loss. The transition matrix contains all the losses of species i in row i, the gains in column i:

$$P = \begin{bmatrix} p_{11} & \cdots & \downarrow & \cdots \\ \cdots & \cdots & \downarrow & \cdots \\ \rightarrow & \rightarrow & p_{ii} & \rightarrow \\ \cdots & \cdots & \downarrow & \cdots \end{bmatrix} \quad \text{with} \downarrow = \text{gain}, \rightarrow = \text{loss of pecies i}$$

The diagonal elements contain the final proportions of space each species covers, i.e. $x_{i,t+1}$. The gains of species i at the expense of species h as well as the losses of species i compared with species h are given by the equation

$$\mathrm{Dev}(i,h) = |\mathrm{Diff}(i)| \frac{x_{h,t+1}}{\sum_i x_{i,t+1}} \tag{4}$$

This means that gains and losses occur proportionally to the resource (i.e. the space) each species occupies at the end of the current time step. Processing the first species in our example, it can be seen that it loses 20% of the total ground from t to t+1 (Diff(1) = - 20). The new diagonal element is 40. The losses, a portion of 0.2 of 35% cover (second element) and 25% cover (third element), respectively, are noted in the first row:

$$P(t_1;t_2;\ 1) = \begin{bmatrix} 40 & 7 & 5 \\ 0 & 0 & 0 \\ 0 & 0 & 0 \end{bmatrix}$$

Species 2 exhibits a win of 10% (factor 0.1), to be added to column no. 2. This is again proportional to the covers of the species at time t+1:

$$P(t_1;t_2;\ 1{+}2) = \begin{bmatrix} 40 & 7+4 & 5 \\ 0 & 35 & 0 \\ 0 & 2.5 & 0 \end{bmatrix}$$

The same procedure can be applied to species no. 3:

$$P(t_1;t_2) = \begin{bmatrix} 40 & 7+4 & 5+4 \\ 0 & 35 & 3.5 \\ 0 & 2.5 & 25 \end{bmatrix} = \begin{bmatrix} 40 & 11 & 9 \\ 0 & 35 & 3.5 \\ 0 & 2.5 & 25 \end{bmatrix}$$

After normalising the rows (the sum adjusted to 1) the transition matrix is

$$P'(t_1;t_2) = \begin{bmatrix} 0.667 & 0.183 & 0.150 \\ 0 & 0.909 & 0.091 \\ 0 & 0.091 & 0.909 \end{bmatrix}$$

For time steps t=2 to t=3 the matrix is

$$P(t_2;t_3) = \begin{bmatrix} 10 & 18.50 & 11.50 \\ 0 & 55 & 5.50 \\ 0 & 7.0 & 35 \end{bmatrix}$$

For all time steps, the average of all gains and losses is taken and the same adjustments are applied:

$$\overline{P} = \begin{bmatrix} 0.500 & 0.2950 & 0.2050 \\ 0 & 0.9091 & 0.0909 \\ 0 & 0.1367 & 0.8633 \end{bmatrix}$$

Through a simple matrix operation according to formula (1) the simulated relevés are derived:

	$x_{t=1}$	$x_{t=2}$	$x_{t=3}$
species 1	60	30	15
species 2	25	42.5	51.23
species 3	15	27.52	33.77

There, the first relevé is identical with the field data. It represents the initial state of the dynamic system.
Orlóci et al. (1993) present a time series documenting recovery of a heathland after fire. As explained before, the data (Figure 1) have originated from Lippe et al. (1985). From the ordination trajectory (explained later in this chapter), it can be seen that in the first few years there is a directed change (Figure 4). After about 8 years, an equilibrium state is reached in which merely random variations occur. The objective of the analysis is to determine exactly the equilibrium state. For this, the Markov model is derived:
The average transition matrix is calculated as shown above from the 19 states of the system, i.e. it is the mean of 18 matrices. Then, beginning with the first field observation, 18 Markov relevés are calculated. They are shown in the lower graph in Figure 1. Obviously, after this time period, the model depicts an equilibrium state.
In the present example, the model fits the field data almost perfectly. However only the deterministic, directed part of the variation is reflected by the simulated time series. The local fluctuations are completely suppressed.

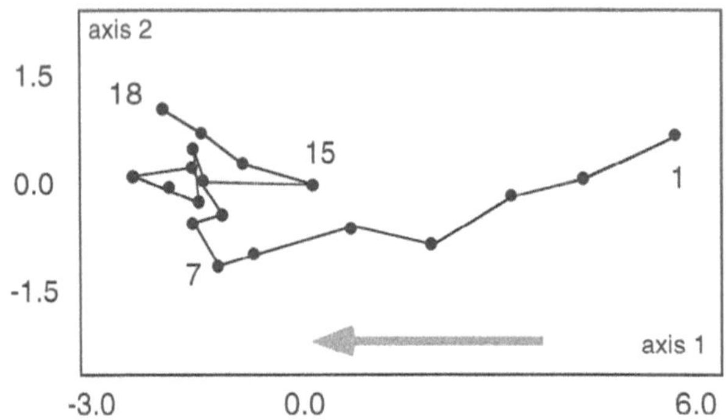

Figure 4. Ordination of the relevés of Lippe et al. (1985). Succession proceeds from right to left.

A Markov chain of the type shown above simulates a classical succession towards a mono-climax: the vegetation reaches an equilibrium state inherent in the model, from which it does not escape. In systems theory, this is a point attractor. In other systems, however, it can be expected that a cycle is reached as proposed by Watt (1947) (i.e. the system has a cyclic attractor) or that there are "random" fluctuations like in the data of Lippe et al. (1985). It is still unknown whether there is also a "strange" attractor present. This would mean that the final state was strongly dependent on the initial conditions and could not be predicted.

5. Multivariate pattern analysis

5.1. ORDINATION AND INTERPRETATION OF TRAJECTORIES

Species vectors represent only a part of the pattern of a multi-species assemblage. Most likely, the change of performance of one species affects one or several others. Interactions occur. Multivariate pattern analysis is used to analyse both, the vectors and their interaction.
Multivariate analysis operates in what we call resemblance space (Orlóci 1978). The properties of this space depend on how resemblance is defined. There are many ways to do this. The most common functions are related to Euclidean space, variance, chi-squares or to measures from information theory. It is the easiest way to explain the Euclidean view, i.e. the geometric approach in the present context (Orlóci 1973). The following artificial data set should be used as an example:

relevé	P_1	P_2	P_3
species 1 (x_1)	2	1	1
species 2 (x_2)	0.5	2.5	3

The three relevés P_1 to P_3 are described by two species. For the geometric representation of these data see Figure 5. In this type of graph, the axes are species (or species combinations). Here, species 1 is used for axis x_1 and species 2 for axis x_2. The species cover values are the co-ordinates for the relevés. The relevés appear as points in the graph. This kind of graph is called ordination. The distance between P_1 and P_2 is a measure of dissimilarity (distance) between the two relevés. Line a represents the most obvious definition of distance. It is called the Euclidean distance. If a is long, then the composition of the two relevés differs. If a is short, the relevés have more common features. A different way of measuring distance is b, which consists of two segments along the two axes. This is the manhattan distance. Line c, the chord distance, is yet another concept. It assumes that the species vectors are transformed to unit length.

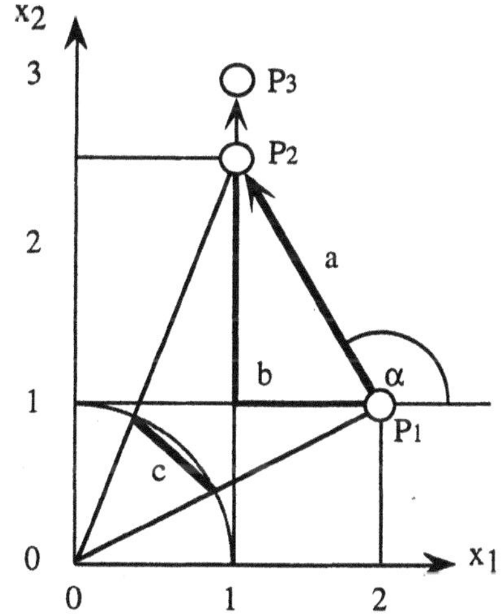

Figure 5. The Euclidean representation of two relevés in two-dimensional space. The arrows are trajectories.

In time series data, P_1 could be a relevé at time step 1 and P_2 the same at time step 2. In this case, distance a would be a measure of change in multivariate space. It would represent a temporal trajectory pointing from P_1 to P_2. The trajectory would have length a and also a direction, α. Since vegetation always varies over time, the direction could just be the result of a random event. In order to find out if a change persists, more time steps need to be known. In the present example, the trajectory continues towards P_3. A succession could be in progress. To evaluate this question more precisely, a statistical test may be necessary, like the one shown in the section about Markov chains.

Ordinating is of course more than just plotting data. The operations performed with many methods include several steps, illustrated in the example shown in Figure 6. There, a series of 6 relevés is characterised by two species whose cover percentages increase and decrease over time just as it can be observed in reality. Using the species as axes, an ordination with 6 relevé points results. In Figure 7, these are presented as circles. The trajectory formed by the arrows resembles a horse shoe, and the phenomenon is therefore known as horse-shoe effect. This is not a deficit of the ordination - as often believed - but the consequence of the bell-shaped response of species in natural systems.

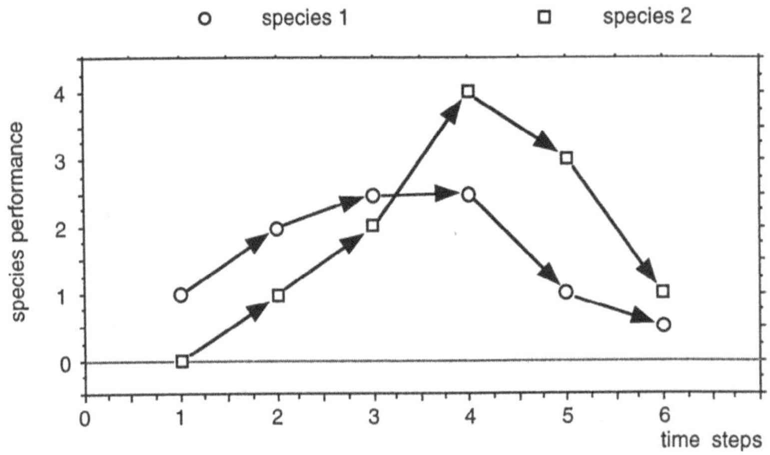

Figure 6. An artificial time series with changing performance of two species.

The geometric configuration in Figure 7 is nothing else than a graphical representation of the data. Many ordination methods, such as Principal Components Analysis (PCA), proceed as follows:

1. A shift of the ordination axes moves the origin into the centre of the point cloud.
2. The axes are transformed so that the first axis displays a maximum of the variation between the points, the second axis the maximum of what remains, etc.
3. The points are plotted in 2 or 3 dimensions using the axes which explain as much variation as possible.

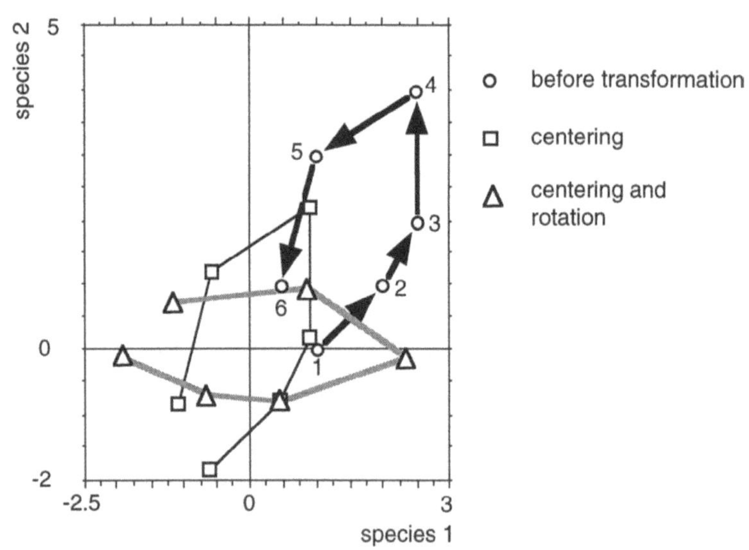

Figure 7. Plot of relevés from Figure 6 in species space and transformations involved in ordination.

Step 1 in Figure 7 generates the squared points, step 2 the triangles. Step 3 in this case is superfluous, because there are no more axes needed to describe the entire variation. The resulting trajectory remains strongly curved. Nevertheless, it represents a logical sequence, so that the process can be qualified as clearly trended.

Using ordination to reveal the trajectory of a time gradient is most effective when the change in time is complex. As an example for a real investigation, the data of Dierschke (1996) are shown in Figure 8. He investigated the vegetation of a river bank in Germany that emerged in 1981 after a heavy flood. In the first few years a type of primary succession occurred. But then, after 1986, several events changed the environment every few years. This has led to a complex type of development and even in recent times there was no stable state to be expected. The ordination also depicts that there was a phase of high species diversity from 1987 to 1989 which will eventually recur in 1996, assuming no further disturbance. Dierschke's data (1996) document extreme fluctuations, but there is also a long-term trend from 1981 to 1995.

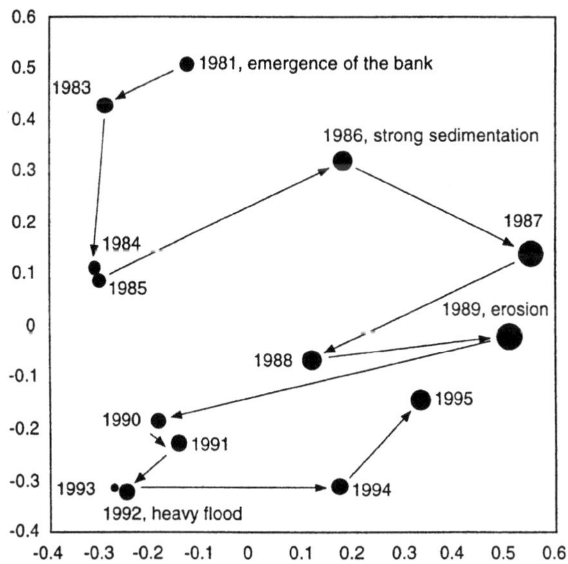

Figure 8. Development of a permanent plot at the shore of the Oder river in Germany (Dierschke 1996). The diameter of the dots in the ordination is proportional to the number of species found. Arrows indicate the time trajectory.

Trend and noise are properties of great interest to plant ecology. While the former is widely discussed in the context of gradient analysis, there is a "low level of interest attached to noise" (Gauch 1982). The first observation usually made on inspection of time series data is the fluctuation in species performance over time. As can be seen in Figure 8, this may sometimes hamper the recognition of a trend. A method to distinguish the two components is to manipulate the data at the level of individual species performances (Wildi 1988; Wagner & Wildi 1997). As long as the change is sufficiently linear, the curve can be approximated by a straight line (Figure 9). Instead of using the observed scores (circles in Figure 9), the expected ones are taken (squares). This is done for all species in a set of relevés. When ordinating the manipulated data, the time trajectory becomes smooth and fluctuations disappear. Figure 10 illustrates the application of the procedure to the data of Dierschke (1996). The points along the straight line come from the derived relevés. Including these artificial data influences the result of the ordination: the configuration of the real relevés is seen from a slightly different direction and is therefore not exactly the same as in Figure 9.

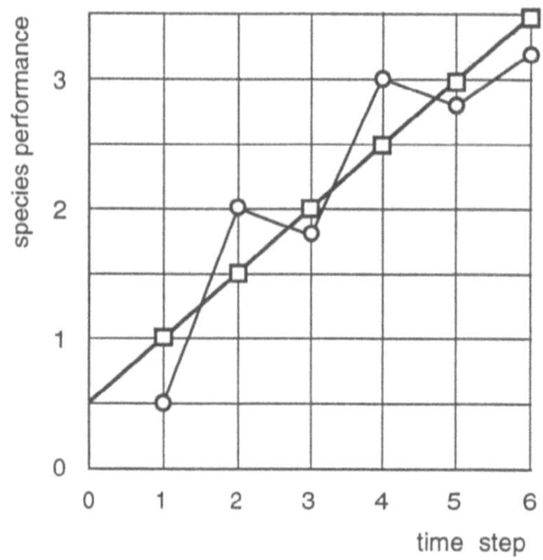

Figure 9. Replacing the observed species performance (circles) by the expected one (squares).

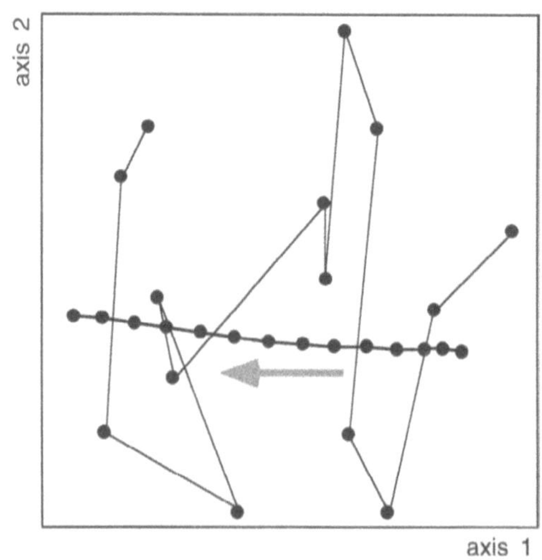

Figure 10. Including relevés without fluctuation into the time series of Dierschke (1996). They form the straight line of data points in the centre of the ordination. The arrow indicates the direction of time.

5.2. TEMPORAL AUTOCORRELATION

The term autocorrelation is used when observations in a statistical sample exhibit dependence either on space or on time. In the case of a succession process, such a dependence must exist. In a normal permanent plot, some species will disappear from one time step to the next while others will persist. It is likely that relevés made within a short interval will be more similar than relevés made in a long interval. To measure this is what has been proposed by Sokal (1986) and Legendre & Fortin (1989). Computational details are explained in Wildi & Orlóci (1996). The method answers the following questions:

1. Is there a general time dependence in the similarity structure of a series of relevés? How strong is this dependence?
2. What is the time range within which a time dependence exists?
3. How long is the time span after which there is complete independence of time?

The first question is answered by the overall autocorrelation, expressed by Moran's I. For this, the compositional dissimilarity of the relevés is correlated with the time elapsed between the sampling events. Formally, I has the same range as a product moment correlation coefficient:

$$-1.0 \leq I \leq +1.0$$

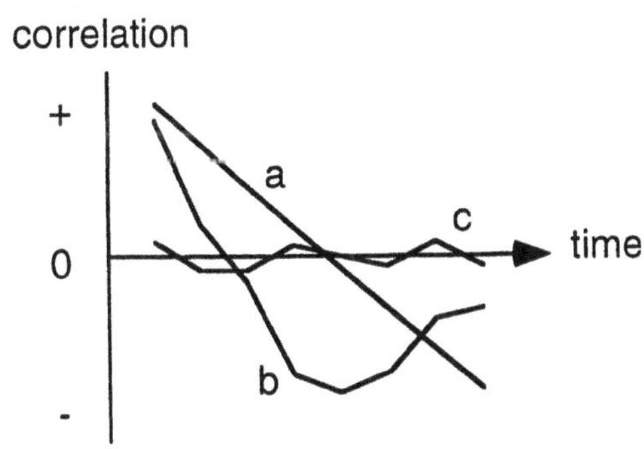

Figure 11. Shapes of correlograms.

If there is time dependence, small differences in composition will occur when the time ranges considered are also small. The I value will therefore exceed zero. Even under ideal conditions, it will hardly reach the maximum of 1.0. The confidence limits of the correlation coefficient found in statistical books are not valid for I, therefore the mantel-test, a Monte Carlo simulation, has to be used (Mantel 1967; Legendre & Fortin 1989). The second and third questions are answered by a correlogram. There, the time variable is divided into a small number of time intervals. For each time interval, a correlation is calculated expressing the dependence between the relevés belonging to the same time category or to all the others, respectively. Again, the resulting coefficients cannot be tested using ordinary statistical tests, but rather by Monte Carlo simulation. When the I-

values are plotted as a function of time steps, then a correlogram is obtained. As shown in Figure 11, the shape is characteristic for the type of process:
A perfect dependence over the entire time of investigation theoretically results in a straight line (line a in Figure 11). This does not happen in practice. As explained in the previous section, non-linearity in vegetation data will reduce correlation at long time intervals.
More often, correlation is strong at the beginning, then decreases and eventually vanishes. Line b is such an example. After some time, it may even return to the zero line. There, random fluctuations may prevail. If there is no dependence at all, the line will fluctuate near the zero line. This is illustrated by line c.
The data of Lippe et al. (1985) are again an interesting example. In this case, the floristic similarity is related to the 19 time steps. The overall autocorrelation, measured in terms of Moran's I is

$$I = -0.4568$$

and

$$I^2 = 0.2087.$$

The mantel-test yields an error probability of $p = 0.0260$, i.e. the probability that $I^2 = 0.2087$ is obtained is 2.6 percent if autocorrelation is zero. In Figure 12 the correlogram for time steps of 1 year is shown. Time dependence is obvious for about the first 9 years. Then dependence starts to decrease. The end of the curve is unreliable: the number of data elements (N) involved in the calculations is too low for a sound interpretation.

```
TIME STEPS                                        FROM     TO    N CORREL.
    -0.5                 0.0                 0.5
      .---.---.---.---.---0---.---.---.---.---.
 1               *        .                        0.0    1.0   18  -0.25
 2                 *      .                        1.0    2.0   17  -0.19
 3                   *    .                        2.0    3.0   16  -0.13
 4                      * .                        3.0    4.0   15  -0.07
 5                       *.                        4.0    5.0   14  -0.03
 6                        *                        5.0    6.0   13  -0.02
 7                        . *                      6.0    7.0   12   0.04
 8                        .   *                    7.0    8.0   11   0.09
 9                        .    *                   8.0    9.0   10   0.11
10                        .    *                   9.0   10.0    9   0.11
11                        .    *                  10.0   11.0    8   0.12
12                        .   *                   11.0   12.0    7   0.09
13                        .   *                   12.0   13.0    6   0.09
14                        .  *                    13.0   14.1    5   0.07
15                        .  *                    14.0   15.0    4   0.07
16                        .     *                 15.0   16.0    3   0.13
17                        .                       16.0   17.0    2
```

Figure 12. Correlogram of the data of Lippe et al. (1985).

5.3. FUZZY ORDINATION: FITTING RELEVÉS TO A TIME VECTOR

The analysis of long-term vegetation data is nothing else but fitting a multivariate data set - the vegetation data - to a continuous variable - the time gradient. Roberts (1986) published an ordination method that precisely fulfils this task. For this reason it can be recommended for application in time series as well. The following questions can be answered:

- Is there a vegetation gradient related to a continuous environmental variable measured in the field?
- How good is the fit between the environmental variable and the vegetation gradient?
- In case of correspondence, are there relevés which do not follow the trend?

The computations rely on the fuzzy set theory (Feoli & Zuccarello 1988). There, the basic concept is the degree of belonging. The following steps are to be performed (Roberts 1986):
The site factor x (in the present case the time variable) is "fuzzyfied", i.e. transformed in order to range from 0 to 1:

$$y(i) = \frac{x(i) - x(\min)}{x(\max) - x(\min)} \quad (5)$$

The vector y now contains the degree of belonging of the relevés to high values of x. The lowest value is 0, the highest is 1. Then, the complement is taken:

$$y'(i) = 1 - y(i) \quad (6)$$

y'(i) is the degree of belonging of relevé i to low values of the environmental factor. An example is given in the following data matrix:

year	1	2	3	4
y	0	0.33	0.66	1
y'	1	0.66	0.33	0
species 1	1	1	0	0
species 2	0	1	1	1

Using the biological data (species 1 and 2 in this case), the similarities among the relevés are calculated. When using the similarity ratio (Maarel et al. 1978), a range of 0 to 1 is obtained as required for fuzzy operations. The matrix comparing the four relevés is:

$$S = \begin{matrix} 1 & & & \\ 0.5 & 1 & & \\ 0 & 0.5 & 1 & \\ 0 & 0.5 & 1 & 1 \end{matrix}$$

Then the similarity between each relevé and relevés having high values as time variable, z, is calculated. For relevé m, this is the average similarity to all remaining relevés, weighted by their degree of belonging to high values:

$$z(m) = \frac{\sum_{i \neq m} S(m,i) \cdot y(i)}{\sum_{i \neq m} y(i)} \tag{7}$$

In an analogous way the similarity of each relevé with relevés of low values in the time vector, z', is calculated:

$$z'(m) = \frac{\sum_{i \neq m} S(m,i) \cdot y'(i)}{\sum_{i \neq m} y'(i)} \tag{8}$$

Then, the two vectors, z and z', are projected onto one axis. For this, the anticommutative difference is determined (Roberts 1986) according to:

$$u(i) = \frac{1}{2}\left[1 + \left(z(i)^2 - z'(i)^2\right)\right] \tag{9}$$

Here, u is the apparent site factor and y the real or measured one. These two variables are used for plotting the relevés into an ordination. An example is shown in Figure 13, where the data of Lippe et al. (1985) are analysed. On the x-axis, there is the "apparent time", which in fact is an expression of relevé similarity. On the y-axis, there is the real time scale, adjusted to range from 0 to 1. In a first phase (a) the change in time coincides with a change in similarity of the relevés. In a second phase (b), fluctuation is independent of time. This confirms the previous findings concerning this data set.

6. Concluding remarks

Measuring change in time under different initial conditions with or without varying external impact is essential for a better understanding of the functioning of natural systems. The present contribution concentrates on the analysis of time series data. However, data collection is always preceded by parameter selection. This selection expresses the way in which the systems are perceived. At the stage of interpretation, the results of data analysis are therefore confined by the selected parameters. The most efficient path in the analysis of vegetation change may start with a formulated hypothesis about succession, followed by an appropriately worked out sampling design and subsequent data collection. The results of data analysis then confirm or question the initial idea. There is no statistical test available which could be used to accept or reject a hypothesis in a strict sense under such complex conditions.

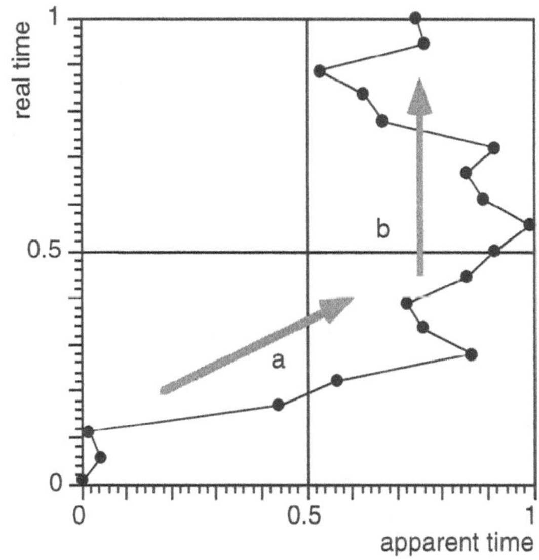

Figure 13. Fuzzy ordination (Roberts 1988) of the data of Lippe et al. (1985).

There are circumstances where changes are expected to occur rarely. Environmental monitoring is such a case. Sometimes, a hypothesis about the succession process can only be stated after a change has occurred already. The only way to proceed in such a situation may be to have "a closer look" at the data. Multivariate pattern analysis offers a wide selection of methods for this purpose and it may therefore be the initial or even the only possible path to follow.

Acknowledgements

The author expresses his thanks to Peter Longatti and Dr. John Innes for comments, corrections and suggestions.

References

Anand, M. 1997. Towards a unifying theory of vegetation dynamics. The University of Western Ontario, Canada, PhD thesis. (Manuscript.)

Austin, M.P., Nicholls, A.O., Doherty, M.D. & Meyers, J.A. 1994. Determining species response functions to an environmental gradient by means of a ß-function. J. Veg. Sci. 5: 215-228.

Begon, M., Harper, J.L. & Townsend, C.R. 1996. Ecology. 3rd ed. Blackwell Science, Oxford.

Clements, F.E. 1916. Plant Succession: Analysis of the Development of Vegetation. Carnegie Institute of Washington Publication No. 242. Washington DC.

Connell, J.H. & Slatyer, R.O. 1977. Mechanisms of succession in natural communities and their role in community stability and organisation. American Naturalist 111: 1119-1144.

Dale, M.B. 1980. A syntactic basis of classification. Vegetatio 42: 93-98.

De Patta Pillar, V. & Orlóci, L. 1993. Taxonomy and perception in vegetation analysis. Coenoses 8(1): 53-66.

De Wit, C.T. & Goudriaan, J. 1974. Simulation of ecological process. Wageningen, Centre for Agricultural Publishing and Documentation.

Dierschke, H. 1996. Sukzession, Fluktuation und Stabilität von Flussufer-Gesellschaften. Ergebnisse 15-jähriger Dauerflächen-Untersuchungen an der Oder (Harz-Vorland). Braunschweiger Kolloquium zur Ufervegetation von Flüssen. Braunschweiger Geobotanische Arbeiten 4: 93-116.

Ellenberg, H., Weber, H.E., Düll, R., Wirth, V., Werner, W. & Paulissen, D. 1991. Zeigerwerte von Pflanzen in Mitteleuropa. Scripta Geobotanica XVIII: 1-248.

Feoli, E. & Zuccarello, V. 1988. Syntaxonomy: A source of useful sets for environmental analysis? Coenoses 3: 141-147.

Gauch, G.H. 1982. Multivariate analysis in community ecology. Cambridge University Press, Cambridge.

Gleason, H.A. 1939. The individualistic concept of the plant association. Amer. Midl. Nat. 21: 92-110.

Green, R.H. 1979. Sampling Design and Statistical Methods for Environmental Biologists. Wiley & Sons, New York, Chichester, Brisbane, Toronto.

Grime, J.P. 1979. Plant Strategies and Vegetation Process. Wiley, Chichester.

Harcombe, P.A., Palmer, M. & Mucina, L. 1997. The importance of spatial and temporal perspectives for understanding vegetation pattern and process: Introduction. J. Veg. Sci. 8: 1-162.

Landolt, E. 1997. Ökologische Zeigerwerte zur Schweizer Flora. Veröff. Geobot. Inst. ETH, Stiftung Rübel, Zürich 64: 1-208.

Legendre, P. & Fortin, M.-J. 1989. Spatial pattern and ecological analysis. Vegetatio 80: 107-138.

Lippe, E., De Smidt, J.T. & Glenn-Lewin, D.C. 1985. Markov models and succession: A test from a heathland in the Netherlands. J. Ecol. 73: 775-791.

Maarel, van der, E., Janssen, J.G.M. & Louppen, J.M.W. 1978. Tabord, a program for clustering phytosociological tables. Vegetatio 38: 143-156.

Mantel, N. 1967. The detection of disease clustering and a generalized regression approach. Cancer Res. 27: 209-220.

Marsili-Libelli, S. 1989. Fuzzy clustering of ecological data. Coenoses 4(2): 95-106.

Moral, R. & Bliss, J.A. 1993. Mechanisms of primary succession: insights resulting from the eruption of Mount St. Helens. Advances in Ecological Research 24: 1-66.

Moral, R. 1983. Initial recovery of subalpine vegetation on Mount St. Helens, Washington. Am. Midl. Nat. 109: 72-79.

Orlóci, L. 1978. Multivariate Analysis in Vegetation Research. 2nd ed. Junk, The Hague.

Orlóci, L., Anand, M. & He, X. 1993. Markov chain: a realistic model for temporal coenosere? Biom. Praxim. 33: 7-26.

Podani, J. 1984. Spatial processes in the analysis of vegetation: theory and review. Acta Botanica Hungarica 30: 75-118.

Roberts, D.W. 1986. Ordination on the basis of fuzzy set theory. Vegetatio 66: 123-131.

Scott Urquhart, N., Scott Overton, W. & Birkes, D.S. 1993. Comparing sampling designs for monitoring ecological status and trends: Impact of temporal patterns. pp. 72-84. In: Barnett, V. & Turkman, K.F. (eds), Statistics for the Environment. Wiley & Sons.

Sokal, R.R. 1986. Spatial data analysis and historical processes. pp. 29-43. In: Diday, E. et al. (eds), Data analysis and informatics, IV. Proceedings of the Fourth International Symposium on Data Analysis and Informatics, Versailles, France, 1985. North-Holland, Amsterdam.

Ter Braak, C.F.J. & Looman, C.W.N. 1987. Regression. pp. 29-77. In: Jongman, R.H.G., ter Braak, C.J.F. & van Tongeren, O.F.R. (eds), Data analysis in community and landscape ecology. Pudoc, Wageningen.

Tilman, D. 1988. Plant Strategies and the Dynamics and Structure of Plant Communities. Princeton University Press, Princeton, NJ.

Usher, M.B. 1981. Modelling ecological succession, with particular reference to Markovian model. Vegetatio 46: 11-18.

Wagner, H.H. & Wildi, O. 1997. Markov chains and vegetation monitoring. Student 2(1): 13-26.

Watt, A.S. 1947. Pattern and process in the plant community. J. Ecol. 35: 1-22.

Whittaker, R.H. 1953. A consideration of climax theory: the climax as a population and pattern. Ecological Monographs 23: 41-78.

Whittaker, R.H. 1967. Gradient analysis of vegetation. Biol. Rev. 42: 207-264.

Whittaker, R.H. 1978. Direct gradient analysis. pp. 7-50 In: Whittaker, R.H. (ed), Ordination of Plant Communities. Junk, The Hague.

Wildi, O. 1977. Beschreibung exzentrischer Hochmoore mit Hilfe quantitativer Methoden. Veröff. Geobot. Inst. ETH, Stiftung Rübel 60: 1-128.

Wildi, O. 1988. Linear trend and noise in multi-species time series. Vegetatio 77: 51-56.

Wildi, O. 1990. A multiple scale sampling design for long term monitoring. Global Natural Resource Monitoring and Assessments: Preparing for the 21st Century. Proceedings of the International Conference and Workshop, Vol. 2. Bethesda, U.S.A.: 975-982.

Wildi, O. & Orlóci, L. 1996. Numerical exploration of community patterns. A guide to the use of MULVA-5. 2nd ed. SPB Academic Publishing bv, The Hague.

Winkler, E. & Klotz, S. 1997. Long-term control of species abundances in a dry grassland: a spatially explicit model. J. Veg. Sci. 8: 189-198.

Zimmermann, H.J. 1985. Fuzzy Set Theory and Its Applications. Kluwer-Nijmhoff Publishing, Boston.

ADMINISTRATION LEVELS AND TASKS OF NATURE CONSERVATION EFFICIENCY CONTROL

MARKUS LABASCH & ANNETTE OTTE

Landscape Ecology and Landscape Planning, Heinrich-Buff-Ring 26 - 32, D-35392 Gießen, Germany

Keywords: Expenditure control, evaluation procedure, implementation control, effect control, goal definition, condition control

Keywords: Aufwandskontrolle, Bewertungsverfahren, Umsetzungskontrolle, Wirkungskontrolle, Zieldefinition, Zustandskontrolle

Abstract

The application possibilities of, and the methodology for the development of various nature conservation control procedures are outlined and systematized for practical application; the goal framework for the future development of landscapes (and parts thereof) is stipulated in nature conservation models. These goals are formulated within the area of responsibility of landscape planning. Controls differentiated according to scale (implementation control, condition control, effect control, expense control) are intended for the review of effective nature conservation measures on various administrative levels (implementation level, condition level, effect level, expense level) and for evaluating these as to the fulfilment of the set goals, as previously stipulated.

Kurzfassung

Die Einsatzmöglichkeiten und die Vorgehensweise zur Entwicklung verschiedener naturschutzfachlich wirksamer Kontrollverfahren werden geschildert und für den Anwendungsbereich systematisiert: In naturschutzfachlichen Leitbildern ist der Zielrahmen für die zukünftige Entwicklung von Landschaften (und Teilen davon) festgelegt. Formuliert werden diese Ziele innerhalb des Aufgabengebietes der Landschaftsplanung. Maßstäblich gestaffelte Kontrollfahrten (Umsetzungskontrolle, Zustandskontrolle, Wirkungskontrolle, Aufwandskontrolle) haben die Aufgabe, durchgeführte naturschutzfachlich wirksame Maßnahmen auf verschiedenen administrativen Handlungsebenen (Umsetzungsebene, Zustandsebene, Wirkungsebene, Aufwandsebene) zu überprüfen und hinsichtlich der vorher festgelegten Zielerreichung zu bewerten.

Burga & Kratochwil (eds.), BIOMONITORING, 41-60

1. Introduction

Whenever the national economic situation worsens, public service tasks are made subject to review. Thereby, political issues and goals which were formerly very highly valued - even if not generally - are called into question, at least with respect to their existing extent. These issues include environmental and nature conservation, which first became incorporated as a national goal in the Basic Law for the Federal Republic of Germany (Article 20a) only a few years ago.
From the necessity of political goals related to nature conservation - and hence to justify the expenditure for nature conservation measures while employing the declining financial resources as usefully as possible - comes the need to **review goal priorities and the efficiency of nature conservation measures**. This applies not only to public institutions, but also to private organizations if these receive public grants or carry out public contract work.
Arguments for, and terms and areas of responsibility of efficiency control procedures are presented in the present paper. The necessity of the development of regional models and object-related goal definitions is discussed, as well as issues of evaluation and the measurement of success of nature conservation measures.

2. Control procedures

2.1. DEFINITIONS

Goal: Definition of terms: efficiency control, (bio-)monitoring and continuous observation

The observation by Goldsmith (1991), "monitoring has become fashionable", is also valid for the terms success control and efficiency control. This has resulted in a very variable usage of these terms; a clarification of this usage is therefore necessary.
Hellawell (1991) makes a clear distinction between **monitoring** and survey or surveillance. He states that 'survey' is a temporally restricted study, while 'surveillance' is an 'extended programme of surveys' for the analysis of temporal variations in values or objects (e.g., to distinguish between short-term fluctuations in the vegetation cover and long-term successional trends). By contrast, Hellawell (1991) defines "monitoring" as: "Intermittent (regular or irregular) surveillance carried out in order to ascertain the extent of compliance with a predetermined standard or the degree of deviation from an expected norm."
As a consequence, monitoring is distinguished in that, before the commencement of a study, a norm or standard must be defined with which the study results can be compared (for the setting of norms and standardization in nature conservation cf. 3.2). Similarly, a monitoring procedure requires that there exists at least an approximate conception of the results to be achieved. According to Hellawell (1991), monitoring is goal-orientated, i.e., in the case of the goal being unattained, alternative or additional measures to be taken for the realization of the goal are foreseen. The statement by Hellawell (1991) is also important: "One activity for which monitoring is not relevant is research, although one may encounter examples where this is claimed." The following example illustrates

this point: when a land use in an area changes, research into the unknown or at least uncertain consequences of the change cannot be regarded as monitoring. Moreover, within the scope of a research project, the previous land use would have to be sustained within a part of the area in order to be able to relate the observed effect back to the changed land use. When however the land use is altered in order to produce a certain, known or at least probable effect, the control of the development towards this result (the norm or standard) may be regarded as monitoring.
Monitoring is often wrongly equated with **continuous observation** (Plachter 1991; Reich 1994) or with **long-term observation** (Mühlenberg 1993). Here, mention is often made of continuous monitoring, especially for the control of the stand development of chosen species. According to Goldsmith (1991) and Hellawell (1991), however, these forms of 'monitoring' should be called 'survey' or 'surveillance', insofar as 'inventory taking' or 'non-goal-orientated', 'pure' acquisition of knowledge are to the fore (generally non-recurring, but frequently taking place over a period of several years). These studies, e.g., on the population biology of individual species, nevertheless often represent the basis for being able to set the necessary standards or norms for monitoring. It is the knowledge of regular stand fluctuations which initially permits the setting of limiting and comparative values, with which the (generally not continuously recorded) data of a monitoring program must be compared in order to distinguish between short-term fluctuations and long-term trends.

Long-term observation can be distinguished from continuous observation as follows:

a) **Long-term observation:** repeated observation of objects (populations/individuals) over a longer period of time (temporal continuity), e.g., locations of animals.

b) **Continuous observation:** repeated observation of occurrences (processes/ developments) at a fixed location (spatial continuity), e.g., botanical continuous sampling plots.

The terms success and efficiency are outlined below. The economic term **efficiency** denotes the wise use of scarce factors. "If no resources are to be wasted, there must, according to the 'economic principle', be either a given set of goals with minimal costs, or otherwise a maximal degree of goal realization must be attained with a given level of expenditure" (Hampicke 1991).

Because the means for implementing nature conservation tasks are limited, priorities must be set in order to employ them efficiently. To this end, the following conditions must be fulfilled:

- the goal to be attained must be clear,
- knowledge of suitable means and measures for goal attainment must exist, and
- the costs of the respective measures must be calculated.

"To operate efficiently does not mean that costs must be avoided unconditionally; rather, it means weighing costs and benefits against one another for all measures," since costs can be understood as missed benefits or the non-realization of other goals (Hampicke 1991).

For the calculation of efficiency in politico-economic terms, a cost-benefit analysis is necessary. According to Hampicke, however, its realization within the scope of nature conservation and agriculture is fraught with difficulties, because there are often no efficiency costs known. The costs involved are not scarcity costs, but are instead subject to a multitude of distortions, such as state subventions. The uncertainty of a possible future value (option value) is a further difficulty in the economic evaluation of natural resources.
A measure can be considered successful when the set goal is attained to a high degree, i.e., success can be measured by the **degree of goal realization**! Hence, the review of goal attainment may be called **success control**. A successful measure can, however, be associated with a very high expenditure (financial and/or personal). If this expenditure is unreasonable, or if the goal can be attained just as well or even better with less expensive measures, the measure is inefficient. In using the term **efficiency control**, it is therefore decisive that the attained achievements (realized goals, aims and objectives) are proportional to the costs involved. Consequently, measures which only attain the goal to a limited degree, but which are also only associated with a small expenditure, can likewise be regarded as efficient as more expensive measures with a correspondingly higher degree of goal attainment, although the latter are the more successful.
From the above reflections, it cannot however be concluded that the primary task of efficiency controls should be to minimize expenditure for nature conservation. "The practical goal should not be to maximize efficiency to the utmost, but rather to reduce inefficiency where this is easiest!" (Hampicke 1994). In this context, by far the most effective method is a ban on "activities detrimental to nature which, quite apart from ecological considerations, are uneconomic" (Hampicke 1994).
In official practice however, other definitions of the above-named terms have developed. An example is that of the federal state North Rhine-Westphalia, where the 'Landesanstalt für Ökologie, Bodenordnung und Forsten/Landesamt für Agrarordnung' ('State Institution for Ecology, Soil Utilization and Forests/State Agency for Agricultural Utilization NRW' [LÖBF]) is responsible for tasks of state-wide biomonitoring and efficiency control. According to Schmidt (1996), biomonitoring serves the recording and documentation of environmental changes and pressures with long-term studies. According to the definition by Hellawell (1991), this form of monitoring must be called long-term observation (surveillance) because no comparison is made with either a norm or a standard. These can, however, be derived from the results of such long-term observations. "Within the scope of efficiency controls, measures of nature conservation and landscape planning, the implementation of these, and the expenditure required are analyzed and, if appropriate, suggestions for improvement are derived from the study results" (Schmidt 1996). The term efficiency control is used in the LÖBF as a collective term "for all studies and analyses on effects and degrees of effectiveness of nature conservation; there is no differentiation between success, effectiveness and efficiency" (Weiss 1996).

Studies carried out prior to planning or implementation can be classified as follows:

a) Site analysis: 'Non-specific' study of biotic and abiotic site factors as the first step of object-related goal development (cf. 2.3). With respect to goal development and subsequent planning of measures, site analysis gives an indication of what is possible under the present site conditions. However, it should also state the possibilities of, or necessities for site change, should the desired goal be unattainable under present circumstances.

b) **Documentation:** When the object-related goals, the masterplan (containing measures) and the monitoring concept are set up (cf. 2.3), documentation of the status quo or the previous state of the object follows, i.e., documentation of the year zero of the monitoring program. Here, only those parameters required for the assessment of goal attainment are recorded, because these and only these are necessary for later goal analyses.

2.2. EFFICIENCY CONTROLS - WHY?

Goal: Necessities of, and argumentation for efficiency controls: scarcity of means and inadequate knowledge of the effects of measures.

The demands for efficiency controls for nature conservation measures can be related back to two issues (Blab & Völkl 1994):

a) Do the public means spent on nature conservation result in demonstrable success with justifiable expenditure? → Review of the efficiency of means employment.

b) How do the nature conservation 'impacts' (measures) act on the environment and on natural resources? → Review of the effectiveness of the measures.

The efficiency of means employment can be determined at the levels of **administration** (planning expenditure, organizational structure, etc.), **protected** objects (protection, development, management and maintenance measures as well as goal definitions) or **political economics** (cost-benefit analyses of programs and strategies). Only efficiency control at the object level is considered below.

According to Weiss (1996), efficiency controls can serve the following goals:

a) to guarantee goal attainment,

b) to guarantee a high degree of effectiveness,

c) to guarantee successes,

d) to optimize goal finding and priority setting in nature conservation.

According to Maas & Pfadenhauer (1994), efficiency controls have "two priority goals:

- Success control: proof of the expediency of the measures
- Gain of experience: comparison of the conditions of the area with the goal conception."

While the former goal is related more to the control of the measures implemented, the latter is primarily concerned with the review of goal conceptions, also with respect to external influences independent of the measures.

2.3. AREAS OF RESPONSIBILITY OF NATURE CONSERVATION EFFICIENCY CONTROLS

Goal: Levels and forms of efficiency controls are presented.

A variety of tasks are associated with the term 'efficiency control' in practical nature conservation. Wey (1994) distinguishes between four areas of responsibility in major federal nature conservation projects. Weiss (1994) also classifies the forms of efficiency controls carried out by the LÖBF into four similar areas, for which both authors have used the following terms (Weiss 1994; Wey 1994):

a) implementation control, or control of measures

b) condition or stand control,

c) effect control (both authors), and

d) expenditure or economic control.

These areas of responsibility are discussed in the following (synonyms were taken largely from Eekhoff et al. [1984]):

Implementation control/control of measures
Control of planning execution, determination of planning deficits and grounds for deviations from planning stipulations. Predominantly administrative procedures are assessed according to type, extent, costs, and time frame (Wey 1994). "When the ecological effectiveness of a usual measure is sufficiently well documented, the review of its complete and appropriate implementation in the sense of a labour-economic reduction of expenditure is sufficient in order to demonstrate efficiency" (Weiss 1996). For example, when a weir is to be re-flooded, the review of the functionality of the weir in the sense of an implementation control is sufficient where no other goals, e.g., concerning the restoration of certain plant communities, have been specified. The assessment is conducted, in the case of (e.g.) Scheible (1996), in a normative fashion using the three categories 'complete', 'partly implemented' or 'not implemented' on the basis of a comparison with the measures contained in the management and maintenance plans.
The following terms are used in this connection:
Approval control, control of independent variables, execution control, guideline control, implementation control, input control, instrumentation control and means control.

Condition controls/stand control
Here, the condition of an object is recorded on the basis of evident, predominantly structural and physiographic, but also simpler biological characteristics. The result is compared with a reference condition defined as a goal, or with a general minimal quality condition. Evaluation categories include, e.g., 'without deficiencies', 'with minor deficiencies', 'with clear deficiencies', and 'with severe deficiencies' (Behlehrt & Weiss 1996). When the measures have been largely carried out (implementation control) and their effectiveness or otherwise is evident (and proven in other similar cases), the condition control is sufficient for the evaluation of the measures (Schütz & Behlert

1996). Wey (1994) designates this type as stand control and regards it as the biological and landscape-ecological evaluation of the measures carried out, which are associated with impacts on the environment. This also includes the clarification of the connection between the determined developments and the impacts made (mainly on the basis of literature evaluations). This definition is somewhat wider and leads to the next type. Because of the goal definition, condition control may be regarded as monitoring in the sense of Hellawell (1991). Where the attainment of a certain standard plays only a minor role, and the primary goal of the study is the clarification of cause-effect relationships, one must, according to HELLAWELL, speak of surveillance. The goal-targeted orientation of measures distinguishes monitoring from surveillance, which serves as research orientation for the acquisition of knowledge.

Synonymous terms:
Control of dependent variables, goal attainment control, goal standard control and success control in the strictest sense.

Effect control
If the direction of effects of the measures implemented and/or their effectiveness in comparison to alternative measures are unclear, more extensive studies of the abiotic and biotic parameters most important for assessment of goal realization are necessary. Effect controls are "thus the central instrument of nature conservation efficiency control" (Weiss 1996).

Similarly used terms for effect control are:
Cause analysis, effect analysis, effectiveness analysis, effectiveness control, impact analysis and impact assessment. The control of exogenous influencing factors may also be called circumstances control, control of the (internal and external) framing conditions or 'monitoring' of exogenous influences.

Expenditure control/economy control
Both monetary as well as non-monetary comparisons of variably costly measures serve the optimization of nature conservation, both on the level of the measure as well as on the level of goal finding and priority setting (Weiss 1996). No politico-economic cost-benefit analyses are conducted by the LÖBF; nor are they included in the efficiency control proposed by Wey (1994) for major nature conservation projects.

Synonym: efficiency control in the strictest sense.

Administrative levels of the control procedures
The four above-named areas of responsibility may be classified into four administrative levels. The use of different terms for the administrative levels and for the different control procedures - which nevertheless complement one another - is necessary in order to be able to arrive at clear statements. A proposal will be made in the following, in consideration of Chapter 2.1. (Figure 1).
In the case of condition control (= success control), the implementation level is supplemented by the condition level, where the degree of goal attainment is reviewed (stand or success control). Because the term 'success control' can easily be

misunderstood in a wider sense, the terms 'goal attainment' or 'goal realization control' are used in the following.
Should in addition the effects of the measures be analyzed (effect control), the procedure can also be called effectiveness control.
Expenditure control is the most comprehensive control procedure in this system, which in addition to the previously described administrative levels (areas of responsibility) also comprises the expenditure level (efficiency or economy control).

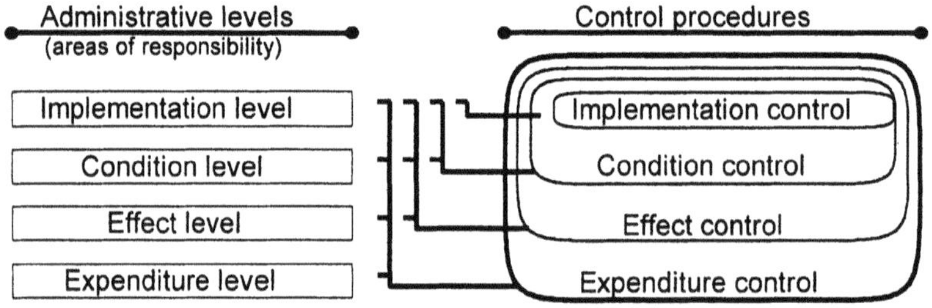

Figure 1. Administrative levels of the various control procedures. On the implementation level, only the control of implementation (= execution) of a measure takes place (implementation or measure control).

3. Goal finding for the administrative levels of the control procedures: the development of nature conservation models

3.1. GENERAL GOALS OF NATURE CONSERVATION

Goal: Goals of statutory nature conservation, higher goal systems and problems of general goal definition.

According to the Bundesnaturschutzgesetz (Federal Nature Conservation Law) Paragraph 1, the goals of nature conservation and of landscape planning are the sustainment of:

- the functionality of the environment,
- the usability of natural resources,
- the animal and plant kingdoms, as well as
- the diversity, distinctive character and beauty of nature and landscape.

Of the above resources, "the functionality of the environment is the fundamental resource encompassing all others" (Gassner 1995). The resources are to be protected, managed and maintained as the basis of the livelihood of humanity.

The 'skeleton' goals of nature conservation legislation are nevertheless so generalized that it is still not possible to derive definite, object-related goals or models from them

(Dierssen 1994; Plachter 1991). Dierssen names the following essential (more specific) aims of nature conservation:

- "the establishment of a representative reserve system, as well as the sustainment of as many ecosystem types as possible, where the functions of these have not been so severely impaired (poly- to metahemerobic systems) that the stands, as typical of a particular landscape region, have suffered severe losses with respect to structures, habitats and species, and impair neighbouring systems,
- the containment of regional species losses,
- the lasting protection of the abiotic resources soil, water and air, as well as
- the preservation and carefully regulated, resource-protecting development of the landscape."

In order to implement these goals for the entire landscape, Dierssen (1994) proposes a graduated goal system. According to this system,

- oligohemerobic habitats should be left undisturbed,
- "semi-natural" *Kulturlandschaften* ('cultural landscapes') representative of former land use periods should be preserved by sustained/appropriate management and maintenance,
- impaired ecosystems should be restored, or in the case of irreversible impacts be transformed into a new, in landscape-ecological terms, 'unobjectionable' condition, and,
- outside these areas, a resource-protecting management form should be practised, without long-term impairment of priority areas for nature conservation.

These goals are still too non-committal for an efficiency control, and can hardly be made subject to review. The goals fulfil their function as higher models, towards which the goal definitions on the lower levels must be orientated.

3.2. MODEL DEVELOPMENT

Goal: Definition of the terms model, guideline, quality objective, and quality standard. Procedures of model development.

The term model (in German, *Leitbild*) is used in various disciplines (according to Wiegleb 1997, it was originally a psychological term), and is correspondingly defined in different ways. Table 1 contains an overview on the basis of the definitions by Jessel (1994a), Kiemstedt (1991), Marzelli (1994) and Wiegleb (1997).
The exactness of goal definition increases in the sequence **models → guidelines → environmental quality objectives → environmental quality standards**; however, the area of validity (spatial, temporal and disciplinary) becomes increasingly restricted. This increased specificity is essential, because it is for example not possible to realize all of the (sometimes contradictory) goals in a relatively small area.

Wiegleb (1997) sees the great advantage of the model method in that "the model itself (regardless of how exactly or inexactly - or statically or dynamically - it is formulated) retreats into the background and the model development as social process gains in significance". Behind this statement is the realization that models cannot be stipulated by experts on the basis of their knowledge, but that they need to be legitimized by public discussion between all participating or concerned persons and institutions. **The distinction between knowledge and values is decisive** (Jessel 1994b), as well as that **between scientific perceptions and social norm setting**. "Science can only say how something can be protected, i.e., it can show the best way of attaining a defined environmental goal. The decision on exactly what and how much of nature should be protected is, however, ultimately a value decision to be made by society, one which cannot in fact be made objectively" (Jessel 1994a).

Table 1. Model, guideline, environmental quality objectives and environmental quality standards (after Kiemstedt 1991; Jessel 1994a; Marzelli 1994 and Wiegleb 1997).

Term	Definition	Examples
Model	General (graphic) goal conceptions, on which there should be largely agreement; without spatial, disciplinary (sectoral) or temporal definition.	Restoration of the *Kulturlandschaft* of around 1850 (nevertheless controversial).
Guideline	Spatially or sectorally defined administrative principles.	Reintroduction of salmon into the Rhine.
Environmental quality objectives	Particular sectorally, spatially and (if appropriate) temporally defined qualities of resources, potentials or functions which can be preserved or developed in specific situations (also called developmental objectives).	Minimal requirements of near-natural water: Water Body Resource Class II as well as passableness for migratory fish.
Environmental quality standards	Quantified environmental quality objectives, very precise evaluation methods of varying liability (limiting, guideline, orientation and discussion values) and function (protection = ban on deterioration, provision = order of minimization).	50 mg/l nitrate as limiting value for drinking water; a longer tradition exists in abiotic resource conservation, because such values are easier to stipulate!

Dierssen refers to a further problem in model development:
"**Moreover**, in the development of models and goal hierarchies, **criteria for the assessment of areas** (e.g., the presence of rare and/or endangered species, species diversity or near-naturalness of an area) **or instruments for goal implementation** (e.g., newly established small water bodies as [potential] habitats and newly established linear networking structures as [potential] dispersal areas for species] **become ends in themselves**" (Dierssen 1994; author's own emphasis).

This problem in model development of an inadequate distinction between goals, measures, facts and values leads to the question of nature conservation evaluation, in which the distinction made between facts and values is likewise often inadequate (Wiegleb 1997). As examples of facts which are used as evaluation criteria without expressed attribution of values, Wiegleb lists (amongst others) species number, diversity of habitats, area size, rarity, and number of endangered species.

3.3. EVALUATION

Goal: Differentiation between 'ecological assessment' and 'nature conservation evaluation'. Classification of evaluation procedures.

The difference between knowledge and values, important for model development, must be considered in nature conservation evaluation. To this end, Erz (1986) and Plachter (1991) argue for a differentiation between ecology as a natural science and "nature conservation as an applied discipline", because the latter has a value dimension.

According to Wiegleb (1997), there are four meanings behind the term 'evaluation':

- 'Analysis' = evaluation of data,
- 'Assessment' = assessment of data without explicit value setting,
- 'Ranking' = relative comparison on the basis of value-giving criteria,
- 'Target-Status Quo Comparison' = evaluation in a stricter sense (e.g., standardization).

According to the above, analysis and assessment can be regarded as **'ecological evaluation'**; by contrast, ranking and target-status quo comparison* can be considered as **'nature conservation evaluation'**. In order to be better able to differentiate between these two 'evaluations', the term **'ecological evaluation'** instead of **'ecological assessment'** is used in the following. This is the prerequisite for nature conservation evaluation.

According to Wiegleb (1997), 'ecological assessment' can sometimes result in value judgements being derived from untransformed facts, so that a 'status quo' can be turned into a 'target' (naturalistic fallacy). The "hit list of evaluation" compiled by Usher (1994) contains numerous criteria, e.g., species diversity, which as "factual-ecological realities (...) must first be transformed into values" (Wiegleb 1997).

A logical evaluation should be guaranteed by a formal, and - with respect to content - structured procedure, which theoretically "should be so operationalized that it can be programmed in the form of an expert system" (Wiegleb1997).

A system of nature conservation evaluation procedures has been prepared by Wiegleb(1997). According to this, non-monetary procedures, which fulfil the demands of the model method, must be "based on explicit values and goal values which are defined as accurately as possible" (Wiegleb 1997). These aspects are listed in the left column of Table 2; the opposite terms are listed in the right column.

---

* cf. condition control in Chapter "*Condition Controls/Stand Control*" and target-status quo comparison in Chapter "*The Target-Status Quo Comparison*"

The differentiation into sectoral vs. synthetic, nominally/ordinally scaled vs. numerical, as well as typological vs. object-orientated evaluation procedures is not so easy to make as in the case of the previously named aspects (Table 2). For one thing, the decision as to whether or not the one or other procedure of the model method is appropriate is dependent on the individual case, while there are also transitional and mixed forms (Wiegleb 1997).

Table 2. Dichotomies of nature conservation evaluation procedures (after Wiegleb 1997, listed below in tabular form).

Aspect	Opposite aspect
Definite goal values: Previous stipulation of limiting values, minimal requirements, etc. (environmental quality standards according to Table 1). Example: Minimal population size	***Without definite values*:** These procedures attempt a relative comparison of objects. Example: Application of (a) and (b) (see below) for the evaluation of biotope complexes
Non-monetary scale: Illustration of the value on scales which are derived from the resources themselves and their interrelationships.	***Monetary scales:*** Illustration of the value on an external, generally accepted scale (monetary value). Example: Compensation tax in Hesse (a)
Explicit attribution of value: The facts are explicitly transformed into values (transformation rule). Example: Condition-value relationship (see below)	***Implicit attribution of value:*** Implicit values are attributed to scientific facts. Example: Meadow/pasture evaluation method according to KAULE (1991) (b)
Sectoral assessment: Because this procedure can only be valid for the assessed resource, it does not permit a comprehensive evaluation of the chosen study area.	***Synthetic (global) procedures:*** Combination of various (sectoral) part evaluations to give a total value with 'global' validity.
Nominal or ordinal scales: Rough classificatory (good-medium-poor), nominal (near-natural, semi-natural, minimally natural, unnatural) and ordinal (1-2-3-4) evaluation. Individual values may not however - in a strict mathematical sense - be combined with one another (addition/multiplication).	***Cardinal scales:*** Numerical procedures permit mathematical combination; they can however feign an accuracy which is in fact absent from the method or the data.
Type evaluation: Uniform evaluation of all objects of a type, independent of their specific character, e.g., definite biotope values which are always valid, and problematic if they are to be valid for different landscape regions in the same way.	***Object evaluation:*** Evaluation of the characteristics of objects (e.g., the number of Red Data Book species), independent of the value of the type.

The 'synoptic evaluation procedure' proposed by Plachter (1994) can, in the sense of Wiegleb, be classified as a non-monetary, explicit procedure with definite goal values (in which case type parameters are modified by object parameters), which uses both cardinal and (on practical grounds) nominal/ordinal value attributions. It is a quasi-'multi-sectoral' procedure, which can however be transformed into a synthetic one if required (e.g., impact-compensation procedure) (Plachter1994).
Plachter (1994) expressly distinguishes between type level and object level; this is particularly important for the logic of the evaluation. All objects (e.g., biotopes) of the same type are assigned a basic value for a defined study area (**type value = T**) which **reflects the relative value specified in the models** for the considered study area (e.g., regional landscape unit). This basic value is independent of the expression of the studied object (Plachter 1994).
On the object level, the **object value (= O)** is determined by a comparison of the specific expression of a single natural element with the expression of all objects of the same type (in the respective study area). This expression should be recorded **using indicators** since, in practice, a 'holistic' evaluation fails due to the complexity (and sometimes the dynamics) of the objects (e.g., biotopes, landscapes) (Plachter 1994).
Because of the necessity of a clear distinction between type and object level, Plachter (1992, 1994) also suggests the use of different terms for type and object parameters, e.g., the type parameter 'rarity' corresponds to the object parameter 'abundance'. He designates the object parameters as 'value-determining criteria', too. This designation however is problematic given its strict distinction between type value and object value; here the wrong impression might be gained that the (total) value of the natural elements can only be obtained from their expression (object level).

3.4. LEVELS OF SUCCESS MEASUREMENT FOR IMPLEMENTATION, CONDITION, EFFECT, EXPENDITURE AND THEIR PROCEDURES

Goal: Different measures of success (especially those of regional planning) and their suitability for nature conservation goal control.

Kötter (1989) proposes various measures of success for the evaluation of village renewal measures. He suggests that the effects of the measures should be assessed using indicators, and subsequently be transformed into dimensionless goal fulfilment grades. It is then possible to give percentage values of goal realization and hence to compare the success of different measures.

The Target-Status Quo Comparison
This procedure requires a clear definition of nature conservation goals for the specific object, and the subsequent choice of measures with which these goals or aims can most probably, and with acceptable costs, be attained. Following the goal finding, the criteria by which the success of the measures can be assessed must be stipulated. The definition of limiting and reference values for the chosen success criteria should already be made during the planning of the measures in order to prevent a later 'softening' of the set goals (Wey 1994).

This **"normative measure of success"** (Kötter & Schäfer 1990) is designated by Marti & Stutz (1993) as a **"simple measure of success"**, since there are "still no statements regarding what proportion of the implemented measures have permitted success or failure", because the effect of external influences is not taken into account. If success control is only carried out using a goal attainment control (target-status quo comparison), "measures can be praised as highly successful, without having in fact led to the observed development. Their ineffectiveness would probably show itself in other applications under different circumstances" (Marti & Stutz 1993). An example is that of species conservation-orientated management or maintenance measures directed only towards the promotion of a target species (a practice which frequently has high public effectiveness); cf. examples from dry/rough grassland management (Quonger et al. 1994).

Both authors repeatedly emphasize the necessity of clearly defined, object-related goals and their 'operationalization' in assessable (in an optimum case, exactly measurable) parameters in order to be able to check on the goals at all.

The before-after comparison

The conditions before and after implementation of the measures are compared with one another. According to Reich (1994), this is the usual measure of success in practical nature conservation. Development tendencies in the desired direction are often already considered a success. No definite goals or success criteria need to be defined before the implementation of the measures for a 'before-after comparison'. The former are simply unnecessary, while the latter can be chosen even after evaluation of the after-condition (Wey 1994). This means that discussion on 'what constitutes success?' can be held between the concerned parties **after the measures have been realized**, instead of setting the criteria already before the planning of the measures.

The with-without comparison

For this procedure, 'null plots' are established (in an optimum case) on which no measures are carried out (Wey 1994). Success can be judged by a comparison of the development on the 'null plot' with that on the 'measure plot', on which the measures to check are implemented. As an alternative to 'null plots', Wey suggests making the comparison "by considering the hypothetical development of an area without nature conservation measures".

The with-without comparison is, especially in the case of the establishment of 'null plots', suitable for recording the actual effect of the measures carried out, and is thus an essential component of effect control.

Comparison and evaluation of measures of success

For regional planning, Eekhoff et al. (1984) regard the with-without comparison as the only plausible measure of success, "because on the one hand, the target value aspired to can arbitrarily be set high or low, and because on the other hand, it is unknown which influencing factors have affected the actual goal level, the status quo value. Therefore, the target-status quo comparison can at best be the impetus for effect control". The before-after comparison also fails to clarify which effects the actual measure has had; instead it measures 'only' the actual change, which can go into a desirable or an undesirable direction.

"One of the named measures of success is nevertheless insufficient on its own for an accurate description of the success or failure of a measure, because the statements are orientated either towards the goal norm only, or towards the actual trend" (Kötter 1989). A plausible measure of success is attained only after comparison of the actual change on the 'measure-plot' with the target conditions and the conditions without measures in the 'null-plot'.

4. Proposal for a procedure

Marti & Stutz (1993) propose a fundamental structure for an efficiency control* (Figure 2) on the basis of the analysis of numerous efficiency controls, both from conceptional research and from case studies. A fundamental distinction is made between the control of measure implementation (= implementation control) and that of the effects of the measures (condition, effect and expenditure control), as similarly proposed by Blab & Völkl (1994).

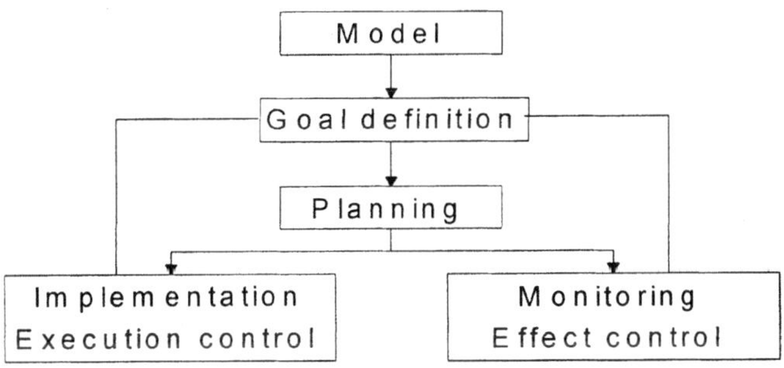

Figure 2. Modules of efficiency control and their interconnection (after Marti & Stutz 1993, modified).

This model was developed for nature conservation efficiency controls in the sense of impacts (or omissions), not however for the assessment of administrative structures or instrumentations in nature conservation politics. All forms of control (execution, condition, ecological effects and expenditure) and a goal analysis of individual objects of nature conservation measures are applied. The individual modules of an efficiency control according to Marti & Stutz (1993) are presented below (Figure 3).

* The Swiss authors use the term success control instead; this is however largely identical to the expression efficiency control, as used in the present paper. Marti & Stutz themselves note that the term efficiency control is preferred in the Federal Republic of Germany. They only use this term for expenditure control (efficiency control in the strictest sense).

4.1. MODEL, OBJECT-RELATED GOAL DEFINITION AND CO-ORDINATION

"A model should indicate the fundamental vision for a landscape region and hence a framework in which the individual object can be incorporated and assessed with respect to its significance and status" (Marti & Stutz 1993). In most cases, however, models do not come from public discussion.

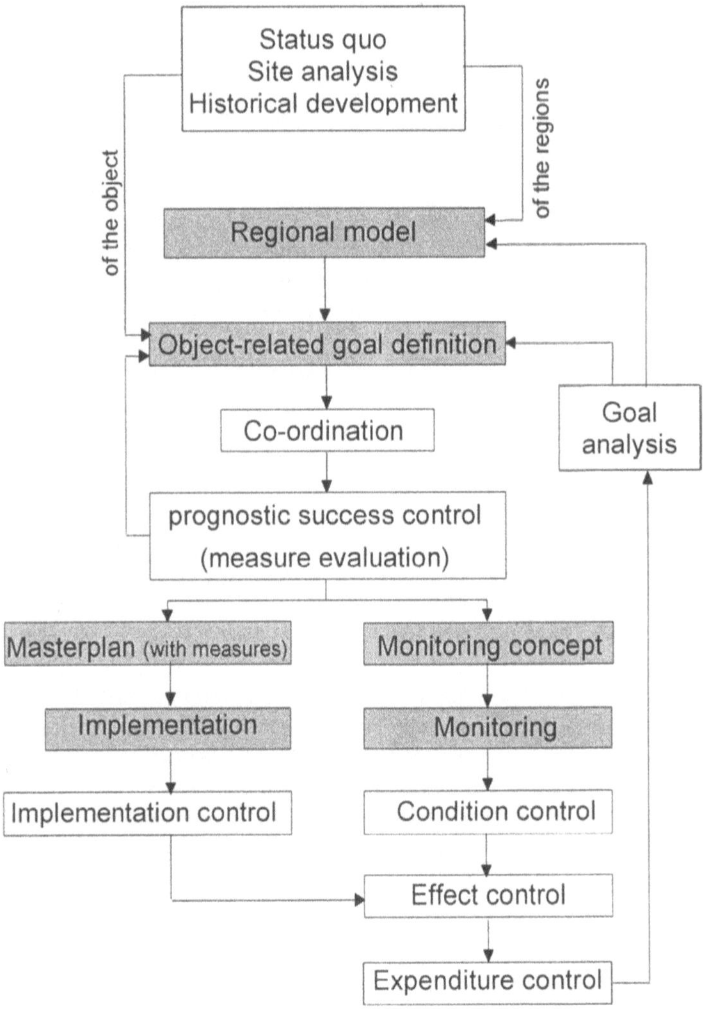

Figure 3. Development of nature conservation models and assessment of goal attainment: the modules and their procedural steps (after Marti & Stutz 1993, modified and expanded).

Still unclarified is the matter of the spatial or political level at which this should take place. One possibility would be the level of the district, although districts often do not correspond to regional landscape units.
Because of these problems, efficiency controls should initially be conducted without a model. Instead, "emphasis should be placed on a comprehensive assessment of the success or effects of individual measures" (Marti & Stutz 1993).
The most critical point however is the object-related goal definition, which can be derived from the measures and goal parameters. The goal parameters are preferentially chosen according to their suitability for assessment of the measures. The target values stipulated for the goal parameters can be regarded as environmental quality standards, while the object-related goals can be regarded as environmental quality objectives. The goals are often inadequately formulated, so that they give no **directions for administration**, and/or they can lead to 'standardized landscapes' due to a **lack of regionalization**. Also, the **time frame** in which the attainable (!) goals are to be realized should be defined, so that failure is not diagnosed too soon. Moreover, a distinction should be made between higher management goals and the objectives of the individual measures (Marti & Stutz 1993). These requirements should be fulfilled through the establishment of a goal attainment hierarchy. A practical example is the concept by Kötter & Schäfer (1990) for village renewal measures. The purpose of co-ordination is to modify the principal procedure of an efficiency control so that it is appropriate for the project; in other words, the expenditure for the control procedure must be reasonable in proportion to the expenditure for the actual measures (cf. Marti & Stutz 1993).

4.2. PROGNOSTIC EFFICIENCY CONTROL

In the planning stage, an **evaluation of the measures** should take place through a prognosis of the positive and potentially negative effects, as well as of the success prospects. With inclusion of costs, an environmental impact assessment should effectively be made for each individual measure, because these must be assessed just as critically as other impacts on the environment. "The fact that a measure has been proposed or implemented by nature conservationists is in no way a guarantee of its success" (Marti & Stutz 1993).

4.3. PLANNING OF MEASURES AND MONITORING CONCEPT

The masterplan (with measures) should specify the extent, time of implementation as well as the estimated costs of the planned measures, as defined in maintenance plans for nature conservation areas, but also in a way which is suited to the implementation control to be carried out.
The goal parameters to be studied, derived from the goals, should be stipulated in the monitoring concept, including the method of recording, data management, and evaluation methodology. A decision must also be made regarding reference areas outside the object to be recorded, or as to whether or not comparison plots ('null plots') should be established within the actual area for measures.

4.4. IMPLEMENTATION OF MEASURES AND IMPLEMENTATION CONTROL

Because it is often not possible to realize measures exactly as they were planned, the implementation of all planned and unplanned impacts (measures) must be documented as accurately as possible with respect to time, location, and procedure (work documentation). Only then does a credible effect control become possible. The implementation control measures the degree of correspondence between the masterplan and the implementation of the measures, but in no way permits an assessment of the success or efficiency of the measures.

4.5. CONDITION, EFFECT AND EXPENDITURE CONTROL

The degree of goal attainment is calculated with a target-status quo comparison (cf. 3.4).
In calculating the economicalness of the measures, the specific effect of the measures must be stipulated in relation to associated costs. In general, only an assessment of the expense of individual measures is possible. On the generally problematic nature of efficiency in nature conservation, cf. once again Hampicke (1991). These politico-economic aspects in particular are normally not considered within the scope of an efficiency control of individual nature conservation measures.

4.6. GOAL ANALYSIS

Using the results of the implementation, condition, effect and expenditure controls, both the object-related goals and the regional model must be reviewed at regular intervals.
"On this basis, it is possible to decide whether the goals were set too high or too low, whether or not the measures were inadequate, or whether or not hitherto unobserved effects occurred. In the sense of a renewed reflection on goal definition however, the further bases included must also be reviewed for changes. This concerns the current state of environmental research, the prevailing conditions with respect to abiotic environmental pressures as well as social, political and economic circumstances. As Dierssen (1992) emphasizes, it is social acceptance which determines whether or not goals and measures of nature conservation are realistic" (Marti & Stutz 1993).
However, developments in regional planning, agriculture and forestry, as well as in other areas relevant to nature conservation, must also be considered for the goal analysis. As far as possible, their impacts on nature and landscape should be recorded and quantified within the scope of **general environmental monitoring** (pollutant or biomonitoring). Particular value should be set on controls of the prevailing conditions necessary for the realization of the developed models. For example, high nitrogen deposition by air and/or water can make the conservation of ombrogenous bogs as oligotrophic habitats quite dubious. These **'external effects'** can thus be of considerable importance for model development.

The experiences gained and (possibly corrected) models form the basis for goal optimization in the case of the considered object, as well as for the planning of future nature conservation measures, including the further development of efficiency control itself.

References

Behlert, R. & Weiss, J. 1996. Landesweite Effizienzkontrolle von Kleingewässern. Eine Vorstudie. LÖBF-Mitteilungen 2: 49-55.

Blab, J. & Völkl, W. 1994. Voraussetzungen und Möglichkeiten für eine wirksame Effizienzkontrolle im Naturschutz. Schriftenr. f. Landschaftspfl. Naturschutz 40: 291-300.

Dierssen, K. 1994. Was ist Erfolg im Naturschutz? Schriftenr. f. Landschaftspfl. Naturschutz 40: 9-23.

Eekhoff, J., Fischer, K., Hellstern, G.-M., Hübler, K.H. & Wollmann, H. 1984. Begriffe und Funktionen der Evaluierung räumlich relevanter Sachverhalte. pp. 29-40. In: Veröffentlichungen der Akademie für Raumforschung und Landesplanung: Wirkungsanalysen und Erfolgskontrollen in der Raumplanung. Forschungs- u. Sitzungsberichte 154.

Erz, W. 1986. Ökologie oder Naturschutz? Überlegungen zur terminologischen Trennung und Zusammenführung. Ber. Akad. Natursch. u. Landschaftspfl. 10: 11-17.

Gassner, E. 1996. Das Recht der Landschaft. Neumann Verlag, Radebeul.

Goldsmith, B. (ed) 1991. Monitoring for Conservation and Ecology. Chapman & Hall, London.

Hampicke, U. 1991. Naturschutz-Ökonomie. Ulmer, Stuttgart.

Hampicke, U. 1994. Die Effizienz von Naturschutzmaßnahmen in ökonomischer Sicht. Schriftenr. Landschaftspfl. Naturschutz 40: 269-290.

Hellawell, J. M. 1991. Development of rationale for monitoring. pp. 1-14. In: Goldsmith, B.: Monitoring for Conservation and Ecology. Chapman & Hall, London: 1-14.

Jessel, B. 1994a. Leitbilder - Umweltqualitätsziele - Umweltstandards - Einführung in die Themenstellung und Ergebnisse. Laufener Seminarbeiträge 4: 5-10.

Jessel, B. 1994b. Methodische Einbindung von Leitbildern und naturschutzfachlichen Zielvorstellungen im Rahmen planerischer Beurteilungen. Laufener Seminarbeiträge 4: 53-64.

Kiemstedt, H. 1991. Leitlinien und Qualitätsziele in Naturschutz und Landschaftspflege. pp. 338-342. In: Henle, K. & Kaule, G. (eds), Arten- und Biotopschutzforschung für Deutschland. Berichte aus der ökologischen Forschung 4.

Kötter, T. 1989. Wirkungen und Erfolge der Dorferneuerung. Ein Konzept zur Bewertung von Dorferneuerungsmaßnahmen auf der Grundlage eines systemaren Dorfmodells und seine praktische Anwendung. Schriftenreihe des Inst. f. Städtebau, Bodenordnung und Kulturtechnik der Rheinischen Friedrichs-Wilhelms-Universität Bonn 10: 105-121.

Kötter, T. & Schäfer, G. 1990. Effizienz der Dorferneuerung - Anwendungsfälle. Schriftenreihe des Bundesministers für Ernährung, Landwirtschaft und Forsten, Reihe B: Flurbereinigung 77.

Maas, D. & Pfadenhauer, J. 1994. Effizienzkontrolle von Naturschutzmaßnahmen - fachliche Anforderungen im vegetationskundlichen Bereich. Schriftenr. f. Landschaftspfl. Naturschutz 40: 25-50.

Marti, F. & Stutz, H.-P.B. 1993. Zur Erfolgskontrolle im Naturschutz. Literaturgrundlagen und Vorschläge für ein Rahmenkonzept. Ber. Eidgenöss. Forsch.anst. Wald Schnee Landsch. Birmensdorf.

Marzelli, S. 1994. Zur Relevanz von Leitbildern und Standards für die ökologische Planung. Laufener Seminarbeiträge 4: 11-23.

Mühlenberg, M. 1993. Freilandökologie. 3rd ed. Quelle & Meyer, Heidelberg.

Plachter, H. 1991. Naturschutz. Fischer, Stuttgart.

Plachter, H. 1992. Grundzüge der naturschutzfachlichen Bewertung. Veröff. Naturschutz Landschaftspflege Bad.-Württ. 67: 9-48.

Plachter, H. 1994. Methodische Rahmenbedingungen für synoptische Bewertungsverfahren im Naturschutz. Z. Ökol. und Naturschutz (ZÖN) 3: 87-106.

Quinger, B., Bräu, M. & Kornprobst, M. 1994. Lebensraumtyp Kalkmagerrasen - 1. Teilband. Landschaftspflegekonzept Bayern, Band II.1. Bayerisches Staatsministerium für Landesentwicklung und Umweltfragen (StMLU) und Bayerische Akademie für Naturschutz und Landschaftspflege (ANL) (eds). München.

Reich, M. 1994. Dauerbeobachtung, Leitbilder und Zielarten - Instrumente für Effizienzkontrollen des Naturschutzes. Schriftenr. Landschaftspfl. Naturschutz 40: 103-111.

Scheible, A. 1996. Effizienzkontrolle im Waldnaturschutz. LÖBF-Mitteilungen 2: 63-65.

Schmidt, A. 1996. Neuer LÖBF-Schwerpunkt - Landesweite Effizienzkontrolle im Naturschutz. LÖBF-Mitteilungen 2: 1-10.

Schütz, P. & Behlert, R. 1996. Effizienzkontrolle von Biotoppflege und -entwicklungsplänen. LÖBF-Mitteilungen 2: 55-63.

Usher, M. 1994. Erfassen und Bewerten von Lebensräumen: Merkmale, Kriterien, Werte. In: Usher, M. & ERZ, W. (eds), Erfassen und Bewerten im Naturschutz. Quelle & Meyer, Heidelberg.

Weiss, J. 1996. Landesweite Effizienzkontrolle in Naturschutz und Landschaftspflege. LÖBF-Mitteilungen 2: 11-16.

Wey, H. 1994. Effizienzkontrollen bei Naturschutzgroßprojekten des Bundes. - Schriftenr. Landschaftspfl. Naturschutz 40: 187-197.

Wiegleb, G. 1997. Leitbildmethode und naturschutzfachliche Bewertung. Zeitschrift für Ökologie und Naturschutz 1: 43-62.

CHAPTER B

EXAMPLES OF APPLIED BIOMONITORING IN GERMANY AND SWITZERLAND

MONITORING RECENT VEGETATION CHANGES IN NUTRIENT-RICH BEECHWOODS IN CENTRAL GERMANY

WILFRIED HAKES

Langenbeckstr. 29, D-34121 Kassel, Germany

Keywords: Forest decline, basalt, fluctuation, redundancy analysis, N-supply, *Fagus sylvatica*

Abstract

Recent changes in the vegetation structure and site conditions of nutrient-rich beech woods were studied in a set of permanent plots uninfluenced by silviculture. Within the nine-year observation period, there has been a remarkable temporal variation in vegetation structure reflecting significant changes in soil reaction and nitrogen supply of the top soil. By means of redundancy analysis part of the spatial variation in the data set could be partialled out. A temporary increase of nitratophilous species in the herbaceous layer is accompanied by the development of a shrub layer formed largely by *Sambucus sp.* and *Rubus idaeus*. This process is induced obviously by considerable defoliation of the overstorey tree crowns as a consequence of forest decline. The increase in light availability and nitrogen supply as expressed by the increase of nitratophilous species in the ground vegetation and the shrub layer turns out to be more or less reversible. The suppressed trees of the understorey profit by the increased light level as well and show an increase in foliation, whereas the overstorey trees develop adventitious twigs and branches in the lower trunk area. As an outcome the light availability in the shrub and herbaceous layer decreases again from the middle of the observation period onward and the vegetation structure returns back close to the original situation. However, the increase in the pH of the top soil appears to be more permanent.

Burga & Kratochwil (eds.), BIOMONITORING, 61–71

1. Introduction

For some time now slow, more or less inconspicuous changes have been taking place in the vegetation structure of Central European beech woods. Many case studies have shown that essentially the entire spectrum of different types of beech woods are affected in one form or another. The causes most often cited are nitrogen emission and soil acidification. At the same time there is a direct link between this phenomenon and the still very high level of forest decline. As an example, the average leaf loss for older beech trees in Hessen has been 30% for many years. In this paper I will present the results of a nine-year monitoring case study of nutrient-rich beech woods in northern Hessen. Examples from this area (Hakes 1991) were also used by Ellenberg in his latest edition of "Vegetation Mitteleuropas mit den Alpen" (Ellenberg 1996) to direct attention to this phenomenon. The aim of this study is to find answers to the following questions:

1. Is there a directional shift in the vegetation structure of the beech stands (succession)?
2. Which site factors can explain these changes?
3. How fast are these changes?

2. Methods

The data set consists of nine permanent plots in nutrient-rich beech woods in the Habichtswald range near Kassel (north Hessen, central Germany), which were recorded four times (1988, 1993, 1995, 1997). The stands, including a buffer zone, had been uninfluenced by silviculture for several years before the observation began. The kind of silvicultural regime had been the same for all of the stands. The age of the stands ranges from 110 to 140 years. The original relevés recorded in 1988 are part of a data set which has been studied several times (see Hakes 1991, 1994, for details). The geological substratum consists of loess loam over basalt and in one case loess loam over limestone. The sample plot size was 300-400 m^2. Recording of the relevés followed the Braun-Blanquet method. Synsystematically the beech stands belong to different types of the *Galio odorati-Fagenion* (see Hakes 1991 for details). The cover-abundance values were transformed into an ordinal scale. In addition to the recording of the vegetation structure, in each relevé 6-10 soil samples were taken from the top soil (0-8 cm) repeatedly. The following site measurements were obtained: pH(H_2O), Ca-saturation, base saturation, utilizable water capacity and elevation. Weighted indicator values after Ellenberg et al. (1991) were calculated for moisture, light, nitrogen and soil reaction.
The nomenclature follows Oberdorfer (1990).
Among the site parameters recorded were elevation and the utilizable water capacity, both of which do not vary over time (variation in "space", see Borcard et al. 1992). The variations based on these variables were excluded from the study in order to concentrate on the more interesting variation over time, by analyzing the data set by means of a partial redundancy analysis (see Borcard et al. 1992; Œkland & Eilertsen 1994; Ter Braak & Verdonschot 1995 for details). The Wilcoxon-Signed-Rank-Test was used to calculate the significance of changes for the site factors.

3. Results

Despite the relatively short observation period a notable temporal variation in vegetation structure can be observed in the data set. Between 1988 and 1993 most of the relevés shift more or less to the top (-right) of the ordination diagram (Figure 1). In several plots this shift takes place between 1993 and 1995 (for example, the plot symbolized by a filled square, see Figure 2). Several relevés are already moving in the opposite direction during this time (e.g. filled circle). Between 1995 and 1997 an underlying backward trend can be seen for most of the plots (Figure 3). Generally a directional forward and backward shift can be ascertained. Moreover all of the plots show a tendency of moving to the right in the ordination diagram throughout the observation period.

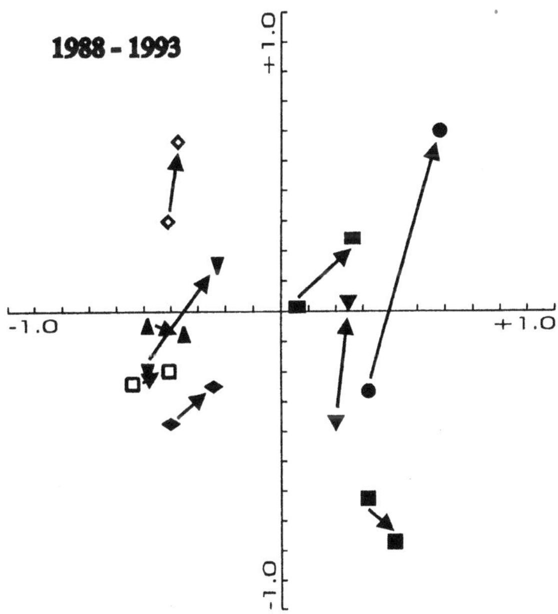

Figure 1. Shift of plots from 1988 to 1993 in the Ordination diagram of the Redundancy analysis.

The partial redundancy analysis shows the following results (Table 1): a variance of 26.0 % can be attributed to the covariables utilizable water capacity and elevation combined. The remaining environmental variables make up another 37.5 % of the variance. In other words, nearly two-thirds of the variance in the data set can be explained.

Which species are responsible for the change? Figure 4 shows the plot for the syndynamically most important species. These species are almost exclusively nitratophilous (nitrogen indicator values > 8, according to Ellenberg et al. 1991) with a high need for light (light indicator values > 6). The shifts of the relevés can be clearly related to these species with their vectors pointing in the same direction. Especially the two nitratophilous *Sambucus* species (*S. nigra* and *S. racemosa*) temporarily form a shrub layer on many of the plots. Along with these shrub species with a high degree of coverage several new herbaceous species appear temporarily which only reach low degrees of coverage (therefore they are not represented in the ordination). Examples are typical clear-cut species, such as *Calamagrostis epigejos* and *Epilobium angustifolium*, or in some cases nitrogen indicators and/or indicators of disturbance, such as *Galeopsis tetrahit, Galium aparine* and *Alliaria petiolata*, of which the latter was never found in beech woods before (Ellenberg 1996). The results of the ordination show a lower variation for *Sambucus* when compared to *Urtica* because the former varies more strongly in relation to elevation than the latter. As a result, the exclusion of elevation from the analysis is more noticeable for *Sambucus*.

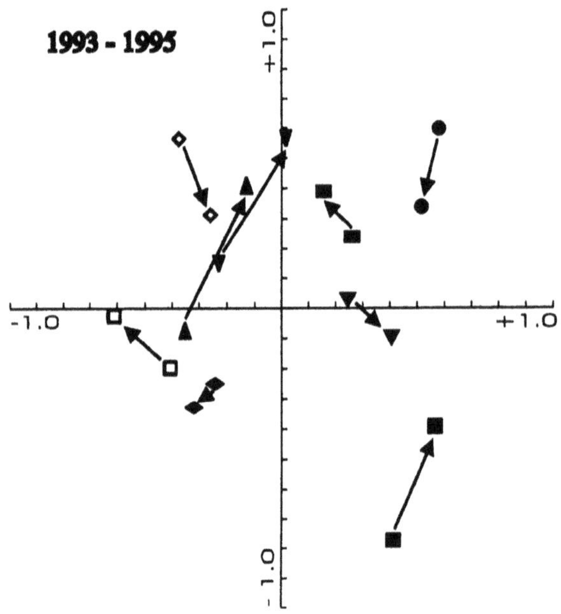

Figure 2. Shift of plots from 1993 to 1995 in the Ordination diagram of the Redundancy analysis.

The fact that the forward and backward movement of the plots followed the same direction as the vectors of the nitratophilous species is an indication of a (temporary) change in the species composition due to a change in the availability of light and the resulting nitrogen availability for ground vegetation. These indications of changes in the site conditions are further supported by the results of the environmental variables: the movement to the top right and the return movement down are indicative, also in the context of environmental variables, of a corresponding change. The first axis corresponds to a gradient of soil reaction (Figure 5) where the pH of the upper soil and the reaction figure (mR) according to Spearman show a correlation of $r=0.65$ ($p < 0.001$). A more significant correlation may be prevented by the well-known time lag in the reaction of the vegetation structure to changes in the soil chemistry (see Wilmanns 1989). The high negative correlation of the acid indicator *Luzula luzuloides* with the first axis points in the same direction (Figure 4). Thus, the shift to the right of most of the plots indicates a pH increase in the top soil.

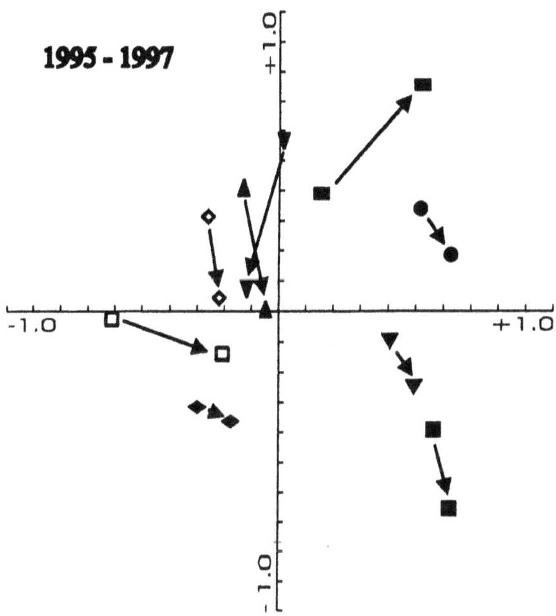

Figure 3. Shift of plots from 1995 to 1997 in the Ordination diagram of the Redundancy analysis.

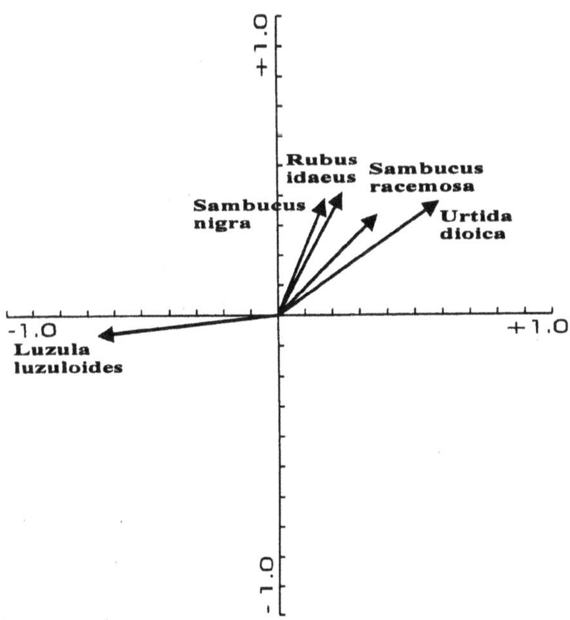

Figure 4. Syndynamically important species in the Ordination diagram of the Redundancy analysis.

The available mineralized nitrogen (as expressed by mN) also shows a distinct correlation with the first axis, while the light penetration for the field layer (as expressed by mL) is strongly correlated to the second axis. The variation of the moisture figure is not a temporal variation (see the results of the indicator value analysis below), but rather a spatial variation between sample sites which continues to exist despite exclusion of utilizable water capacity from the analysis.

Table 1. Variance explained by site variables in the Redundancy analysis.

Variable	Utilizable water capacity + Elevation (Covariables)	Other site factors: pH, Ca-ex, Base saturation, mL, mR, mF, mN	Total
Variance explained (%)	26.0	37.5	63.5

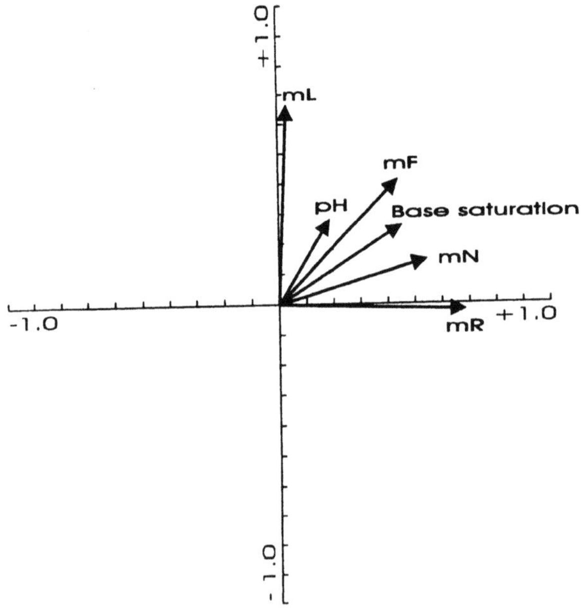

Figure 5. Site variables in the Ordination diagram of the Redundancy analysis.

pH, Ca-ex and base saturation increase steadily during the study period (Figure 6 and Figure 7). The base saturation increases from about 40% to over 50%. The decisive cation is Ca which increases by nearly 10%. The pH increases by about 0.2 points. Despite the small number of samples these increments are found to be significant according to the Wilcoxon-Test (Table 2). Possible explanations are a decrease in acidic input from the atmosphere or an input of basic cations into the upper soil by litter fall and/or leaching from leaves. In addition, there could be an influence resulting from the soil N metabolism (cf. Liu 1988).

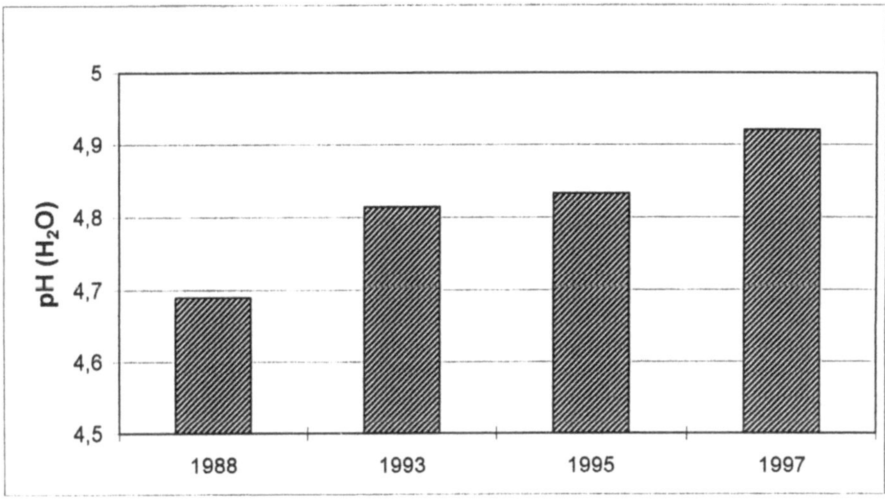

Figure 6. Change of mean pH(H_2O) from 1988 to 1997.

The reaction figure (mR) increases significantly by half a unit throughout the study period as well and is therefore indicative of a distinct change (see Böcker et al. 1983), while the light figure (mL) increases noticeably until 1993 and then decreases again (Figure 8). The nitrogen figure exhibits a similar behaviour. The moisture figure points towards an unchanged soil moisture throughout the observation period.

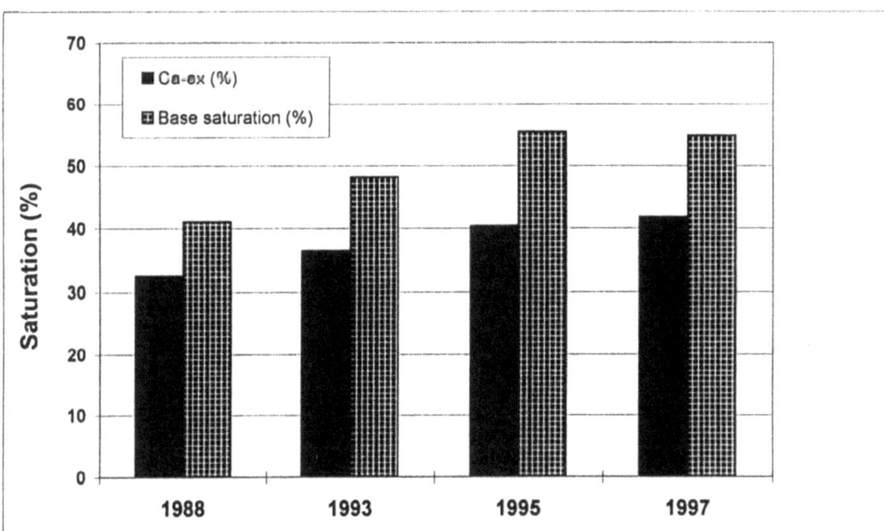

Figure 7. Changes of mean Ca-saturation and base saturation from 1988 to 1997.

Table 2. Significance of increase of parameters as calculated by the Wilcoxon-Signed-Rank-Test (Significance levels: + = $P < 0.05$; ; ++ = $P < 0.01$; +++ = $P < 0.001$; n. s. = not significant).

Parameter	1988 - 1997
pH (H_2O)	+
Ca-ex (%)	++
Base saturation (%)	++
mF	n. s.
mL	n. s.
mR	+++
mN	n. s.

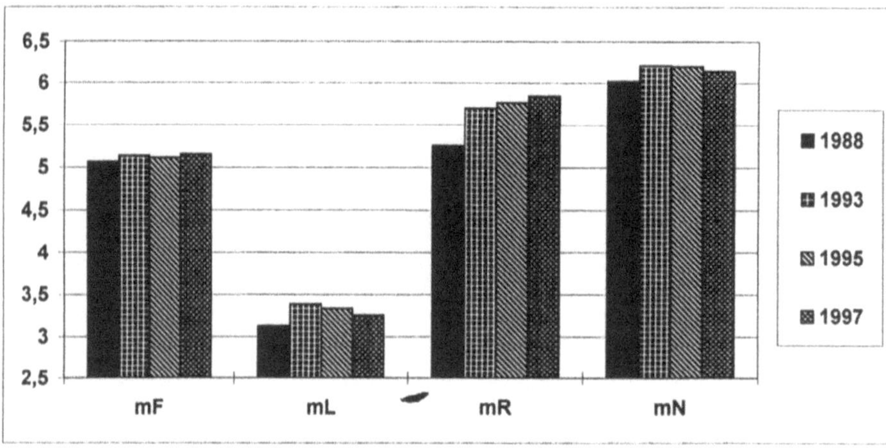

Figure 8. Changes of mean weighted indicator values after Ellenberg et al. (1991) for moisture, light, nitrogen and soil reaction from 1988 to 1997.

So what was the primary cause for the described structural changes? An answer can be found by looking at the vertical structure of the beech woods over a longer period of time. Before the appearance of the phenomenon of forest decline (let us say before 1980) mature beech woods which were not cleared or strongly thinned had the following structural characteristics (see Ellenberg 1996): The canopy was dense, resulting in a relative light intensity at ground level of about 2-4% and the lack of a shrub layer (Figure 9, top). Understorey trees had more or less died out or were dying while sub-dominant trees were severely pressured by dominant trees ("cathedral"). Canopy damage (for example, an average of 30% foliage loss for older beech trees over the last years in Hessen is reported, see Waldschadensbericht Hessen 1997 [Report on forest damage in Hessen 1997]) leads to an increase of the relative light intensity at ground level of up to 10%. The result is an atypical situation for beech woods with the occurrence of a shrub layer (Figure 9, middle). In addition, a revitalization of the formerly strongly suppressed trees of the understorey, as well as the sprouting of dormant buds in the lower trunk region of overstorey trees and the development of adventitious branches could be observed. Due to this increase of foliage mainly in the understorey, the shrub layer dies back and the vegetation structure of the field layer returns back close to the original situation (Figure 9, bottom).

Figure 9. Recent changes in the (vertical) structure of beech woods (further explanation in the text): Structure prior to crown defoliation without shrub layer (top), structure after crown defoliation; a shrub layer has developed (middle), structure after increase of foliation of the understorey trees and spring-off of dormant buds in the lower stem area of overstorey trees; the shrub layer has died back (bottom).

4. Discussion

Numerous studies in different regions of Central Europe have shown changes in the vegetation structure of nutrient-rich beech woods in recent years (e.g. Bürger-Arndt 1994). But the results of these studies are very heterogeneous. This is due mainly to the different types of beech woods investigated and the different manner in which they had been managed in the past. In most former coppice forests, for example, an increase in canopy density has been observed (e.g. Kuhn et al. 1987), leading to a definite increase in nitrogen supply. Generally this is an autogenic succession which can be enhanced by the input of atmospheric nitrogen. These changes are not dealt with here. Instead this paper deals with changes in high forest beech woods which generally have shown a reduced canopy density for many years due to forest decline. There are not many monitoring studies about this phenomenon (Gertzmann 1985; quoted from Bürger-Arndt 1994; Bürger 1991). Bürger (1991) points to a weak increase in nitrogen supply (indicated by the nitrogen figure) in submontane beech woods of the Black Forest, whereas the reaction figure decreases. The study of Gertzmann (1985) leads to the same conclusions with respect to the nitrogen supply, but the reaction figure did not change. In most studies from high forest beech woods the stands were thinned during the observation period, an influence having similar effects on the ground vegetation as the phenomenon described here (temporary increase of light intensity and nitrogen mineralization). These will not be dealt with here at any length. Almost all of these studies have found a definite increase in the nitrogen supply (e.g. Thimonier et al. 1994).

Whether the increase in available nitrogen is allogenic (from emissions), autogenic or both in nature cannot be decided for this study. The beech stands observed are generally very well supplied with soil nitrogen and therefore it is probable that it originated from within the ecosystem itself. The change in the soil nitrogen status was inferred from vegetation data only. Nevertheless this interpretation seems reliable because the validity of Ellenberg's indicator values has been verified in many studies. On the other hand, the increase of soil reaction has been proven both by means of measurements and of bioindication. The increase in the reaction figure has been shown in many studies but this alone may not be proof for an increase of the soil pH (e.g. Röder et al. 1996).

Acknowledgements

I wish to thank the Forestry Department of Hessen, especially Mr. FOR T. Arend of the State Forestry Department Kassel for issuing the permits needed for this study and for his interest in the project.

References

Böcker, R., Kowarik, I. & Bornkamm, R. 1983. Untersuchungen zur Anwendung der Zeigerwerte nach Ellenberg. Verh. Ges. Ökol. 11: 35-56.

Borcard, D., Legendre, P. & Drapeau, P. 1992. Partialling out the spatial component of ecological variation. Ecology 73: 1045-1055.

Bürger, R. 1991. Immissionen und Kronenverlichtung als Ursache für Veränderungen der Waldbodenvegetation im Schwarzwald. Tuexenia 11: 407-424.

Bürger-Arndt, R. 1994. Zur Bedeutung von Stickstoffeinträgen für naturnahe Vegetationseinheiten in Mitteleuropa. Diss. Bot. 220. Cramer, Berlin, Stuttgart.

Økland, R.H. & Eilertsen, O. 1994. Canonical Correspondence Analysis with variation partitioning: some comments and an application. J. Veg. Sci. 5: 117-126.

Ellenberg, H. 1996. Vegetation Mitteleuropas mit den Alpen. 5th ed. Ulmer, Stuttgart.

Ellenberg H., Weber, H.E., Düll, R., Wirth, V., Werner, W. & Paulissen, D. 1991. Zeigerwerte von Pflanzen in Mitteleuropa. Scripta Geobot. 18.

Gertzmann, Ch. 1985. Veränderungen der Bodenvegetation in immissionsgeschädigten Beständen an der Löwenburg im Siebengebirge. Dipl.-Arb. Forstl. Fak. Univ. Freiburg.

Hakes, W. 1991. Das Galio odorati-Fagenion im Habichtswald bei Kassel - Untersuchungen zur ökologischen Feingliederung. Tuexenia 11: 381-406.

Hakes, W. 1994. On the predictive power of numerical and Braun-Blanquet classification: an example from beechwoods. J. Veg. Sci. 5: 153-160.

Hessisches Ministerium des Innern und für Landwirtschaft, Forsten und Naturschutz 1997. Waldschadensbericht '97: 1-28.

Kuhn, N., Amiet, R. & Hufschmid, N. 1987. Veränderungen in der Waldvegetation der Schweiz infolge Nährstoffanreicherungen aus der Atmosphäre. Allg. Forst- u. Jagdz. 158: 77-84.

Liu, J. 1988. Ertragskundliche Auswertung von diagnostischen Düngungsversuchen in Fichtenbeständen (Picea abies Karst.) Südwestdeutschlands. Freiburger Bodenkdl. Abh. 21: 1-193.

Oberdorfer, E. 1990. Pflanzensoziologische Exkursionsflora. 6th ed. Ulmer, Stuttgart.

Röder, H., Fischer, A. & Klöck, W. 1996. Waldentwicklung auf Quasi-Dauerflächen im Luzulo-Fagetum der Buntsandsteinrhön (Forstamt Mittelsinn) zwischen 1950 und 1990. Forstw. Cbl. 115: 312-335.

Ter Braak, C.J.F. & Verdonschot, P.F.M. 1995. Canonical correspondence analysis and related multivariate methods in aquatic ecology. Aquatic Sciences 55(4): 1-35.

Thimonier, A., Dupouey, J.L., Bost, F. & Becker, M. 1994. Simultaneous eutrophication and acidification of a forest ecosystem in North-East France. New Phytol. 126: 533-539.

Wilmanns, O. 1989. Zur Frage der Reaktion der Waldboden-Vegetation auf Stoffeintrag durch Regen - eine Studie auf der Schwäbischen Alb. Allg. Forst- u. Jagdz. 160(8): 165-175.

BIOMONITORING - EVALUATION AND ASSESSMENT OF HEAVY METAL CONCENTRATIONS FROM TWO GERMAN MOSS MONITORING SURVEYS

UWE HERPIN[1], ULRICH SIEWERS[1], KURT KREIMES[2] & BERND MARKERT[3]

[1]Federal Institute for Geosciences and Natural Resources (Bundesanstalt für Geowissenschaften und Rohstoffe [BGR]), Stilleweg 2, D-30655 Hannover, Germany; [2]Institute for Environmental Protection Baden-Württemberg (Landesanstalt für Umweltschutz [LfU]), Griesbachstr. 1, D-76185 Karlsruhe, Germany; [3]International Graduate School (Internationales Hochschulinstitut [IHI]) Zittau, Markt 23, D-02763 Zittau, Germany

Keywords: Biomonitoring, bioindication, mosses, air-borne pollution, metals, ecological quality, evaluation model, assessment

Abstract

On the basis of two national moss monitoring programmes carried out in 1990/91 and 1995/96 this paper shows the geographical and time-related changes in lead concentrations in mosses (this element being used as an example). It proves that the lead concentrations fell considerably over large areas between 1990/91 and 1995/96. This is attributed to the continuing trend towards low-lead petrol and the closing or modernization of industrial firing systems, especially in eastern Germany.

Such bioindication techniques for environmental monitoring require more sophisticated methods of evaluating the results. This paper describes the methods available for evaluating bioindication data. "Normal values" and deviations are calculated for all elements as a basis for evaluation and shown in the form of a map, using lead as an example. With these normal values and a specific evaluation model an integrated evaluation of all the elements analyzed is performed to determine the quality of the air. With the aid of the evaluation model the data thus obtained can be included as a partial result in an overall rating of environmental quality. The objective is a combined evaluation of different ecosystems and sub-models covering all media and their representation in an "overall ecological quality map".

1. Introduction

Emissions of heavy metals cause these substances to enter ecosystems, where they may have negative effects on the organisms in relation to the concentrations of specific elements. The effects take the form of an accumulation of elements, and in the case of mercury and cadmium, for example, they may result in severe damage to the food chain.

Burga & Kratochwil (eds.), BIOMONITORING, 73-95

The accumulation of heavy metals over long periods and large areas necessitates careful monitoring of their input, movements and effects (Ernst 1974; Steubing & Jäger 1982; Adriano 1992; Kabata-Pendias & Pendias 1992; Fiedler & Rösler 1993; Klein & Paulus 1995).

In Germany, environmental monitoring is at present based largely on measuring networks in which the quality of the environment is determined by chemico-physical measuring techniques and evaluated with the aid of standards and limits. But the use of instrumental measuring techniques alone for environmental monitoring has the disadvantage that the equipment does not supply information on either the bio-availability of pollutants or their biological effects. Moreover, because of analytical difficulties the instrumental measuring techniques tend to neglect some pollutants – including many trace elements – that occur in very low concentrations in the environment. This may mean that potential effects are underestimated, as some elements are toxic even at very low concentrations.

Biological methods make a valuable contribution to investigating air-borne pollution. The biological effects of such pollution are only determined by measurements on and in the organism itself. Plants, especially, are useful indicator organisms (bioindicators/biomonitors). Certain species lend themselves to use as indicator organisms if they show measurable reactions (in the form of sensitivity or accumulation) to concentrations and the bioavailability of pollutants (Goodmann & Roberts 1971; Little & Martin 1974; Martin & Coughtrey 1982; Schubert 1985; Arndt et al. 1987; Hertz 1991; Zimmermann 1992; Markert 1993; Markert 1996; Oehlmann et al. 1996; Herpin et al. 1997; Bruns et al. 1997; Nimis & Cislaghi 1997; Schüürmann & Markert 1998). Moreover, the use of bioindicators is cheaper than instrumental methods, and this makes it possible to establish a close measuring network allowing observations throughout an area.

Mosses have a wide range of applications as indicator organisms because of their special physiological and morphological characteristics, and they have been known as heavy metal indicators for many years (Rühling & Tyler 1968, 1969). Since the 1970s their use in environmental monitoring has been systematically extended (Rühling & Tyler 1971, 1973; Ellison et al. 1976; Callaghan et al. 1978; Maschke 1981; Loetschert & Wandter 1982; Steinnes 1984; Markert & Weckert 1989; Ross 1990; Herrmann 1990; Steinnes 1993; Markert 1993; Schmidt-Grob et al. 1993; Grodzinska et al. 1993; Markert & Weckert 1994; Bruns et al. 1995; Markert et al. 1996a, 1996b; Herpin et al. 1996; Berlekamp et al. 1998; Reimann et al. 1998). In 1985 the Scandinavian countries carried out the first large-scale joint monitoring programme for assessing heavy metal inputs with the aid of mosses, and it has been repeated every five years since then (Rühling et al. 1987, 1992; Rühling & Steinnes 1998). Taking this programme as a model, Germany then carried out a similar project with the assistance of the Federal Environmental Protection Agency in 1990 (Herpin et al. 1994); in 1995 it was repeated on a national scale with the aim of showing geographic differences in heavy metal pollution and the development of these over time (trend cadaster).

But the more established methods of this kind become, the greater are the requirements in respect of evaluation and presentation of the results (Kreimes 1996). A comprehensive collection of standards and limits for the media water, soil and air is available for evaluating chemico-physical measurements in the context of environmental monitoring, but there is an almost total lack of such data for the living environment. The natural complexity of biological systems, the abiotic and anthropogenic factors acting on them and also the different modes of action of mixtures of harmful substances make it much more difficult to draw conclusions from the results of bioindication methods.

We shall use the German moss monitoring programme as an example to show what methods are available for representing and evaluating bioindication data. The basis is a bioindication technique that primarily investigates the accumulation of heavy metals within a passive monitoring system.

As examples of this method the following heavy metals in mosses are shown and evaluated both singly and in an integrated form as sub-models within an ecosystem.

Table 1. Selected sub-models (SM 1-12) from an ecosystem.

Sub-model (SM)	Sample type	Part of plant sampled	Sections sampled	Parameter
SM 1	moss	leaves/stem	green/greenish-brown	As
SM 2	moss	leaves/stem	green/greenish-brown	Cd
SM 3	moss	leaves/stem	green/greenish-brown	Cr
SM 4	moss	leaves/stem	green/greenish-brown	Cu
SM 5	moss	leaves/stem	green/greenish-brown	Fe
SM 6	moss	leaves/stem	green/greenish-brown	Hg
SM 7	moss	leaves/stem	green/greenish-brown	Ni
SM 8	moss	leaves/stem	green/greenish-brown	Pb
SM 9	moss	leaves/stem	green/greenish-brown	Sb
SM 10	moss	leaves/stem	green/greenish-brown	Ti
SM 11	moss	leaves/stem	green/greenish-brown	V
SM 12	moss	leaves/stem	green/greenish-brown	Zn

The aim of this technique is to combine various bioindication methods in a measuring system for all media and to incorporate the sub-models investigated in an integrated evaluation with the aid of the representation and evaluation systems described. An "ecological quality map" derived from this could be used to document developments in the state of the environment and the situation of various ecosystems in respect of pollution on a space/time basis.

2. Material and methods

The German and European moss monitoring programmes of 1990 and 1995 were based on the Scandinavian guidelines (Rühling 1989). The experience of sampling gained in the pilot study of 1990/91 was used in the 1995/96 programme, for example in planning the sampling period and the density of the measuring network.

2.1. MOSS SPECIES USED

The main moss species used in both monitoring programmes was *Pleurozium schreberi* (P.s.). In the absence of this species, samples of *Scleropodium purum* (S.p.), *Hypnum cupressiforme* (H.c.) and *Hylocomium splendens* (H.s.) were taken instead. In 1995/96 species such as *Rhytidiadelphus squarrosus, Brachythecium rutabulum, Brachythecium albicans, Eurhynchium praelongum, Abitinella abitinella, Hypnum jutlandicum* and *Plagiothecium undulatum* were sampled at some locations to permit a comparison and included in the evaluation there. Table 2 gives an overview of the main moss species sampled in the individual German states.

Table 2. Number of samples of the moss species ***Pleurozium schreberi*** (P.s.), ***Scleropodium purum*** (S.p.), ***Hypnum cupressiforme*** (H.c.), ***Hylocomium splendens*** (H.s.) and other species taken in the individual German states during the moss monitoring projects of 1990/91 and 1995/96.

	Moss monitoring 90/91				**Moss monitoring 95/96**				
	P.s.	**S.p.**	**H.c.**	**H.s.**	**P.s.**	**S.p.**	**H.c.**	**H.s.**	**others**
Schleswig-Holstein	19	10	12	-	11	18	13	-	4
Hamburg	4	-	-	-	4	-	-	-	-
Lower Saxony	72	6	4	-	83	27	6	-	9
North Rhine-Westphalia	30	20	4	1	33	46	4	1	-
Hesse	11	12	2	1	23	21	7	-	1
Rhineland Palatinate	26	6	-	-	18	9	2	-	2
Baden-Württemberg	7	2	49	1	4	1	69	-	-
Bavaria	98	17	2	1	80	30	5	4	-
Saarland	4	2	-	-	3	4	-	-	-
Berlin	3	-	-	-	3	-	-	-	-
Brandenburg	55	-	-	-	75	42	7	-	2
Mecklenburg-West Pomerania	37	-	-	-	24	71	12	5	1
Saxony	20	7	-	-	40	11	22	-	8
Saxony-Anhalt	20	6	-	-	38	27	3	-	18
Thuringia	10	11	1	-	27	16	24	-	8
Germany	416	99	74	4	466	323	174	10	53

2.2. MEASURING NETWORK

The measuring network for 1995/96 was improved by including the experience gained in the 1990/91 moss monitoring project, and the sub-networks (West German states, northern and southern parts of the East German states) were combined. In the West German states, sampling was carried out on the basis of a 25x25 km^2 grid as in 1990/91. In the East German states the sampling grid was 16x16 km^2.

Table 3 shows the differences in the number of moss monitoring locations between 1990/91 and 1995/96.

Table 3. Number of locations sampled in the individual German states during the moss monitoring projects of 1990/91 and 1995/96.

	Locations 1990/91	**Locations 1995/96**
Schleswig-Holstein	41	46
Hamburg	4	4
Lower Saxony	82	126
North Rhine-Westphalia	55	84
Hesse	26	52
Rhineland Palatinate	31	31
Baden-Württemberg	59	74
Bavaria	118	119
Saarland	6	7
Berlin	3	3
Brandenburg	54	126
Mecklenburg-West Pomerania	37	113
Saxony	27	80
Saxony-Anhalt	26	86
Thuringia	22	75
Germany	591	1026

Figure 1 shows the distribution of the sampling points and the occurrence of the individual species in Germany. The ratio of the number of sampling points to the area results in a density of 1.7 sites/1000 km^2 for the 1990/91 moss monitoring project and a density of 2.9 sites/1000 km^2 for the 1995/96 project (Table 4).

Table 4. Densities of the measuring networks in Germany per 1000 km² in 1990/91 and 1995/96.

	Sampling points per 1000 km^2 90/91	**Sampling points per 1000 km^2 95/96**
Germany	1.7	2.9
West Germany	1.8	2.2
East Germany	1.5	4.6

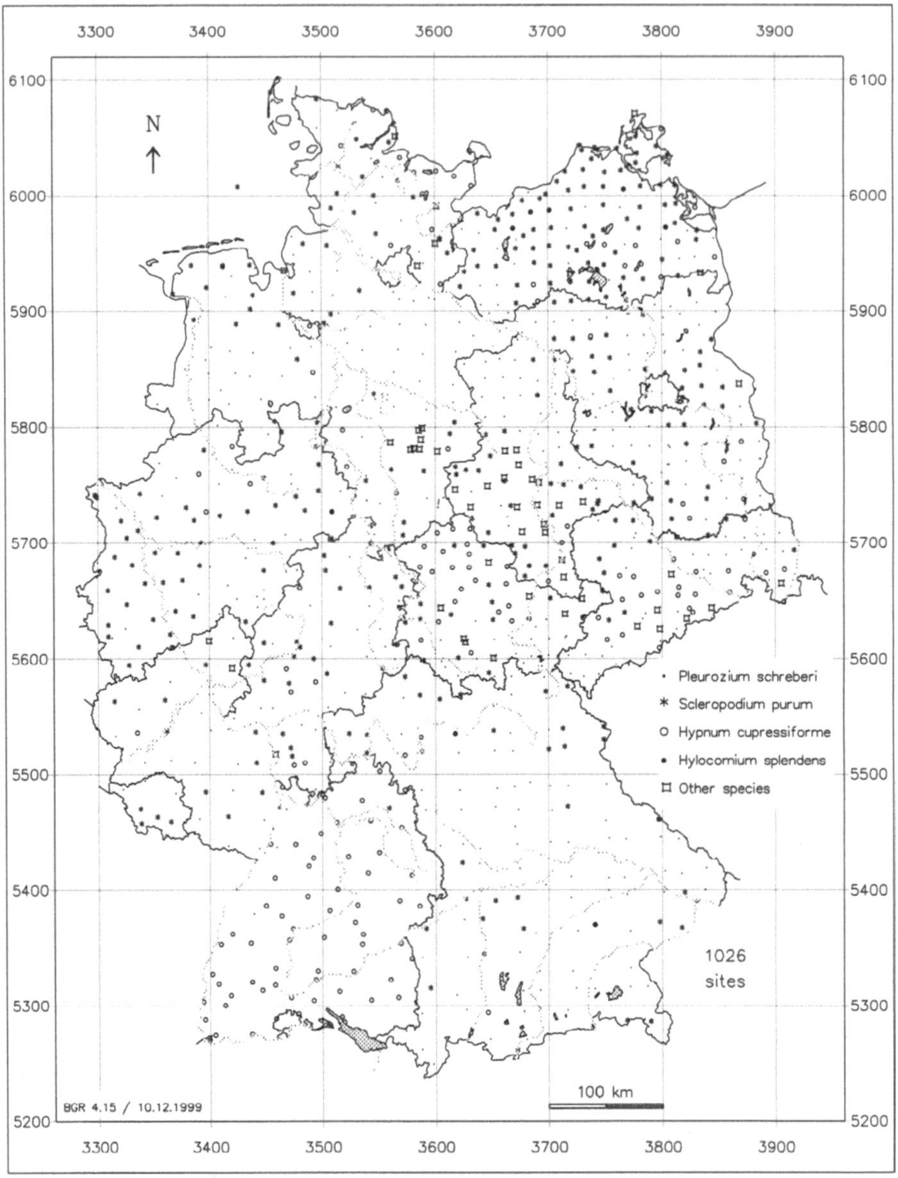

Figure 1. Distribution of the sampling points (n = 1026) and moss species sampled in Germany in the 1995/96 moss monitoring programme.
· *Pleurozium schreberi*, * *Scleropodium purum*, ° *Hypnum cupressiforme*,
• *Hylocomium splendens*, □ *Other species*

2.3. SAMPLING PERIOD

Sampling for the 1990/91 moss monitoring project was carried out from September 1991 to April 1992 in the West German states and the southern part of the East German states and from 1989 to 1990 in the north-eastern parts of the East German states. In the 1995/96 project sampling was carried out from September to December 1995.

2.4. ELEMENTS ANALYZED

As in the earlier project, the elements analyzed in 1995/96 were the standard elements arsenic (As), cadmium (Cd), chromium (Cr), copper (Cu), nickel (Ni), iron (Fe), lead (Pb), titanium (Ti), vanadium (V) and zinc (Zn). In 1995/96 the elements mercury (Hg) and antimony (Sb) were added. Analyses of 28 further elements were also performed.

2.5. SAMPLE PREPARATION AND ANALYSIS

Preparation

In both projects, any foreign material (leaves, roots, humus particles etc.) adhering to the samples was removed by hand using plastic tweezers and plastic gloves; the samples were not washed. In the analysis the green and greenish-brown parts of the plants were used. These represent a period of about 2-3 years (Pakarinen & Rinne 1979).

The samples were dried to constant weight at 40°C/72h in a drying cabinet and ground in an agate mill.

Digestion of samples and instrumental measurement

In the 1990/91 moss monitoring project (University of Osnabruck) 400 mg of the sample material was weighed into quartz vessels and mixed with 5.0 ml HNO_3; the vessels were closed with a screw cap. The material was digested under pressure (≈ 10 bar) for 2.5 hours at 150 °C. Then the samples were made up to 40 ml with bidistilled water without filtration.

In the 1995/96 moss monitoring project (Federal Institute for Geosciences and Natural Resources, Hannover) 250 mg of the sample material was weighed into Teflon vessels, and mixed with 5.0 ml HNO_3 (Merck suprapur) and 2.0 ml H_2O_2. The samples were then digested in high-pressure Teflon containers (≈ 100 bar) in a microwave oven (MLS-1200) and made up to 200.0 g with bidistilled water without filtration.

In 1990/91 the instrumental measurement of the samples was carried out using the following methods, depending on the element: AAS (hydride system: As), AAS (graphite tube: Cd, Pb) and ICP-OES (Cr, Cu, Fe, Ni, Ti, V, Zn). In the 1995/96 programme all samples were analyzed by ICP-MS.

To check the comparability of the old and new data (different digestion and measuring techniques), homogenized residue samples from 1990/91 were digested again in Hannover by the above microwave method and analyzed by ICP-MS. The results yielded by the same sample material from 1990/91 and 1995/96 were compared by means of regression analysis. The results of the analysis ensure relatively good comparability of the data. Figure 2 shows the element lead as an example.

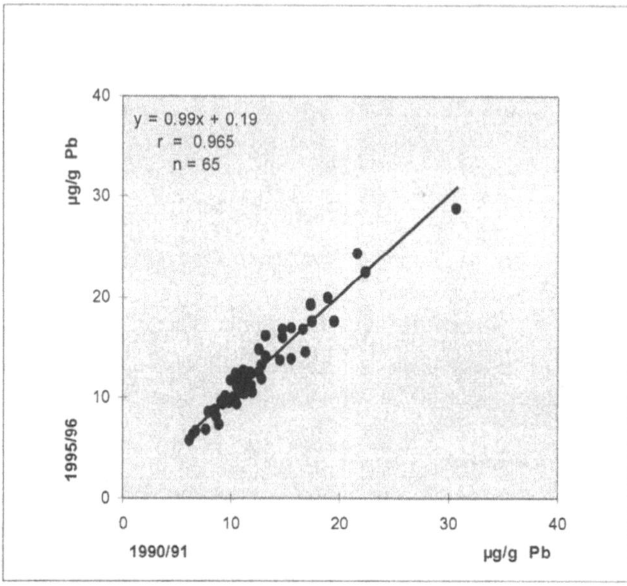

Figure 2. Lead concentrations (µg/g) in moss samples from the 1990/91 programme. 1990/91: digestion in quartz vessels with a screw cap, analysis by AAS (graphite tube). 1995/96: digestion in a microwave oven, analysis by ICP/MS.

Quality control

The quality of the results was ensured by blank analyses and analyses for reproducibility and accuracy. The reference materials used were SRM 1572 and SRM 1573 in 1990/91 and BCR 60 and BCR 61 in 1995/96.

2.6. REPRESENTATION AND EVALUATION METHODS

Representation methods

- Contour maps and dot charts

At the Federal Institute for Geosciences and Natural Resources (BGR) in Hannover the heavy metal analyses from the moss monitoring projects of 1990/91 and 1995/96 were fed into a DEC/VAX computer together with the coordinates of the relevant sampling points (Gauss-Krüger) so that coloured contour maps could be made with the colour raster programme UNIRAS (Figures 6-9).

The basic step in making contour maps is to transform the irregularly (statistically) distributed measuring points in the field into a regular grid (e.g. 3x3 km²) covering the whole of Germany. This requires interpolation to transpose the measurements from the original measuring points to the corresponding corners of the squares in the grid, so that each square has a calculated value. Within a specified search radius (e.g. 20 km) the method calculates the interpolated value for the centre of the circle from the values of all measuring points within the radius. Each value is given a weighting factor equal to the inversely proportional relationship of the square on the distance of the measured value from the centre of the circle ($1 \cdot d^{-2}$). This system gives more weight to the values close to the centre of the circle than to those farther away. It produces mean values weighted according to distance for the corresponding squares, which in turn serve as a basis on which to prepare the contour maps. The white areas in the contour maps mean that no interpolation was carried out because of inadequate sampling density.

Evaluation methods
In principle the following methods can be used for evaluating the distribution of the element concentrations over the area to permit a classification of pollution:

- Evaluation on the basis of binding standards and limits

As mentioned in the Introduction, the statutes and ordinances contain no binding standards and limits for use in bioindication. Standards and limits for biological systems such as food and animal feed that are similar to the bioindicator are often used as yardsticks for evaluation.

- Evaluation on the basis of specialist literature (expertise)

The literature contains data on the effects of pollutants on organisms. This information can be compared with one's own data to evaluate the results of bioindication methods.

- Evaluation on the basis of the existing data

The procedure for determining the normal value described by Erhardt et al. (1996) is suitable for the results of bioindication methods used in investigating the accumulation of pollutants. It will be explained and enlarged upon in the following sections.

Individual evaluation of heavy metal distribution (lead as a sub-model). The **normal value** (NV) for individual sub-models (Table 1) is calculated by the method described by Erhardt et al. (1996). This method is based on the observation that regional surveys yield numerous values that are fairly low and within narrow margins. The remaining measurements cover a wide range of higher concentrations. The method makes it possible to filter out the group of the many rather low measurements that are close together; the mean of these and its standard deviation constitute the "normal range" for a particular element in the area surveyed. The normal range or background may be assumed to have a more or less normal distribution (Lorenz 1988). If higher concentrations occur than those within a normal distribution of the measurements (anomalies, outliers), these are considered to be outside the normal range. They are probably caused by air-borne pollution that is greater than the base level in the area surveyed.
The decision as to which values are within the normal range is made on the basis of the mean value and the standard deviation of all measurements.
In a normal distribution, 97.5 % of the measurements are below the level of the mean plus 1.96 times the standard deviation. Values above this are excluded from the calculation of the standard range. Then the mean and the standard deviation of the remaining measurements are re-calculated. Again, those values are excluded that exceed the mean plus 1.96 times the standard deviation. After several repetitions only those values are left that do not exceed the test level (mean plus 1.96 times the standard

deviation). This procedure yields all measurements whose mean and scatter (plus/minus the single standard deviation) constitute the "normal value, normal range".
For further interpretations the element concentrations are considered significantly higher than the normal value for all measurements if they exceed this by at least three times the standard deviation. With this method it is possible to define classes or grades. This is the explanation of class 5, for example: If a value for lead at a particular site exceeds the corresponding "normal value" plus six times the standard deviation (> 18.6 µg/g) there is a "greatly increased" effect of air-borne pollution (Table 5).
Figure 8 shows this classification (evaluation) in the form of a map.

Table 5. Classification of the lead concentrations (µg/g) in mosses after calculation of the "normal value" (NV) and the standard deviation (1σ, 3σ, 6σ).

	Evaluation	Range	Lead (µg/g)
Grade 1	Slight immission	<NV - 1σ	<4.8
Grade 2	Normal range	NV ±1σ	6.8±2
Grade 3	Increased immission	>NV + 1σ	>8.8
Grade 4	Noticeably increased immission	>NV + 3σ	>12.7
Grade 5	Greatly increased immission	>NV + 6σ	>18.6

Integrated evaluation of several sub-models. Using the above method and evaluation it is possible to represent the results for individual sub-models (SM 1-12) in such a way that they show geographic differences in pollution.
But in addition there is a need to achieve an integrated evaluation and representation of all sub-models and different ecosystem compartments (Kreimes 1996). However, as the number of sub-models increases it becomes more difficult to view them together. The following problems arise when combining several sub-models to generate integrated information:

- The data from the sub-models are not directly comparable with each other.
- There are no objective weighting factors.
- The sub-models analyzed differ in their significance for the ecosystem.
- The results of sub-models correlate very closely. If such sub-models are weighted in the same way, too much emphasis is placed on certain parameters.
- The quality of the data varies according to the method used.

The following model is an attempt to eliminate the above difficulties by specifying a standard procedure for analysis and evaluation.

Evaluation model
For each location to be investigated the evaluation model shows the measured value (x_i) of each sub-model as a sub-model quality (Q_{SM}) (Figure 3). This sub-model quality is

calculated from the product of the standardized measured value (x_s), the ecological weighting (W_e) and the methodological weighting (W_m).

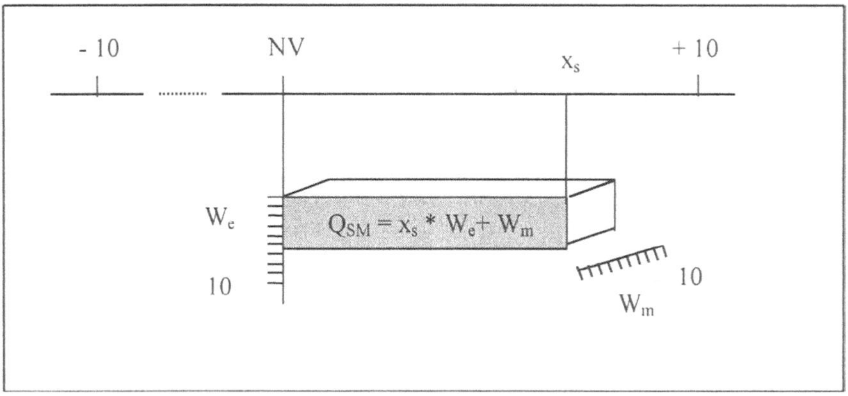

Figure 3. Calculation of sub-model qualities.

The measured values are standardized by setting the normal value (NV) at 0 on a scale from -10 to +10 (Figure 4). The maximum value is -10 and the minimum value +10. In this procedure it is assumed that the maximum value has a negative effect on the bioindicator (e.g. accumulation of pollutants). For the data ranges ≤ NV and > NV the measurements are standardized with the aid of linear functions. The maximum value (x_{max}) and minimum value (x_{min}) are calculated from the same set of data (N) as the normal value.

Ecological weighting (W_e) of the various sub-models is carried out according to their functionality and representativeness of the ecosystem. The maximum weighting factor is 10, and steps of whole numbers are permissible. This weighting has to be carried out according to accepted rules that serve as a basis for making decisions. Such rules are at present being formulated.

The following is an example of the problem of parallel measurements in some sub-models. The accumulation of Pb in various sub-models is to be investigated. The results of three different sub-models correlate very closely. We have to conclude that the accumulations in individual sub-models influence each other or have the same causes. If we were to weight the sub-models thus related as highly as those that have no connection with other sub-models, the parallel sub-models would distort the overall evaluation. This is the reason for the rule that the total weighting of parallel sub-models must not be more than 10.

Methodological weighting (W_m) of the various sub-models can be carried out by the methods already used. The stipulated maximum weighting factor is 10, and steps of whole numbers are permissible. Here, too, it is necessary to establish practical rules for determining a weighting factor for the relevant methods.

The sub-model qualities (Q_{SMi}) are combined to determine an overall quality (ecological quality) (Q_{tot}) by comparing the total of the sub-model qualities (Figure 5) with a maximum quality (Q_{max}), or more usually by calculating a percentage of the maximum quality.

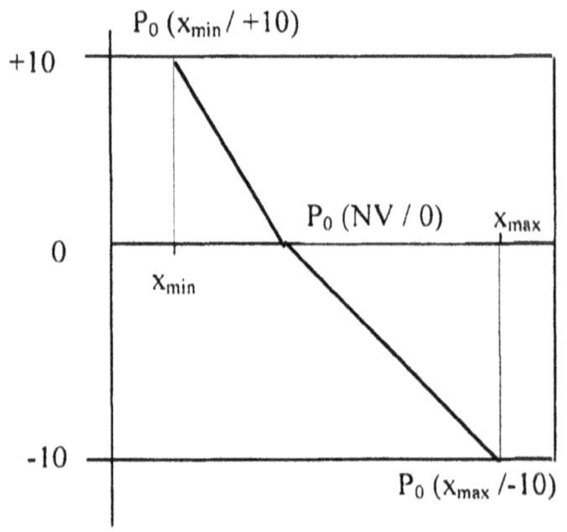

$$x_i \le NV \text{ then } y = -\frac{10}{NV - x_{min}} x + \frac{10NV}{NV - x_{min}}$$

$$x_i > NV \text{ then } y = -\frac{10}{x_{max} - NV} x + \frac{10NV}{x_{max} - NV}$$

Figure 4. Standardization method.

The maximum quality (Q_{max}) is calculated from the sum of all maximum sub-model qualities $Q_{(SMimax)}$. These result from the product of ecological weighting, methodological weighting and x_{max} or x_{min} (since the maximum and minimum values of the sub-models are 10).

$$Q_{max} = \sum_{i=1}^{i=n} Q_{SMi[max]}$$

$$Q_{SM} = \sum_{i=1}^{i=n} Q_{SMi}$$

The sum of the individual sub-model qualities is calculated and the percentage of the maximum quality determined.

$$Q_{tot} = \frac{Q_{SM} \cdot 100}{Q_{max}}$$

This results in quality statements in the form of percentages for all sampling points. Classification is carried out by dividing the absolute amount between the ascertained maximum and minimum quality by the number of quality grades and thus assigning quality grades to the percentages. For the sub-models listed in the Introduction (SM 1-12) the "normal values" (NV) calculated by the method of Erhardt et al. (1996) were used (Table 6). With the aid of these normal values and the evaluation model described above the pollution situation for the listed sub-models (Table 1) was determined from all moss analyses together ("pollution level map") (Figure 9). As no rules yet exist for the weighting factors, these were weighted in the same way for all sub-models.

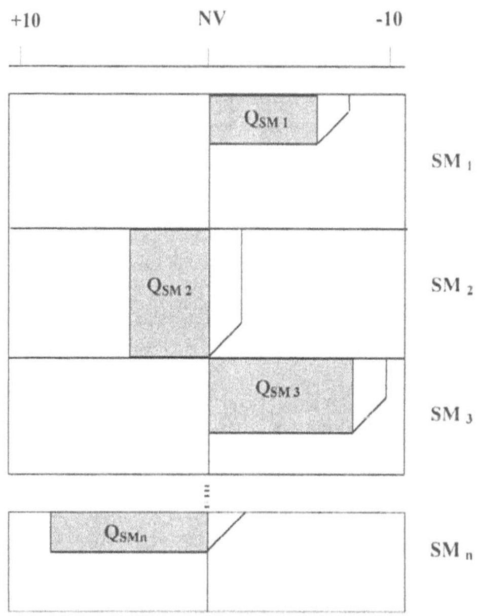

Figure 5. Evaluation model for calculating an overall quality (ecological quality) from n sub-models (SM_n).

3. Results

3.1. LEAD CONCENTRATIONS IN THE MOSS MONITORING PROGRAMMES OF 1990/91 AND 1995/96

The Figures 6 and 7 show the geographical and time-related changes in lead concentrations in Germany. In the 1990/91 moss monitoring project, elevated lead concentrations were found in the densely populated and industrialized western, south-western and eastern areas of Germany. Some parts of Saxony and Brandenburg, in particular, were found to have very high levels of lead.

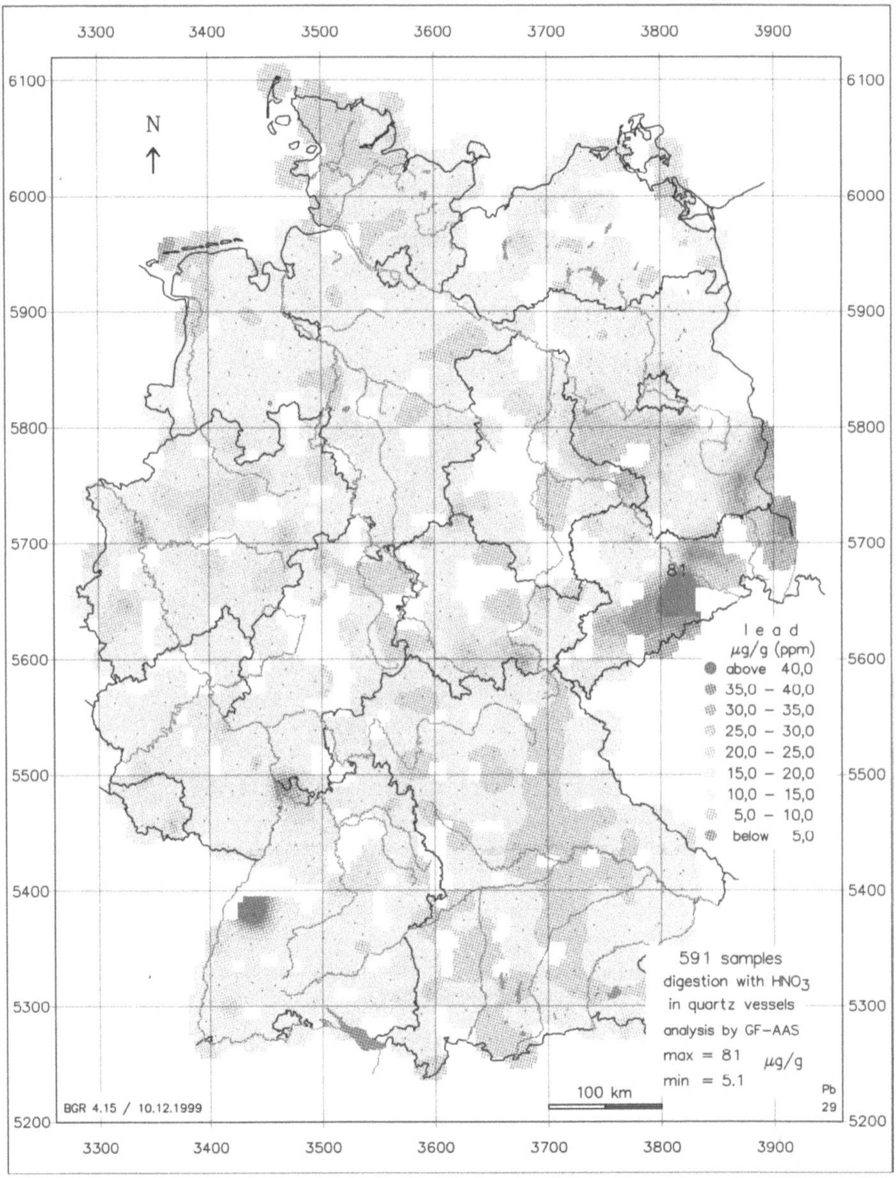

Figure 6. Contour map showing lead concentrations (µg/g) in Germany in the 1990/91 moss monitoring project. Number of samples: 591.

Large areas of the West German states had lower lead concentrations than the East German states in spite of a greater traffic density; this is presumably due to the declining use of leaded petrol (Leaded Petrol Act). In 1990/91 the difference was all the greater because East German petrol still contained large amounts of lead. The high levels of lead in Brandenburg and Saxony also resulted from the industrial and household use of brown coal, which can emit considerable lead quantities. The maximum concentration in Saxony was caused by the non-ferrous metal industry, in particular the smelting of lead-zinc ores.

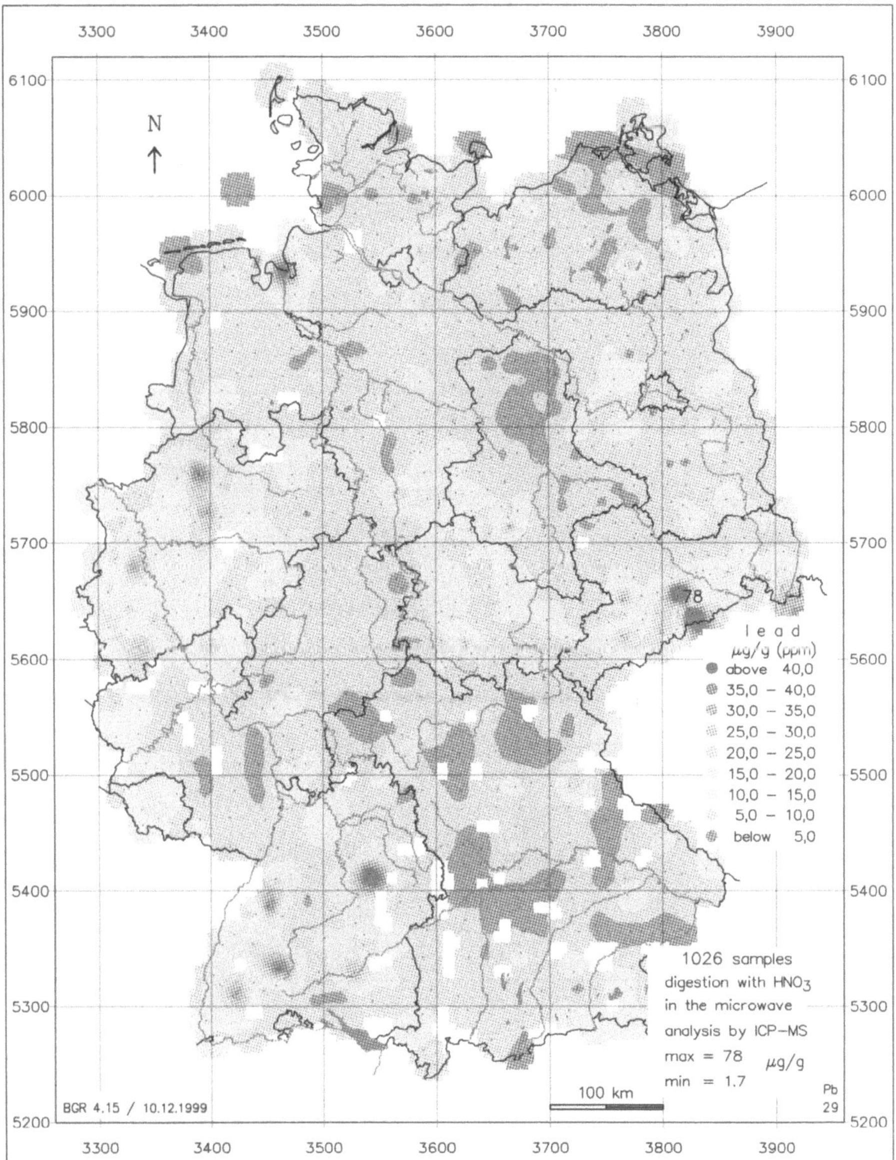

Figure 7. Contour map showing lead concentrations (µg/g) in Germany in the 1995/96 moss monitoring project. Number of samples: 1026.

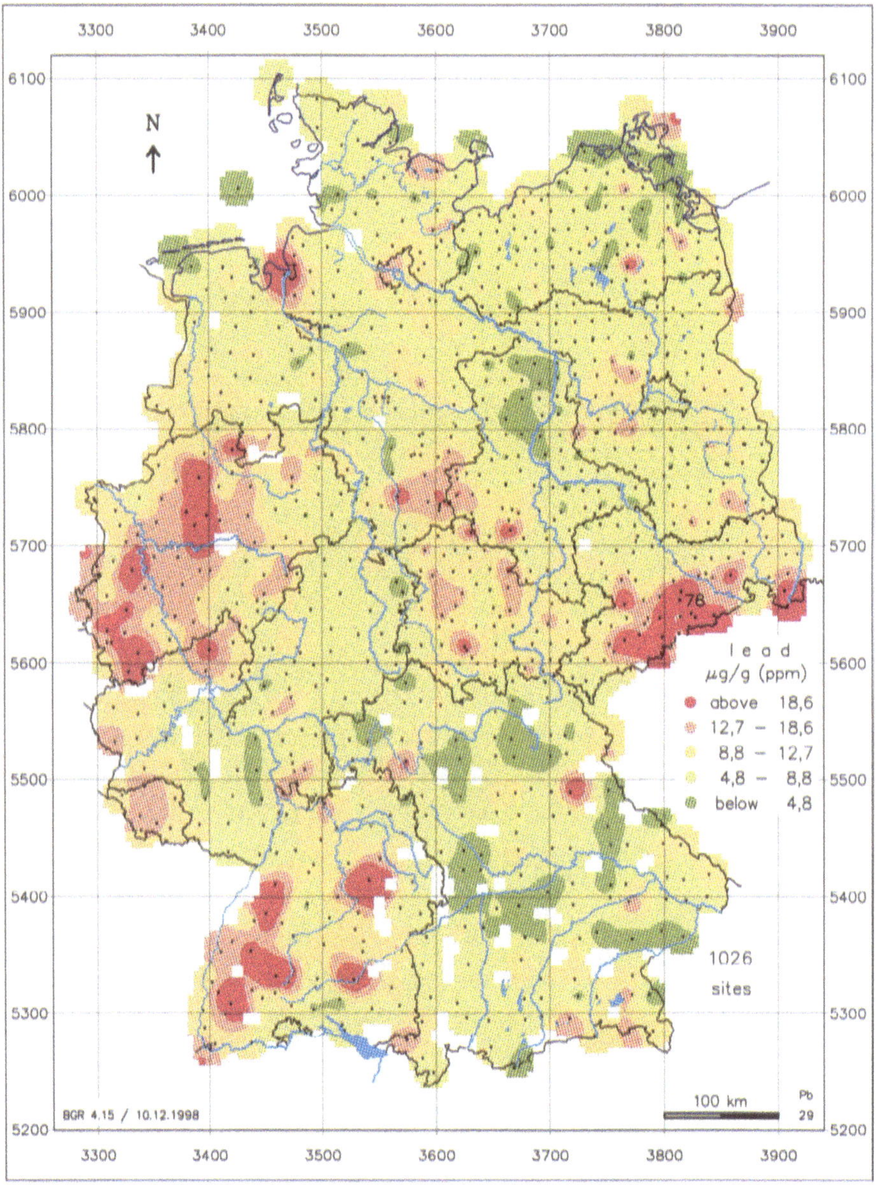

Figure 8. Contour map showing the air-borne pollution classification for lead derived from the normal value and standard deviation (1σ, 3σ, 6σ). Number of locations: n = 1026.
Grade 1: (< 4.8 µg/g); Grade 2: (≥ 4.8 - ≤ 8.8 µg/g); Grade 3: (> 8.8 - ≤ 12.7 µg/g); Grade 4: (> 12.7 ≤ 18.6 µg/g); Grade 5: (> 18.6 µg/g).

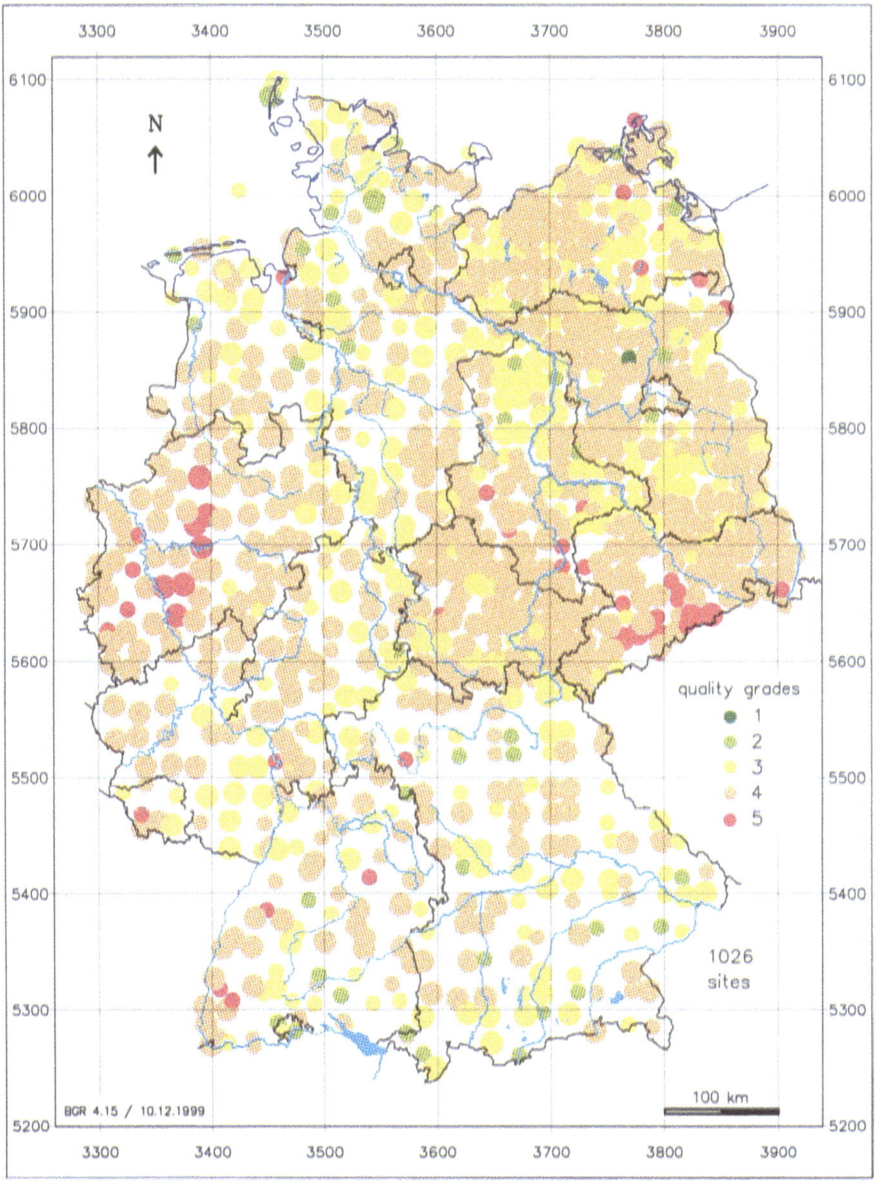

Figure 9. Dot chart showing the summarized and standardized parameters (SM 1-12) as quality grades (1-5) in Germany. Number of locations: n = 1026.

In the West German states the relatively high levels in densely populated and industrialized areas can be explained by the extremely high traffic density; but in some cases no connection is to be found between lead distribution and traffic density. For

example, no reliable interpretations could be given concerning the influence of the industrial regions with dense traffic around Frankfurt and Munich.
As five years earlier, the lead distribution pattern in the moss monitoring project of 1995/96 revealed elevated levels in parts of North Rhine-Westphalia, Baden-Württemberg and the Saarland. In both overviews, Saxony displayed the highest lead concentrations. On the other hand, the high levels in Brandenburg were not found again in the 1995/96 project. When Germany is viewed as a whole, the lead concentrations are seen to have fallen considerably over wide areas. This development is due to the increasing use of low-lead petrol and the closing or modernization of industrial firing systems, especially in East Germany. The 1990/91 lead concentrations range from 5.1 µg/g to 80.5 µg/g with a median of 12.9 µg/g. In 1995/96 the lead measurements ranged from 1.7 µg/g to 78 µg/g with a median of 7.7 µg/g. This represents an overall relative reduction of 40 %.

Table 6. Normal ranges of the selected sub-models (1-12) in µg/g and threshold values for classifying the element concentrations according to the grades "noticeably increased immission" and "greatly increased immission" using the data from the 1995/96 moss monitoring project.

Sub-model (SM)	Sample type	Parameter	**NV ± 1σ** **Normal value**	**NV + 3σ** **Threshold value for** "noticeably increased immission"	**NV + 6σ** **Threshold value for** "greatly increased immission"
SM 1	moss	As	0.201 ± 0.079	0.438	0.674
SM 2	moss	Cd	0.27 ± 0.06	0.45	0.63
SM 3	moss	Cr	1.20 ± 0.4	2.4	3.6
SM 4	moss	Cu	8.8 ± 2.0	14.9	21.1
SM 5	moss	Fe	391 ± 115	735	1078
SM 6	moss	Hg	0.039 ± 0.014	0.080	0.121
SM 7	moss	Ni	1.44 ± 0.38	2.57	3.70
SM 8	moss	Pb	6.8 ± 2.0	12.7	18.6
SM 9	moss	Sb	0.159 ± 0.044	0.291	0.423
SM 10	moss	Ti	18.6 ± 5.7	35.7	52.7
SM 11	moss	V	1.55 ± 0.46	2.92	4.29
SM 12	moss	Zn	49.5 ± 10.6	81	113

3.2. INDIVIDUAL EVALUATION OF LEAD CONCENTRATIONS IN THE 1995/96 MOSS MONITORING PROJECT, AND PRESENTATION OF THE RESULTS

The lead concentrations now have to be represented in the form of contour maps, using the evaluation method described in Chapter "*Individual evaluation of heavy metal distribution*". The classifications obtained with the aid of the "normal value" and its scatter (1σ, 3σ, 6σ) (Table 6) provide a relative yardstick for evaluating air-borne accumulations of lead in mosses and thus go beyond a mere representation of element levels (Figures 6, 7). In particular, by adding the three-fold and six-fold standard deviation to the "normal value" it is possible to calculate threshold values for classifying lead concentrations in the mosses to indicate significant effects of air-borne pollution. In this way it is possible to identify the locations or regions in which the level of air-borne pollution is significantly higher than the "base load" in the area covered by the survey. Table 6 shows the threshold values for "noticeably increased immission" and "greatly increased immission" for all elements (sub-models) analyzed.
Using this method it is possible to classify large areas of North-Rhine Westphalia, Baden-Württemberg and Saxony as having a very high level of air-borne pollution with lead. It must be added that some areas (e.g. Bavaria) are below the normal range (Figure 8, Grade 1). This can be of general importance in biological or ecological investigations, for example when deficiency symptoms are to be identified.

3.3. INTEGRATED EVALUATION AND REPRESENTATION OF ALL SUB-MODELS INVESTIGATED (SM 1-12)

With the evaluation method based on all elements analyzed, which we described in Chapter "*Integrated evaluation of several sub-models*", the quality of the environment in Germany is expressed in the form of quality grades. These grades are an initial integrated measure of quality for the medium "air" (Figure 9).
Some locations in North Rhine-Westphalia and Saxony, for example, have Quality Grade 5; this indicates poor environmental quality in the area due to high atmospheric inputs of metals. In Bavaria, on the other hand, there are locations with Quality Grade 2; these regions have a much better air quality than the locations in North Rhine-Westphalia or Saxony on the basis of the parameters investigated (SM 1-12).

4. Discussion

Because of their physiological and morphological characteristics, and since they occur everywhere, mosses seem to be more suitable than other organisms for regular monitoring of heavy metal inputs, although the moss analyses only document the relative geographical and time-related changes in element concentrations.
When depicting element concentrations in mosses over large areas it has to be taken into account that even mosses are subject to biological effects that may result in inaccurate interpretations of the heavy metal measurements. There is no doubt that the

concentrations determined are not due to the absolute atmospheric inputs alone, but are influenced by other factors as well.

The various species of moss and also genetic variations within one and the same species have to be considered when investigating the uptake of preferred elements (Folkeson 1979; Ross 1990) and exchange processes resulting from the charge density or size of ions. This is reflected in differing degrees of tolerance to heavy metals (Brown & Sidhu 1992) and in element-specific uptake capacity (Rühling 1987; Ross 1990; Berg et al. 1995; Wolterbeek et al. 1995). The growth form of the mosses varies considerably according to location, which is one of the main reasons for the different uptake characteristics of the individual species. *Hypnum cupressiforme*, for example, is smaller and less branching than *Pleurozium schreberi* or *Scleropodium purum* and forms more compact cushions. In particular, close leaf growth and a more compact shape may result in more efficient filtering of the air and thus higher element concentrations. On the other hand, a vigorous increase in biomass may have the effect of diluting the elements taken up. This is the reason for phenological changes and seasonal fluctuations in the element concentrations in some mosses over the year (Markert & Weckert 1989).

A further factor is that certain trace elements (e.g. Cu, Zn) are necessary for the life of plants and therefore present at a natural base level that is not directly related to the atmospheric inputs (Berg et al. 1995). Moreover, soil contamination can have a significant influence on the element concentrations in mosses. Wallace et al. (1980) show that this affects levels of heavy metals, especially in the case of the following elements: Cr > Fe > Co > Mn > Ni > Zn > Cu > Mo > Cd. Washing experiments have revealed contamination with Fe, Al, Si, Ti and Pb by dust and other means. According to Rühling et al. (1987) the erosion of soil material may cause high element concentrations in mosses in areas with sparse vegetation or none at all and with high heavy metal levels in the soil. This in turn results in an over-estimation of the heavy metal inputs caused by man.

But these uncertainties are within a range that is still acceptable for the interpretation of the results, especially as far as the basic idea of the moss method is concerned. This method offers a means of identifying regional differences and long-term trends over wide areas at relatively low cost.

However, it must be emphasized that the use of red and green in the figures should not be interpreted as indicating that an ecosystem is "polluted" or otherwise. When merely element distributions are shown (Figures 6, 7) the hazard can only be estimated. The colours in the contour maps indicate areas or regions in which the heavy metal concentrations found in the mosses were either higher (red) or lower (green) than the average. The natural complexity of biological systems and abiotic factors, and also the way in which mixtures of pollutants act, make it difficult to draw conclusions from the results of bioindication methods.

In the analysis of element concentrations in biological samples the fundamental question arises which levels are "normal" and which are elevated by air pollution. The present paper is an attempt to determine the normal levels of the elements analyzed and thus establish a yardstick for identifying unusual features and deviations (see similar values for lead by Peichl & Köhler 1993 and Markert 1996). These "normal values" from the sub-models (elements) analyzed were incorporated in an integrated evaluation of environmental quality using the evaluation model described and expressed in the form of quality grades. But strictly speaking, only one aspect of environmental quality (air quality) can be deduced in this manner when the above uncertainty factors specific to moss are taken into account. The results obtained here can therefore be incorporated in an overall evaluation directed towards establishing an "overall ecological quality map" representing different ecosystems and sub-models together and covering all media.

Acknowledgements
We are grateful to the Federal Environmental Agency (Umweltbundesamt [UBA]) for supporting this project (F+E 295 830 87/01). Especially we would like to thank Ms. G. Knetsch and Dr. H. Bau for many discussions and help throughout the project.

References

Adriano, D.C. (ed) 1992. Biochemistry of trace metals. Lewis Publisher, Boca Raton, Florida.

Arndt, U., Nobel, W. & Schweitzer, B. 1987. Bioindikatoren, Möglichkeiten, Grenzen und neue Erkenntnisse. Ulmer, Stuttgart.

Berg, T., Royset, O. & Steinnes, E. 1995. Moss (*Hylocomium splendens*) used as a biomonitor of atmospheric trace element deposition. Estimation of uptake efficiencies. Atmos. Environ. 29(3): 353-360.

Berlekamp, J., Herpin, U., Matthies, M., Lieth, H., Markert, B., Weckert, V., Wolterbeek, B., Verburg, T., Zinner, H. J. & Siewers, U. 1998. Geographic classification of heavy metal concentrations in mosses and stream sediments in the Federal Republic of Germany. Water, Air and Soil Pollution 101: 177-195.

Brown, D. H. & Sidhu, M. 1992. Heavy metal uptake, cellular location, and inhibition of moss growth. Crypt. Bot. 3: 82-85.

Bruns, I., Siebert, A., Baumbach, R., Miersch, J., Günther, D., Markert, B. & Krauss, G. J. 1995. Aspects of heavy metals and sulphur-rich compounds in the water moss *Fontinalis antipyretica* L. ex Hedw. Fresenius J. Anal. Chem. 353: 101-104.

Bruns, I., Friese, K., Markert, B. & Krauss, G.-J. 1997. The use of *Fontinalis antipyretica* L. ex Hedw. as a biomonitor for heavy metals. 2. Heavy metal accumulation and physiological reaction of *Fontinalis antipyretica* L. ex Hedw. in active biomonitoring in the River Elbe. The Science of the Total Environment 204: 161-176.

Callaghan, T.V., Collins, N.J. & Callaghan, C.H. 1978. Photosynthesis, growth and reproduction of *Hylocomium splendens* and *Polytrichum commune* in Swedish Lapland. Oikos 31: 73-88.

Erhardt, W., Höpker, K. A. & Fischer, I. 1996. Verfahren zur Bewertung von immissionsbedingten Stoffanreicherungen in standardisierten Graskulturen. Z. Umweltchem. Ökotox. 8(4): 237-240.

Ellison, G., Newham, J., Pinchin, M. J. & Thompson, I. 1976. Heavy metal content of moss in the region of Consett (North East England). Environm. Pollut. 11: 167-174.

Ernst, W.H.O. 1974. Schwermetallvegetation der Erde. Fischer, Stuttgart.

Fiedler, H.J. & Rösler, H.J. (eds) 1993. Spurenelemente in der Umwelt. Enke, Stuttgart.

Folkeson, L. 1979. Interspecies calibration of heavy metal concentrations in nine mosses and lichens. Applicability to deposition measurements. Water, Air and Soil Pollution 11: 253-260.

Goodmann, G.T. & Roberts, T.M. 1971. Plants and soils as indicators of metals in the air. Nature 231: 287-292.

Grodzinska, K., Szarek, G., Godzik, B., Braniewski, S. & Chrzanowska, E. 1993. Air pollution mapping in Poland by heavy metal concentration in moss. Proc. Polish-American Workshop. Climate and atmospheric deposition monitoring studies in forest ecosystems. Nieborow, 6-9 October 1992.

Herpin, U., Berlekamp, J., Markert, B., Wolterbeek, B., Grodzinska, K., Siewers, U., Lieth, H. & Weckert, V. 1996. The distribution of heavy metals in a transect of the three states the Netherlands, Germany and Poland, determined with the aid of moss monitoring. The Science of the Total Environment 187: 185-198.

Herpin, U., Markert, B., Siewers, U. & Lieth, H. 1994. Monitoring der Schwermetallbelastung in der Bundesrepublik Deutschland mit Hilfe von Moosanalysen. Forschungsbericht 108 02 087, UBA - FB 94-125.

Herpin, U., Markert, B., Weckert, V., Berlekamp, J., Siewers, U. & Lieth, H. 1997. Retrospective analysis of heavy metal concentrations at selected locations in the Federal Republic of Germany using moss material from a herbarium. The Science of the Total Environment 205: 1-12.

Herrmann, R. 1990. Biomonitoring of organic and inorganic trace pollutants by means of mosses. pp. 319-333. In: Zinsmeister, H.D. & Mues, R. (eds), Bryophytes. Their chemistry and chemical taxonomy. Proceedings of the Phytochemical Society of Europe 29. Clarendon Press, Oxford.

Hertz, L. 1991. Bioindicators for monitoring heavy metals in the environment. Metals and their compounds in the environment - occurrence, analysis and biological relevance. pp. 221-231. In: Merian, E. (ed), VCH Verlagsgesellschaft mbH, Weinheim, New York, Basel, Cambridge.

Kabata-Pendias, A. & Pendias, H. 1992. Trace elements in soils and plants. CRC Press, Inc., Boca Raton, Florida.

Kreimes, K. 1996. Bewertungs- und Darstellungsverfahren. Verfahren zur Bewertung und Darstellung von Bioindikationsdaten am Beispiel des Ökologischen Wirkungskatasters Baden-Württemberg. Z. Umweltchem. Ökotox. 8(5): 287-292.

Little, P. & Martin, M. H. 1974. Biological monitoring of heavy metal pollution. Environm. Pollution 6: 1-19.

Lorenz, R. 1988. Grundbegriffe der Biometrie. Stuttgart, New York.

Klein, R. & Paulus, M. (eds) 1995. Umweltproben für die Schadstoffanalytik im Biomonitoring. Fischer Verlag, Jena, Stuttgart.

Loetschert, W. & Wandtner, R. 1982. Schwermetallakkumulation im Sphagnetum magellanici aus Hochmooren der Bundesrepublik Deutschland. Ber. Deutsch. Bot. Ges. 95: 341-351.

Markert, B. (ed) 1993. Plants as biomonitors - indicators for heavy metals in the terrestrial environment. VCH Verlagsgesellschaft mbH, Weinheim, New York, Basel, Cambridge.

Markert, B. 1996. Instrumental element and multi-element analysis of plant samples - methods and applications. Wiley & Sons, Chichester, New York, Brisbane, Toronto, Singapore.

Markert, B., Herpin, U., Berlekamp J., Oehlmann, J., Grodzinska, K., Mankovska, B., Suchara, I., Siewers, U., Weckert, V. & Lieth, H. 1996b. A comparison of heavy metal deposition in selected Eastern European countries using the moss monitoring method, with special emphasis on the "Black Triangle". The Science of the Total Environment 193: 85-100.

Markert, B., Herpin, U., Siewers, U., Berlekamp J. & Lieth, H. 1996a. The German heavy metal survey by means of mosses. The Science of the Total Environment 182: 159-168.

Markert, B. & Weckert, V. 1989. Fluctuations of element concentrations during the growing season of *Polytrichum formosum* (Hedw.). Water, Air and Soil Pollution 43: 177-189.

Markert, B. & Weckert, V. 1993. Time-and-site integrated long-term biomonitoring of chemical elements by means of mosses. Toxicological and environmental Chemistry Vol. 40: 43-56.

Markert, B. & Weckert, V. 1994. Higher lead concentrations in the environment of former West Germany after the fall of the Berlin Wall. The Science of the Total Environment 158: 93-96.

Martin, M.H. & Coughtrey, P. J. 1982. Biological monitoring of heavy metal pollution, land and air. Applied Science Publisher, London, New York.

Maschke, J. 1981. Moose als Bioindikatoren von Schwermetallimmissionen. In: Cramer, J. (ed). Bryophytum Bibliotheka 22.

Nimis, P. L. & Cislaghi, C. 1997. Lichens, air pollution and lung cancer. Nature 387: 463-464.

Oehlmann, J., Stroben, E., Schulte-Oehlmann, U., Bauer, B., Fiorini, P. & Markert, B. 1996. Tributyltin biomonitoring using brosobranchs as sentinel organisms. Fresenius J. Anal. Chem. 334: 540-545.

Pakarinen, P. & Rinne, R. J. K. 1979. Growth rates and heavy metal concentrations of five moss species in paludified spruce forests. Lindbergia 5: 77-83.

Peichl, L. & Köhler, J. 1993. Vergleich verschiedener Moosarten als Bioindikatoren für Schwermetalle in Bayern. Abschlußbericht. Bayerisches Landesamt für Umweltschutz, Rosenkavaliersplatz 3, 81925 München, im Auftrag des Bayerischen Staatsministeriums für Landesentwicklung und Umweltfragen. Oktober 1993.

Reimann, C., Äyräs, M., Chekushin, V., Bogatyrev, I., Boyd, R., Caritat, P. de, Dutter, R., Finne, T.E., Halleraker, J.H., Jæger, Ø., Kashulina, G., Lehto, O., Niskavaara, H., Pavlov, V., Räisänen, M.L., Strand, T. & Volden, T. 1998. Environmental geochemical atlas of the central Barents region. Norges geologiske undersøkelse (Geological Survey of Norway), Trondheim.

Ross, H.B. 1990. On the use of mosses (*Hylocomium splendens* and *Pleurozium schreberi)* for estimating atmospheric trace metal deposition. Water, Air and Soil Pollution 50: 63-76.

Rühling, Å., Brumelis, G., Goltsova, N., Kvietkus, K., Kubin, E., Liiv, S., Magnusson, S., Mäkinen, A., Pilegaard, K., Rasmussen, L., Sander, E. & Steinnes, E. 1992. Atmospheric heavy metal deposition in Northern Europe 1990. NORD 1992: 12, Copenhagen.

Rühling, Å., Rasmussen, L., Pilegaard, K., Mäkinen, A. & Steinnes, E. 1987. Survey of atmospheric heavy metal deposition - monitored by moss analyses. NORD 1987: 21, Copenhagen.

Rühling, Å., Rasmussen, L., Mäkinen, A., Pilegaard, K., Steinnes, E. & Nihlgård, B. 1989. Survey of the heavy-metal deposition in Europe using bryophytes as bioindicators. Proposal for an international programme. Steering Body of Environmental Monitoring in the Nordic Countries.

Rühling, Å. & Steinnes, E. (eds). 1998. Atmospheric heavy metal deposition in Europe 1995 - 1996. NORD 1998: 15, Copenhagen.

Rühling, Å. & Tyler, G. 1968. An ecological approach to the lead problem. Bot. Notiser 121: 321-342.

Rühling, Å. & Tyler, G. 1969. Ecology of heavy metals, a regional and historical study. Bot. Notiser 122: 248-259.

Rühling, Å. & Tyler, G. 1971. Regional differences in the deposition of heavy metals over Scandinavia. J. Appl. Ecol. 8: 497-507.

Rühling, Å. & Tyler, G. 1973. Heavy metal deposition in Scandinavia. Water, Air and Soil Pollution 2: 445-455.

Schmidt-Grob, I., Thöni, L. & Hertz, J. 1993. Bestimmung der Deposition von Luftschadstoffen in der Schweiz mit Moosanalysen. Bundesanstalt für Umwelt. Wald und Landschaft (BUWAL) (eds). Schriftenreihe Umwelt 194.

Schubert, R. (ed) 1991. Bioindikation in terrestrischen Ökosystemen. Fischer, Stuttgart.

Schüürmann, G. & Markert, B. (eds) 1998. Ecotoxicology - ecological fundamentals, chemical exposure and biological effects. Wiley & Sons, Chichester, New York, Brisbane, Toronto, Singapore and Spektrum Akademischer Verlag, Heidelberg, Berlin.

Steinnes, E. 1984. Monitoring of trace element distribution by means of mosses. Fresenius Z. Anal. Chem. 317: 350-356.

Steinnes, E. 1993. Some aspects of biomonitoring of air pollutants using mosses, as illustrated by the 1976 Norwegian survey. pp. 341-401. In: Markert, B. (ed). Plants as biomonitors - Indicators for heavy metals in the terrestrial environment. VCH-Verlagsgesellschaft mbH, Weinheim, New York, Basel, Cambridge.

Steubing, L. & Jäger, H. (eds) 1982. Monitoring of trace elements of air pollutants. Methods and problems. T:VS 7. Junk, Den Haag, Boston, London.

Wallace, A., Cha Kinnear, J. W. & Romney, E. M. 1980. Effect of washing procedures on mineral analysis and their cluster analysis for orange leaves. J. Plant Nutr. 2(1+2): 1-9.

Wolterbeek, H.T., Kuik, P., Verburg, U., Herpin, U., Markert, B. & Thöni, L. 1995. Moss interspecies comparisons in trace element concentrations. Environ. Monitoring and Assessment 35: 263-286.

Zimmermann, R. D. 1992. Bioindikation/Wirkungsermittlung - Arbeitskreis der Landesanstalten und -ämter. Konzeption der künftigen Aufgabenbereiche. Z. Umweltchem. Ökotox. 4: 286-287.

DO PHYTOPHENOLOGICAL SERIES CONTRIBUTE TO VEGETATION MONITORING?

CLAUDIO DEFILA

MeteoSwiss, Postfach 514, CH-8044 Zürich, Switzerland

Keywords: Global climate change, phytophenology, Switzerland, time series, trends

Abstract
Phenological observations have been made in Switzerland since 1951. In addition to these observation programmes, there are two very long phenological series in Switzerland. The foliation of horse chestnut trees in Geneva has been observed since 1808, the full flowering of cherry trees in Liestal since 1894. Apart from the presentation of these series, trends for the regions Ticino and Engadine are calculated with national data from 1951 to 1998. The earlier foliation of the horse chestnut trees in Geneva can mainly be put down to the city effect (Warmth Island). This phenomenon was not observed with the cherry tree flowering in Liestal. A tendency towards earlier appearance dates in spring and later appearance dates in autumn could be made out with the data from the national observation network. It must be noted that the different phenophases and plant species react differently to various environmental influences. A general development of the vegetation – computed from data from various observation posts and phenophases for the different phenological seasons – is not suited for the investigation of phenological trends. This essay has shown that it is worthwhile to analyse more data from other regions in Switzerland.

1. Introduction

The aim of the science of phenology is to temporally register the annually recurrent growth and development of plants, as well as to study the influences thereon. The phenological phases – such as foliation, beginning of flowering, full bloom, ripening of fruit, leaf colouring and leaf fall – are observed and the relevant beginning times noted. Besides plant phenology there is also animal phenology that studies the migration of birds. Plant-phenological observations have been made since the Middle Ages, the data can be found in archives. Especially extraordinary phenomena, like the twofold flowering of cherry trees within one vegetation period, were noted (Pfister 1984). The botanist Carl von Linné founded a phenological observation network in Sweden in 1750. In the 19th century, there was a European network with observation posts in Belgium, the Netherlands, Italy, France, Great Britain, and in Switzerland. The Bernese forestry service established a forest-phenological observation network in 1869 that was operational until 1882 (Vassella 1997). During the fifties of this century, various

Burga & Kratochwil (eds.), BIOMONITORING, 97-105

European states initiated systematic phenological observations by setting up observation networks within the national weather services. In this manner a network that covered all the regions and altitudes was installed in Switzerland in 1951 (Primault 1955). The German weather office began its systematic observation in 1953 (Schnelle 1955). In 1957, the well-known German phenologist F. Schnelle founded the International Phenological Gardens (IPG). The idea of the gardens is that identical, i.e. cloned, plants are planted in different countries in Europe and that these plants (bushes and trees) are observed during the year, thus allowing a close comparison of the development of the vegetation in the various climatic regions in Europe. The Geographical Institute of the University of Bern has conducted a phenological observation programme since 1970 (Jeanneret 1996). Across the Jura, the Berner Mittelland ("Bernese midland"), and the Berner Oberland ("Bernese highland") a few phenological phases are observed at selected stations. Observations of individual plants at one place go back a lot further. The oldest phenological observation series world-wide is from Japan. There the dates of the flowering of cherry trees have been noted since 812. Two very old phenological data series in Switzerland have been continued until now. The budding of the leaves of chestnut trees has been observed in Geneva since 1808 and the flowering of cherry trees in Liestal (near Basel) since 1894. The phenological data are applied to integrated plant production, and used for the forecasting of phenophases (Bider & Meyer 1946), for frost warnings, or the creation of phenological maps (Primault 1984 and Schreiber 1977). At the time of the problem of forest damage the question came up whether the phenological timing was influenced by the damaged trees (Hartmann 1991). In recent times the phenological data have been focused upon in connection with the possibility of a climatic global change (Menzel 1997).
To what extent phenological series can be used for vegetation monitoring is to be discussed. Besides being influenced by the length of the day, the phenological appearance dates are mainly induced by meteorological conditions. In spring the rising temperature is an important factor (Defila 1991). A possible man-induced climatic global change is expected to lead to a global warming. Higher temperatures in winter and spring provoke earlier appearance dates of phenological phenomena. Thus, a warming should become evident in the trends of appearance dates of phenological data. Based on some examples, the shift in the phenological appearance dates shall be studied and discussed.

2. Methods

For the study of trends of phenological series in Switzerland, data are available from the national phenological observation network, as well as two old phenological series. The Swiss phenological observation network was founded in 1951 and initially consisted of 70 observation posts. 37 plant species and 70 phenophases were observed. A real-time phenological observation network was introduced in 1986. 40 selected stations report 17 phenological phases spread over the entire vegetation period immediately on their appearance. Based on this information, up-to-date bulletins can be composed which are published on the Internet. The phenological observation programme was slightly modified in 1996. Today the observation posts register 69 phenophases of 26 different plant species. The observation programme focuses mainly on wild plants. This almost

fifty-year observation series is well suited for trend analysis. It needs to be noted that not all observation posts were operational in 1951, and that many of them have meanwhile been abandoned. There are also numerous gaps where the observations of entire cycles or certain phenophases within a year are missing. A further problem arises with the changing of observing personnel. Due to a certain subjectivity in phenological observation, a change in observers can lead to a break in the series. In spite of all these difficulties, some of the series could be used for analysis.
The southern side of the Alps of Switzerland (Ticino) and the Engadine (the Grisons) were evaluated first. The results need further corroboration by analyses from other regions in Switzerland. A linear regression model was used for the trend analysis. The significance was determined with an F-test with the error limits at $P < 0.05$. The long phenological series of chestnut foliation in Geneva since 1808 and the full bloom of cherry trees since 1894, however, are completely documented until 1998. These two series were displayed with smoothed curves (Gauss low pass filter with a period of 20 years). Finally the mean course of the vegetation development of the Schweizerisches Mittelland ("Swiss midland") in spring was studied. For this analysis, 20 phenophases at 29 stations during spring were used. The appearance dates were sorted progressively by phenophase and station and classed into five categories:

- very early (10% of all cases)
- early (15% of all cases)
- normal (50% of all cases)
- late (15% of all cases)
- very late (10% of all cases)

To lend more weight to the extreme occurrences, the values of the two categories 'very early' and 'very late' were doubled. In this way, the phenological spring for the period from 1951 to 1995 could be characterised according to the five categories.

3. Results

In graph 1, the appearance dates of chestnut foliation in Geneva from 1808 to 1998 are shown. The smoothed curve (Gauss low pass filter with a period of 20 years) is drawn beside the curve of the annual values. This graph shows a clear trend towards earlier appearance dates. The trend becomes clear from about 1900 onwards. The latest appearance date was registered in 1816 on 23rd April, the earliest was observed in 1991 on 3rd January. It needs to be noted that it is not the foliation of the chestnuts that is observed (phenophase observed by the national observation network) but the beginning of the budding of the chestnut leaves. This stage is reached as soon as the first green of the leaves is visible, i.e. the leaf tips protrude from the buds. This trend might be due to a climatic global change, higher temperatures in winter and spring. It must be noted that these trees are in central Geneva and therefore influenced by the city climate. The city of Geneva has grown considerably since 1808; the number of houses and industrial plants as well as the amount of traffic have increased greatly. This has led to a greater heat emission (especially caused by domestic heating during winter). Due to this 'city-effect' middle-sized and large cities form heat islands with average temperatures two or three degrees higher than their surroundings. This additional warmth also affects the phenological appearance dates; in our case it causes the earlier beginning of the budding of the chestnuts. As this phenophase occurs very early, the heating periods during winter are bound to play a particularly important role. The predominantly mild winters in the nineties are certainly also a factor to be considered but the main cause is probably the city climate.

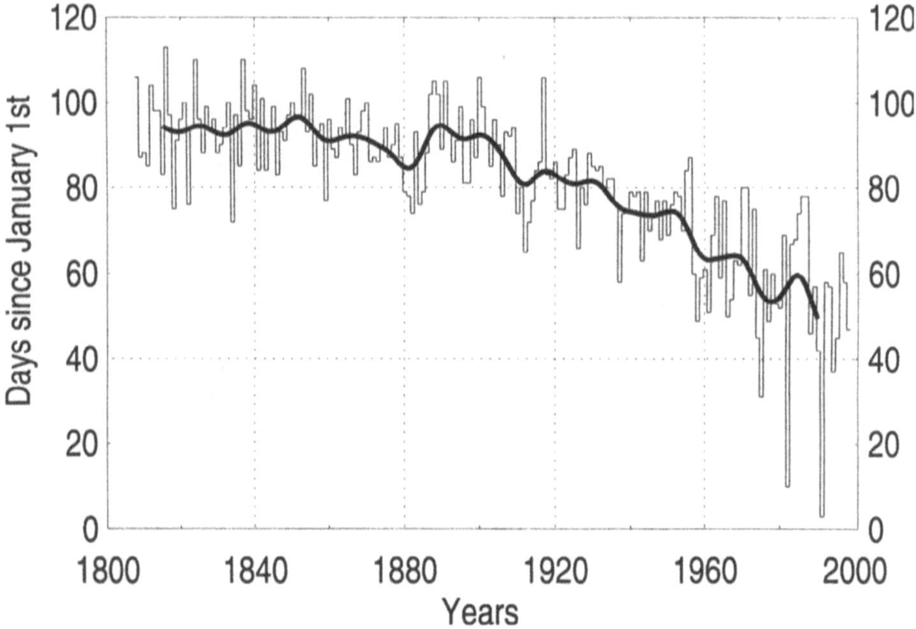

Figure 1. Dates of leaf appearance of the horse chestnut in Geneva, 1808-1998. Smoothing: Gauss low pass filter with a period of 20 years (extended according to Defila 1996).

This assumption is confirmed by the fact that the graph displaying the cherry flowering in rural Liestal (graph 2) does not show a trend. The graph was calculated using the same method as for graph 1 and depicts the appearance dates of cherry blossom in Liestal from 1894 to 1998. The latest appearance was observed on 4th May 1917, the earliest on 16th March 1990. Strictly speaking, the two phenophases are not absolutely comparable as they do not occur during the same period of a year. The early budding of the chestnuts in Geneva is influenced more by the winter temperatures, the later full bloom of the cherry trees in Liestal, in contrast, is more dependent on the spring temperatures. In spite of these restrictions, the city climate of Geneva must be seen as the dominant factor for the budding of the chestnut trees.
The question arises whether the individual phenophases of the national observation network might show trends. Trend analyses were carried out for the regions south of the Alps (Ticino) and Engadine (the Grisons). These two regions are two extreme climate areas of Switzerland. South of the Alps there is a mild Mediterranean climate, while the Engadine – with altitudes ranging from 1000 to 1800 metres above sea level – has an alpine climate with long winters. These climates have a direct influence on the vegetation. It is to be expected that the effects of a climatic global change on the

phenological appearance dates are different in such different climates. The temperatures south of the Alps are optimal while the temperatures in the Engadine are a limiting factor for various plants. Certain plant species no longer grow at these altitudes.

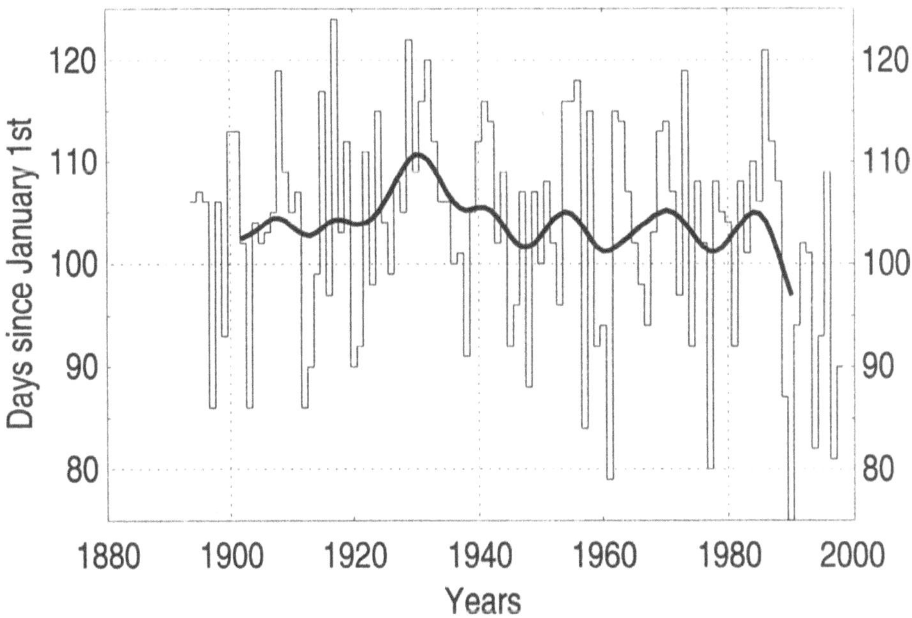

Figure 2. Appearance dates of full flowering of cherry trees in Liestal 1894-1998. Smoothing: Gauss low pass filter with a period of 20 years (extended according to Defila 1996).

Three observation posts were studied for the region south of the Alps: the stations Aurigeno at 350 metres above sea level, Menzonio at 725 metres above sea level, and Prato-Scornico at 750 metres above sea level. 88 phenophases in more or less uninterrupted series could be analysed from these three observation posts. 44% of the cases showed a significant trend ($P > 0.05$) after the F-test, while the remaining 56% were not significant. The trends between earlier and later phenological appearance dates were differentiated. The frequency of earlier and later appearance dates varied depending on the season. A trend towards earlier appearance dates with 55% in spring is evident. In summer the early and the later appearance dates were balanced at 50% each. Autumn only showed a trend towards late appearance dates. This analysis reveals that there is a tendency towards longer vegetation periods (Menzel 1997). It is interesting to note that the later appearance dates occur with foliation as well as during the autumn phases. The trend towards earlier appearance dates occurs solely with flowering phases. It appears that the vegetative phases (foliation) and generative phases (flowering phases) react differently to the climate. A shift of up to 30 days within these 50 years is easily possible.

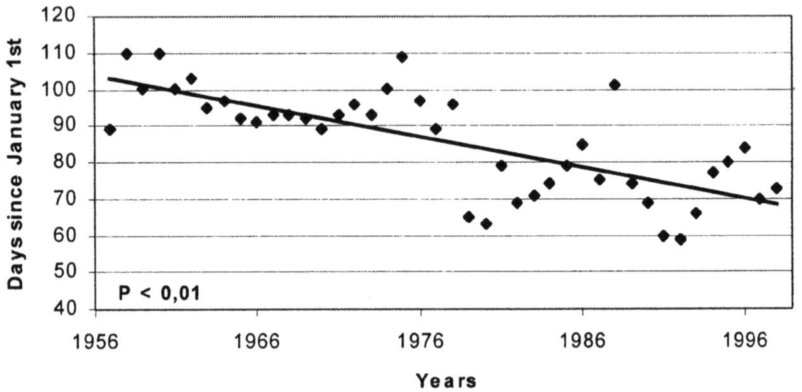

Figure 3. Linear trend of full flowering of the wood anemone in Prato Sornico (Ticino), 1957-1998.

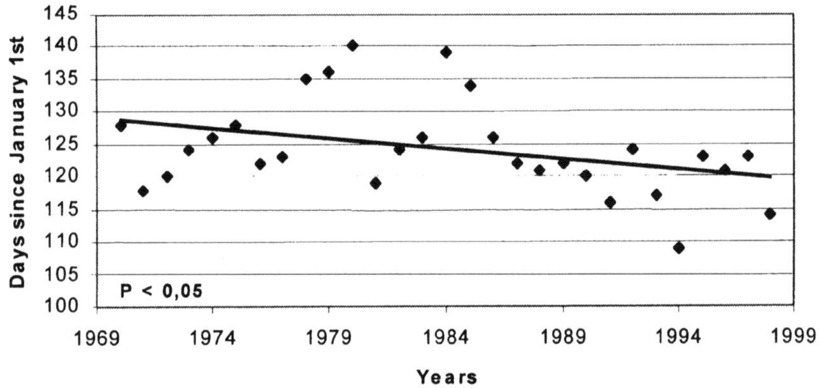

Figure 4. Linear trend of leaf appearance of the larch in Sent (Engadine), 1970-1998.

Unfortunately, the phenological series from the Engadine are not as long. Most of the observation posts were set up in 1970. A total of seven observation posts fulfilled the requirements for the trend analysis: Martina at 1050 metres above sea level, Scuol at 1240 metres above sea level, Sent at 1440 metres above sea level, Pontresina at 1780 metres above sea level, and St. Moritz at 1800 metres above sea level. For the period between 1970 and 1998, 106 phenophases could be analysed. 45% thereof showed a significant trend ($P < 0.05$) while 55% were not significant. These results are surprisingly concurrent with the results from the region south of the Alps. Of the 48 significant series, 28 series are without hiatus or almost (there is at most one missing observation) and the phenophases are such that are not influenced by humans. With one exception, all of the phenophases are spring phases. 26 of the 28 show a trend towards earlier appearance dates (Figure 4). Like south of the Alps, the shift over the 30 years is up to 30 days. In most of the cases, the appearance dates are a mere ten to fifteen days earlier. In contrast to south of the Alps, the trend towards earlier appearance dates occurs for both the vegetative phases (foliation) and the generative phases (flowering phases). The flowering phases are dominant because they occur more often during the observation programme in spring. Of the entirety of the significant trends, a mere nine percent show a tendency towards later appearance dates. Unfortunately the autumn phases are missing in the Engadine as the colouring of horse chestnut and beech tree leaves cannot be observed at these altitudes. Therefore, the trend towards longer vegetation periods is not confirmed by the analysis of the data from the Engadine.

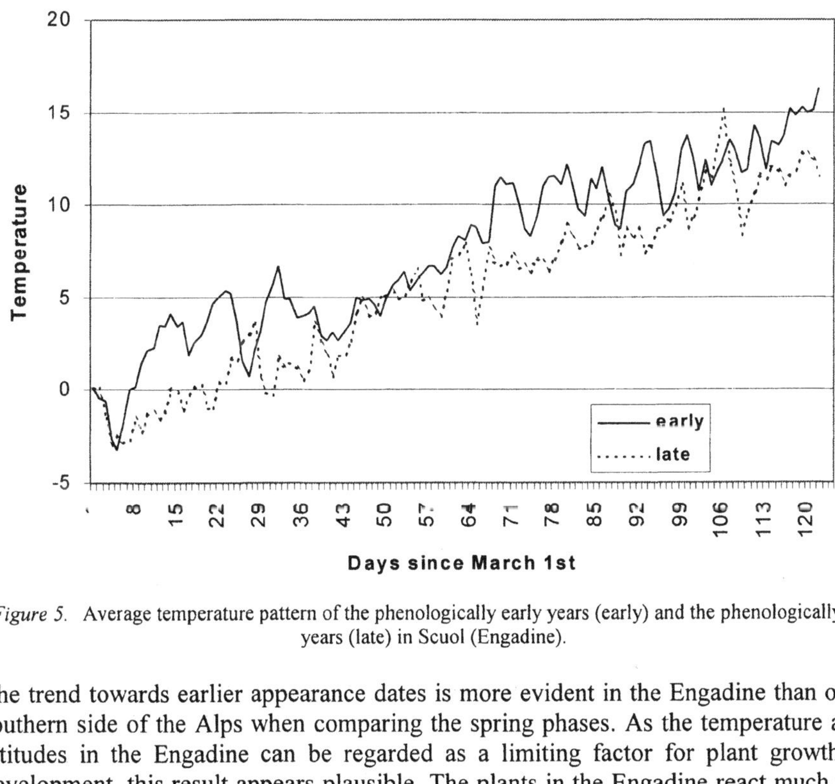

Figure 5. Average temperature pattern of the phenologically early years (early) and the phenologically late years (late) in Scuol (Engadine).

The trend towards earlier appearance dates is more evident in the Engadine than on the southern side of the Alps when comparing the spring phases. As the temperature at the altitudes in the Engadine can be regarded as a limiting factor for plant growth and development, this result appears plausible. The plants in the Engadine react much more strongly to a change in climate (a warming) than south of the Alps where the temperatures are sufficiently high. The strong influence of the air temperature on the phenophases in spring has already been described in Defila (1991). Figure 5 further highlights this discovery. Beginning on 1st March the mean temperatures were recorded at the observation post in Scuol (1240 metres above sea level) for the phenologically early (early) and late (late) springs from 1970 to 1995. It is obvious that the temperature curves of the earlier years are almost all higher than those of the later years. Based on this realisation, the effects of a climatic global change on vegetation need to be studied especially at liminal observation posts. Unfortunately, the phenological series in the Engadine are relatively short. It is hard to discern whether this is a short-term trend. A trend toward earlier appearance dates in spring is shown at relatively many phenophases in the Ticino and in the Engadine. This becomes particularly evident in the Engadine where the temperature is a restrictive factor for plants. South of the Alps, the autumnal phenophases show a definite trend towards later appearance dates. This is an indication of the lengthening of the vegetation period within the last decades that has already been described by Menzel (1997). To substantiate this hypothesis the phenodata from other regions of Switzerland need to be included in the study. The individual plants are subject to many more influences than just meteorological ones. Besides the entire ground complex and the pertinent water supply of the plants, factors such as the age of the plants, the genetic conditions, illnesses, pests, pollution, competition, or anthropogenic sustenance measures can influence the phenological appearance dates of the plants. For this reason, the vegetation development was tested based on various observation posts within a region and different phenophases within a season. By employing this method, the individual characteristics of a plant and the effects of its location are minimised. This analysis is restricted to the spring as the national phenological observation programme studies especially many phenophases in this season. The phenological data from 29 observation posts and 20 phenophases were included in the statistics. The observation posts are at altitudes between 370 metres and 815 metres above sea level. The 20 phenophases are observed on eighteen different plant species; five thereof are grasses and thirteen are trees or bushes. The phenophases consist of fourteen flowering stages and six foliation phases. Figure 6 depicts the vegetation development in the Schweizerisches Mittelland ("Swiss Plateau") in spring. A vague sequence of early, normal, and late years is visible. A trend towards earlier appearance dates cannot be shown with this method.

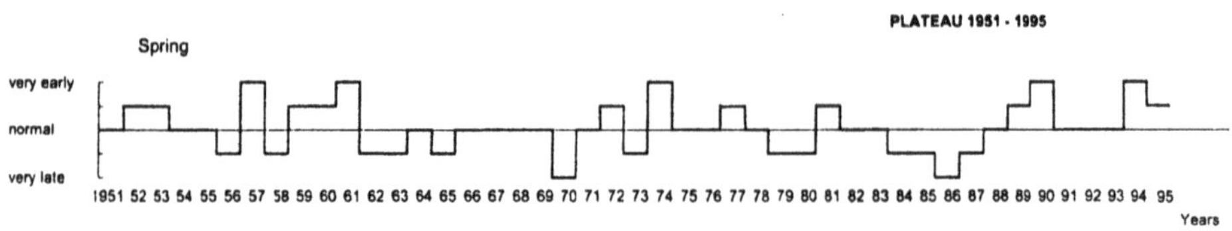

Figure 6. Course of the vegetation development in phenological spring for the Swiss Plateau, 1951-1995.

4. Discussion

The suitability of phenological series for vegetation monitoring cannot yet be confirmed in general. It appears that not all plants and phenophases can be subjected to this method of investigation. The city factor in particular must be taken into account when considering vegetation development. Climate changes affect plants differently. Like plants do not all react in the same way to pollution, it appears that individual plant species and phenophases respond differently to a climatic change (warming). It also needs to be noted that not all climates are equally suited for such studies. The first discoveries from the regions of the Ticino and the Engadine need corroboration and elaboration from other regions and climates. A development of vegetation cannot be shown by data collected from different observation posts and phenophases. Trends do not become evident in phenological series; they are blurred as different plant species react differently to external influences. The danger of trend analysis of individual plants in individual locations is that influences other than meteorological ones – such as ageing or pests – affect the beginning of phenophases or even lead to the appearance of a trend. Whether the vegetative phenophases differ from the generative phenophase in relation to environmental influences needs to be investigated in more detail. Further studies are required to discern which plant species and which phenophases are suitable for vegetation monitoring. Whether there really is a general trend in the phenological series can only be determined with numerous trend analyses which have to be performed at different locations, with various plant species and phenophases. The first results from the Ticino and the Engadine have shown such studies to be promising and worthwhile. Changes in climate should not be studied on the basis of climatological series only. To understand the influence and effects of a rise in temperature on the biosphere and the inherent dangers and opportunities is just as important.

References

Bider, M. & Meyer, A. 1946. Lässt sich der Zeitpunkt der Kirschenernte der Nordwestschweiz vorausbestimmen? Schweiz. Z. f. Obst- u. Weinbau 55(25): 476-483.

Defila, C. 1986. Frostwarnungen im Frühling. Die Grüne 114(17) : 9-14.

Defila, C. 1991. Pflanzenphänologie der Schweiz. Diss. Uni Zürich. Veröff. d. Schweiz. Meteorologischen Anstalt 50: 1-235.

Defila, C. 1996. 45 Years Phytophenological Observations in Switzerland. 1991–1995. Proceedings of the 14th International Congress of Biometeorology, 1-8 September 1996, Ljubljana, Slovenia, Biometeorology 14(2): 175-183.

Hartmann, Ph. 1991. Untersuchungen der Beziehungen zwischen Witterung und der Pflanzenphänologie. Sanasilva-Projekt.

Jeanneret, F. 1996. Phänologie in einem Querschnitt durch Jura, Mittelland und Alpen. Jb. Gg. Ges. BE 59: 159-203.

Menzel, A. 1997. Phänologie von Waldbäumen unter sich ändernden Klimabedingungen – Auswertungen der Beobachtungen in den Internationalen Phänologischen Gärten und Möglichkeiten der Modellierung von Phänodaten. Forstliche Forschungsberichte, München 186: 1-147.

Pfister, Ch. 1984. Klimageschichte der Schweiz 1525 bis 1860 – Das Klima der Schweiz 1525 bis 1860 und seine Bedeutung in der Geschichte von Bevölkerung und Landwirtschaft Band 1. Haupt, Bern.

Primault, B. 1955. Cinq ans d'observations phénologiques systématiques en Suisse. Annalen der Schweiz. Meteorologischen Anstalt 92: 7/4-7/5.

Primault, B. 1984. Phänologie; Blatt 13.1, 13.2; in Kirchhofer, W.: Klimaatlas der Schweiz; 2rd. ed. Verlag des Bundesamtes für Landestopographie, Bern.

Vassella, A. 1997. Phänologie von Waldbäumen. Umwelt-Materialien 73, BUWAL: 9-75.

Schnelle, F. 1955. Pflanzenphänologie. Probleme der Bioklimatologie. 3. Akademische Verlagsgesellschaft, Leipzig.

Schnelle, F. 1985. 25 Jahre phänologische Beobachtungen in den Phänologischen Gärten. Arboreta Phaenologica 29: 1-44.

Schreiber, K.-F. 1977. Wärmegliederung der Schweiz. Grundlagen für die Raumplanung. Eidg. Justiz- und Polizeidepartement. Bern.

SPECIES RESPONSES TO CLIMATIC VARIATION AND LAND-USE CHANGE IN GRASSLANDS OF SOUTHERN SWITZERLAND

ANDREAS STAMPFLI & MICHAELA ZEITER

Institute of Plant Sciences, University of Bern, Altenbergrain 21, CH-3013 Bern, Switzerland

Keywords: Abandonment, drought, restoration, species-rich meadow, seed limitation, succession, variability

Nomenclature: Lauber, K. & Wagner, G. 1996. Flora Helvetica. Haupt, Bern

Abstract
During a period of 10 years, from 1988 to 1997, we monitored the species composition in dry grasslands of high species richness at experimental sites in the 'lower montane zone' of the southern Alps in Switzerland. We examined responses of the abundant herb species to stochastic factors, abandoning, and mowing after abandonment. Although mowing was regularly continued at one meadow site half of the common species showed clear positive or negative frequency trends which are climatically explained. Succession after abandonment resulted in dominance of *Brachypodium pinnatum* and reduction in the number of species at two meadow sites. This process was also affected by spatial heterogeneity and stochastic factors. Resumed mowing in a meadow after 20 years of abandonment resulted in slow shifts of the species composition. *Brachypodium* dominance was unchanged and the establishment success of 'meadow species' was poor. This was not due to unfavourable site conditions but to limited availability of seeds. Seed limitation plays an important role in the dynamics of grasslands. From a conservation perspective this implies that maintenance of still existing meadows of high species richness should be given high priority. A better understanding of regeneration processes and a better knowledge of the regenerative characteristics of grassland herbs would be essential for making better predictions about responses of plant species to altered environmental factors in grasslands.

Kurzfassung
Während 10 Jahren, von 1988-1997, verfolgten wir in der unteren Montanstufe der Schweizer Südalpen die Artenzusammensetzung von artenreichen Halbtrockenrasen. In Feldversuchen untersuchten wir die Reaktionen der häufigen Wiesenpflanzen auf stochastische Faktoren, Vergandung und Mahd nach Vergandung. Die Hälfte aller häufigen Arten einer Wiesen-Versuchsfläche zeigte trotz regelmässig fortgesetzter Mahd bedeutende, klimatisch erklärbare Frequenz-Zunahmen oder Abnahmen. Auf

Burga & Kratochwil (eds.), BIOMONITORING, 107-124

zwei Wiesen-Versuchsflächen führte Vergandung zur Vorherrschaft von *Brachypodium pinnatum* und zu einem Rückgang der Anzahl der Arten. Dieser Prozess wurde durch räumliche Heterogenität und stochastische Faktoren wesentlich mitbeeinflusst. Die wiedereingeführte Mahd bewirkte in einer 20-jährigen Graslandbrache eine langsame Veränderung der Artenzusammensetzung, die *Brachypodium*-Vorherrschaft blieb unverändert, und der Etablierungserfolg von Wiesenarten war gering. Dies war nicht auf ungünstige Standortbedingungen zurückzuführen, sondern auf eine begrenzte Verfügbarkeit von Samen. Letztere spielt in der Graslanddynamik eine wichtige Rolle. Aus der Sicht des Naturschutzes sollte deshalb den noch existierenden artenreichen Wiesen eine hohe Priorität gegeben werden. Damit die Auswirkungen von veränderten Umweltfaktoren auf die Pflanzenarten in artenreicher Graslandvegetation besser vorausgesagt werden können, bedarf es eines besseren Verständnisses der Regenerationsprozesse und vollständiger Kenntnisse der regenerativen Eigenschaften der Graslandarten.

1. Introduction

The fast decline in beautiful meadows as a result of altered land-use practice since the middle of the 20th century has led to advanced conservation initiatives and an increased interest in the processes which cause species abundances to change. In several European countries long-term experiments were started to study the effects of different types of management on grassland vegetation (Schiefer 1981; Krüsi 1981; Willems 1983, 1990; Bakker 1989; Herben et al. 1993). Running several such studies in grasslands at different sites over long periods is necessary and worth-while because the results from any single site are unlikely to be valid for the others and no ecological model can be expected to supply sufficiently accurate between-site predictions. Long-term monitoring provides a basis for constant feedback for refining and restructuring the simple predictive models that are available (Keddy 1991).

In the valleys of southern Switzerland abandonment of traditional small farms and migration of inhabitants to urban centres has been the main cause of vegetation change in unimproved meadows. This process was brought to the public's attention in the 1970s. A decade later a grassland survey showed that only a few hundred hectares of dry meadows and pastures of high species richness remained in the canton of Ticino (Stampfli 1997). Fascinated by the high variety of species assemblages in these dry meadows and the many peculiar management histories which have created these grasslands, one of us (A. Stampfli) started to control the management of experimental plots set up in two dry meadows located in the Blenio valley (Negrentino) and on the slope of Monte Generoso (Pree) and in two dry abandoned grasslands at the slopes of Monte Generoso (Poma) and Monte San Giorgio (Paruscera). Continuous records of the species composition in differently treated plots at annual intervals since 1988 was the main approach by which to recognize the principal processes causing the variation in species composition, and to investigate the effects of different types of management on the responses of species (Stampfli 1993).

The main insight came from an unbroken time series of mown plots at Negrentino; at Poma an understanding of the relevant processes was gained by a seed-addition experiment and subsequent monitoring from 1995 onwards for 2 years. By directing attention to the time dimension in grassland ecology we hope that this review may help to improve the present-day practices of land use and conservation management.

2. Experimental study sites

The three experimental study sites reviewed in this article cover areas of different sizes, *c.* 250 m^2 of a dry meadow at Negrentino, *c.* 350 m^2 of a dry meadow at Pree and *c.* 300 m^2 of an abandoned grassland at Poma. Since 1988 these sites have each been enclosed by a fence to exclude cattle, goats and other domestic or wild animals. For further information on the study sites see Antognoli et al. (1995 p. 164), Stampfli (1992b, 1993, 1995) and Stampfli & Zeiter (1999).

At Negrentino we stopped mowing in 1988 in nine 4.4-m^2 plots which alternate with nine control plots in which mowing, using a scythe, was continued at traditional dates at the end of June and in September. In the mown plots we determined the yearly species composition in June before the first harvest, this cut usually contributing much more to the annual yield than the second one. Yearly sampling took 11-21 days and was started when *Bromus erectus* was in flower and the peak of flowering for most forbs was over. Flowering of other grasses was nearly simultaneous with that of *B. erectus*, except for *Agrostis capillaris* which was usually 2 weeks later. By the time of the first harvest, between 19 June and 2 July in different years, most grasses had completed their life cycles and only a few of the common herbs had not reached the flowering stage: *Pimpinella saxifraga, Prunella vulgaris* and *Carlina acaulis.* We measured the species composition by sampling with the point-frequency technique. This found the percentage of hits per species at 1584 points: 176 points were regularly spaced 10 cm apart in a central 1.76-m^2 area in each of the nine plots. The 'same' points were used in subsequent years and the sampling sequence of points remained the same each year in order to minimize variation due to phenological progress during sampling time. Using this technique we recorded 71 out of 83 herb species which were found in 9 x 1.76-m^2 of our mown plots in 10 years. We selected 43 locally common plants for analysis of responses to climatic variation and excluded 'occasional', and locally rare species which did not reach point-frequency values of 1 % at least once during the 10 years.

From 1988-1993 and in 1997 we determined point frequencies of species in abandoned plots using the same method as in the control plots. Since a *Populus tremula* clone happened to invade our experimental site and we were interested in abundance changes among herbs we had to mow three of our nine experimentally abandoned plots after four years in order to prevent tree establishment. In the analyses on succession we included 52 locally common herbs which came up to average point-frequency values of 1 % at least once during 10 years in the six abandoned or the nine control plots. Thirty-four 'occasional', and locally rare species were excluded.

At Pree we stopped mowing in 1988 in four 20-m^2 plots which represented one of four treatments in an experiment with different management regimes arranged in a 4 x 4 randomized block design. From 1988-1997 we estimated the percentage cover of all the species in central 9-m^2 quadrats of these plots at yearly intervals in early July. In the section on succession we selected 52 locally common herbs of Pree for a comparison with results from the Negrentino site; 28 'occasional', and locally rare herbs which did not come up to 1 % cover at least in one plot or in one year are not presented.

At Poma we resumed regular yearly mowing regimes in two adjacent plots of nearly 100 m^2 in July 1988. One plot was mown twice a year, in early July and late September/early October, the other plot was mown once a year, in early July, and fertilized with 2 kg/m^2 mature manure (containing 17 % dry substance) in November 1994, the seventh year of resumed mowing. The manure originated from a heap of the farm at Pree. Before mowing the aftermath in late September, we recorded the frequency of species in a central strip of 1.2 m x 15.2 m using 456 points per plot, at regular distances of 20 cm (Stampfli & Zeiter 1999).

3. Stochastic variation

Meadows of high species richness are often assumed to be stable assemblages of species as long as mowing is done regularly. In this chapter we ask the question how the 'stable mixture' of about 80 species in an unfertilized dry meadow varies from year to year and we demonstrate the idea that the availability of water during relevant periods of plant growth affects the species composition.

Only in a few cases have fluctuations in the species composition been documented at yearly intervals over a period of 10 years or more in a grassland of high species richness (i.e. Rosén 1985). No such study exists from grasslands of the Alps. In long-term grassland experiments reviewed by van den Berg (1979) fluctuation cycles of plant populations varied from short (one year) to long-lasting (about eight years) but also very slow trends over 20 years occurred. Fluctuations and trends in grassland species composition may be caused by several stochastic factors such as (a) cyclic processes of environmental origin related to climate, soil or hydrology, (b) variation in human activities, (c) cycles of animal activity, mainly insects and voles, (d) variation in the life cycles of some plants and (e) variation in parasite and pathogen reproduction (Rabotnov 1974). Important processes reported by Gigon (1997) from dry meadows of northern Switzerland include external ones such as summer droughts every 5-20 years or occasionally delayed mowing, and endogenous ones such as high vole densities every 5-10 years or half-lives of plant populations which were calculated from short-term observations and varied between 1-45 years for different species (Leutert 1983; Kuhn 1984; Marti 1994).

Climatic effects on the botanical composition are generally presumed to be considerable. In theory climatic variation is suggested to be one of several mechanisms explaining the coexistence of many species (Chesson 1986; Tilman & Pacala 1993). In practice, however, the relationship between plant fluctuations and weather conditions is not well understood yet (Grubb et al. 1982; Bakker 1989). This is due to the scarcity of consecutive long-term series of data series which are truly needed to detect such relationships (Silvertown et al. 1994; Dodd et al. 1995; Dunnett et al. 1998).

In a temperate-humid climate plant growth is periodically limited by low temperatures and occasional snow during the cold season. In dry meadows plant growth is sometimes strongly limited by the availability of water, a variable resource which depends on the local weather. Since mowing was regularly continued and activities of small mammals or extraordinary fluctuations of insects were not apparent at our study site at Negrentino we assume that the species composition was mainly affected by the availability of water during relevant periods of plant growth, in our case between late March and October.

During this period long-term averages of monthly sums of precipitation are above 100 mm and do not indicate any shortage in the water availability. The year-to-year variability of rainfall, however, is high in this part of the Alps. Extended droughts may result from alternating periods of dry anticyclonic weather and cyclonic periods with prevailing dry northern winds. During the past decade such periods of extraordinary drought occurred in autumn 1989, summer 1990, late spring and summer 1991 and in late winter and spring 1997.

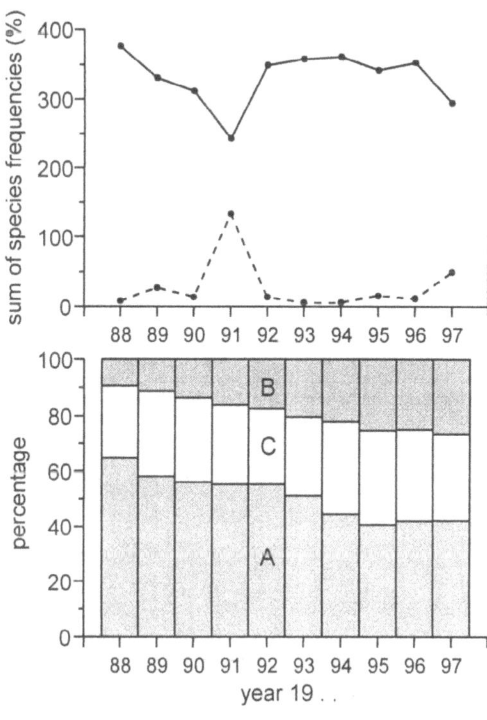

Figure 1. Sum of frequencies (percentages of hits at 1584 points) for all species (solid line), frequency of gaps (dashed line), and proportions (based on frequency sums) for three response groups in nine regularly mown plots at Negrentino: (A) twelve decreasing species, *Bromus erectus, Festuca tenuifolia, Brachypodium pinnatum, Carex caryophyllea, Agrostis capillaris, Trifolium montanum, Lotus corniculatus, Sanguisorba minor, Primula veris, Leontodon hispidus, Achillea millefolium, Dactylis glomerata* (in order of decreasing abundance), (B) ten increasing species, *Helianthemum nummularium, Thymus pulegioides, Scabiosa columbaria, Thalictrum minus, Hypochaeris radicata, Pimpinella saxifraga, Prunella vulgaris, Salvia pratensis, Dianthus carthusianorum, Rumex acetosella* (in order of decreasing abundance), (C) all other species, including 21 common and 28 rare ones.

Exploring plant responses we calculated linear regressions of yearly proportions of point-frequency values of the 43 locally common species on time. We got proportions by dividing point-frequency values of single species by the sum of point-frequency values of all the species.

Almost all the locally common herbs in our investigated area of *c.* 16 m^2 persisted for 10 years. Only one species, *Rumex acetosella,* newly colonized a small part of the area where the proportion of drought-induced gaps after 1991 was high (Stampfli 1995). Although stability of the species composition cannot be tested in a strict sense (Connell & Sousa 1983) because 10 years of observation may not exceed one turnover of all its populations, one might conclude from a qualitative point of view that the meadow was near to compositional equilibrium. However, if emphasis is on species quantities, the idea of a compositional equilibrium cannot be maintained: 22 out of 43 common species showed significant trends (Figure 1).

The variation in species composition observed over 10 years can be explained by similar or different responses of species to variation of climatic humidity. Relative humidity in spring had an important influence on the performance of the four most abundant species *Bromus erectus, Festuca tenuifolia, Brachypodium pinnatum* and *Carex caryophyllea* which showed highly correlated variation with time. Apart from the abundant grasses two different drought-response groups emerged from variation patterns determined for the period 1988-1993 (Stampfli 1995). Some of these patterns in variation dissolved in the long run and species showed individual responses after the strong summer drought in 1991. This can be explained by different modes of regeneration of species, i.e. the clonally growing *Thymus pulegioides* expanded quickly for 1-2 years, while the seed-recruited species *Plantago lanceolata, Scabiosa columbaria, Hypochaeris radicata,* and *Pimpinella saxifraga* responded more slowly; their number increased during a period of 3-5 years.

Different time lags of species responses created by various modes of regeneration but also the fact that the number of plant individuals and plant biomass do not respond in the same way to weather variables (Herben et al. 1995) complicate the efforts to elucidate the relationships between climate and vegetation. A prolonged time series will reliably clarify how the quantitative trends in species composition are related to long-term climatic or oceanic cycles (Willis et al. 1995).

4. Succession in abandoned meadows

Abandonment of meadows of high species richness often results in an increase of a few, and a loss of many, species. Generally species density declines more conspicuously the smaller the scale of observation is. Practically this means that in a few years some species vanish from areas smaller than 1 m^2 (Stampfli & Häfelfinger 1995). The principal mechanism involved in this process is competition for light. Light is altered when a layer of litter is formed (Stöcklin & Gisi 1989). Accumulated litter may prevent seeds from germination and seedlings from getting established (Grime 1979; Kienzle 1979). Moreover, litter was shown to reduce the density of reproductive shoots and the production of seeds of dominant grasses (Grant et al. 1963; Stampfli 1992a; Schlaepfer 1997). Such responses can be explained by decreased light intensity and temperature variation at the soil surface (Zimmermann 1979).

The changes induced by cessation of mowing in the course of years depend on deterministic and stochastic factors (Pickett 1982). The first include genetically fixed characters of the interacting plant species, the latter climatic variability, the activity of herbivores or seed predators and dispersal agents, among others. The rate of change much depends on how these factors interact with each other and it varies from site to site in a physically heterogeneous environment. An investigation of all involved factors and their interaction is an impossible task. Therefore an exact answer to questions like 'after what time does a certain species vanish' or 'when do woody species establish in abandoned meadows' cannot be given. At best such questions could be answered in probabilistic terms. Nevertheless, by analyzing morphological and functional characteristics of species response groups it might be possible to detect the crucial traits which enable plants to cope differently with changed resources.

From a 10-year perspective we complete earlier results on successional trends of species from our abandoning experiment at Negrentino (Stampfli & Häfelfinger 1995) and we add results of species responses in abandoned plots at Pree for comparison. We examine whether plants of different response groups have certain traits in common which may affect their ability to cope with a limited light resource.

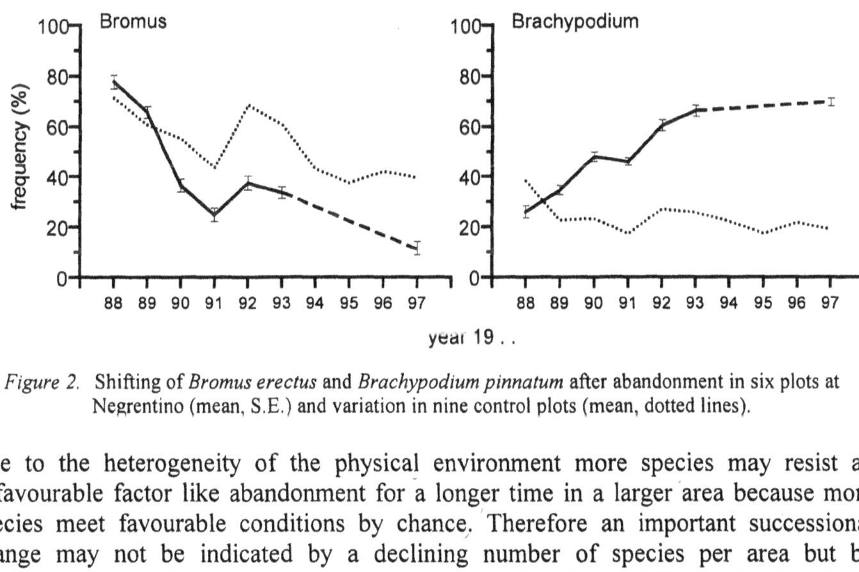

Figure 2. Shifting of *Bromus erectus* and *Brachypodium pinnatum* after abandonment in six plots at Negrentino (mean, S.E.) and variation in nine control plots (mean, dotted lines).

Due to the heterogeneity of the physical environment more species may resist an unfavourable factor like abandonment for a longer time in a larger area because more species meet favourable conditions by chance. Therefore an important successional change may not be indicated by a declining number of species per area but by significant quantitative trends in any measure of abundance. Effects of interactions between abandonment and climate are included and cannot be separated from effects of abandonment. Successional responses of locally common species at Negrentino were classified by various criteria: difference of point-frequency means in abandoned and mown plots after 10 years, linear trend of point-frequency means over 10 years for species with non-zero values in abandoned plots and incidence of a temporary frequency maximum which is significantly different from that ascertained in the control plots, and consistency of change by calculation of difference between frequency values in 1988 and 1997 in single plots taking also error of the sampling method into account (Table 1).

Table 1. Response groups of locally common herbs to abandonment at Negrentino classified by 10-yr trend of yearly point-frequency means, point-frequency difference between abandoned and control plots in the 10th year (1997) or consistency of 10-yr point-frequency change in six single plots. Attributes of species: leaf canopy L, height < 10 cm (1) / 10-30 cm (2) / > 30 cm (3), leaf in a basal rosette or largest leaves towards base (b); principal mode of regeneration R, by seeds (s), vegetative means (v) or mode unclear (.) based on Grime et al. (1988) and own measurements *(in italics)*; preferred habitat H, mown (m), abandoned (a) grassland or no preference (.) based on a survey in southern Switzerland (Häfelfinger et al. 1995). Frequency means: before abandoning 1988 [values missing due to undetermined individuals (?)]; in the 10th year of abandonment 1997; Trend, linear regression over 10 yrs, $P < 0.05$ (pos/neg) or not significant (.), maximum of temporary increase if significantly different from control; indices $^{+/-}$ indicate significant difference between abandoned and control plots (t-test, $P < 0.05$). No. of plots: number of plots in which species shows positive (+), no (~) or negative (–) 10-yr change [values uncertain due to undetermined individuals in 1988 (?)] [10-yr changes were considered significant if calculated differences of observed values were larger than standard errors of sampling multiplied by $t_{0.05} = 2.101$; sampling errors were estimated from errors of 10 repeated samples which were approximated from point-frequency values using Goodall's equation, as in Stampfli (1991)]. Response: classified from experiments at Negrentino (NE) and Pree (PR), positive (+), negative (–), indifferent (~), or not classified (.).

Species Response Groups	**Attributes**			**Frequency Mean**			**No. of Plots**			**Response**	
	L	**R**	**H**	**1988**	**Trend**	**1997**	**+**	**~**	**–**	**NE**	**PR**
Increasing											
Brachypodium pinnatum	3 b	v	a	25.9	pos	69.5 $^{+}$	6	.	.	+	+
Helictotrichon pubescens	3	s	.	0.4	pos	6.0 $^{+}$	6	.	.	+	–
Viola hirta	2 b	vs	.	0.9	pos	2.7 $^{+}$	5	1	.	+	~
Thalictrum minus	3	.	.	1.8	.	11.0	5	.	.	+	.
Galium lucidum	3	.	.	0	.	2.7	1	.	.	+	.
Pimpinella saxifraga	3 b	s	.	1.8	pos	4.2	3	3	.	+	.
Potentilla erecta	3	vs	.	2.4	pos	5.8	3	3	.	+	+
Veronica chamaedrys	2	v	.	0	.	1.9	1	1	.	+	.
Increasing temporarily											
Trisetum flavescens	3	s	m	0.2	1.8 $^{+}$	0.8	2	4	.	+	~
Clinopodium vulgare	3	vs	.	0.3	.	0.6	2	2	.	+	.
Silene nutans	2 b	s	.	5.1	9.0 $^{+}$	3.9	1	2	3	~	–
Primula veris	2 b	.	.	2.4	5.2 $^{+}$	1.8	2	1	3	~	.
Leontodon hispidus	3 b	vs	m	1.5	.	0.9	1	4	1	~	–
Festuca tenuifolia	2 b	vs	m	65.2	neg	42.0	1	1	4	–	~
Rumex acetosa	2 b	s	.	1.9	.	1.3	.	3	3	–	~
Dactylis glomerata	3 b	s	.	1.4	7.7 $^{+}$	1.1	.	4	2	–	–
Phyteuma betonicifolium	2 b	s	.	0.2	2.3 $^{+}$	0	.	4	2	–	~
Decreasing											
Bromus erectus	3 b	s	.	77.6	neg	11.5	.	.	6	–	+
Danthonia decumbens	2 b	s	.	19.4	neg	1.6	.	.	6	–	~
Plantago lanceolata	3 b	vs	m	8.8	neg	0.3	.	.	6	–	.

Table 1 continued

Briza media	2 b	s	.	11.6	neg	0.5^{-}	.	.	6	–	–
Lotus corniculatus	*2*	s	m	7.3	.	0^{-}	.	.	6	–	–
Carex caryophyllea	2 b	v	m	25.1	.	6.5^{-}	.	.	6	–	–
Trifolium montanum	*2*	.	.	12.8	.	2.7^{-}	.	.	6	–	.
Anthyllis vulneraria	2 b	*s*	m	3.8	.	0^{-}	.	.	6	–	~
Hypochaeris radicata	*1* b	vs	m	1.8	.	0^{-}	.	.	6	–	.
Trifolium pratense	*2*	s	m	1.1	.	0^{-}	.	.	5	–	–
Leucanthemum vulg. aggr.	*3*	s	m	1.7	.	0^{-}	.	1	5	–	.
Prunella vulgaris	*1* b	vs	m	2.8	.	0.2^{-}	.	2	4	–	.
Arabis ciliata	*1 b*	*s*	m	0.5	.	0^{-}	.	1	3	–	.
Trifolium repens	2 b	v	m	0.5	.	0^{-}	.	3	3	–	.
Anthoxanthum odoratum	*2*	s	m	1.2	.	0.3^{-}	.	4	2	–	–
Ranunculus bulbosus	2 b	s	m	0.7	.	0.1^{-}	.	4	2	–	–
Luzula campestris	2 b	v	m	7.1	.	1.0	.	.	6	–	–
Potentilla pusilla	*1 b*	.	.	7.1	.	2.7	.	1	5	–	.
Holcus lanatus	*3*	s	.	0.9	3.2^{+}	0.2	.	1	4	–	~
Achillea millefolium	*3* b	v	.	1.2	.	0.3	.	2	3	–	.
Sedum sexangulare	*1 b*	.	.	1.4	neg	0.2	.	.	2	–	.
Hieracium lactucella	*1 b*	.	.	2.5	.	0	.	.	1	–	.
No consistent trend											
Thymus pulegioides	*2*	.	m	6.2	.	8.0	4	.	2	~	~
Sanguisorba minor	2 b	s	m	0.4	.	1.6	4	.	1	~	~
Campanula rotundifolia	2 b	vs	.	1.0	.	0.9	2	2	2	~	.
Salvia pratensis	*2 b*	.	.	4.2	.	4.1	2	2	2	~	~
Koeleria macrantha	*2*	s	.	4.5	.	9.2	2	2	2	~	.
Agrostis capillaris	*3*	vs	.	6.7	.	3.9	2	.	4	~	~
Carlina acaulis	*2 b*	.	.	1.2	.	0.9	1	1	4	~	.
Helianthemum nummular.	*2*	.	.	20.6	.	12.1	1	2	2	~	–
Dianthus carthusianorum	*2 b*	*s*	.	0.1	.	0.1	1	.	1	~	–
Scabiosa columbaria	*1* b	s	m	0.4	.	0.1^{-}	1	2	3	~	–
Hippocrepis comosa	*2 b*	.	.	0.1	.	0.1	.	1	.	~	~
Festuca rubra	2 b	vs	m	?	.	11.4		5?		~	~
Carex ornithopoda	*1 b*	.	.	?	.	0		2?		~	.

Table 2. Response groups of locally common herbs to abandonment at Pree classified by differences in cover between 1989 and 1996/97 (mean) or changed occurrences in the four plots. Attributes of species as in Table 1. Cover: proportion of species as visible to a standing observer, estimated in classes <1 % (1), 1-<2 % (2), 2-<5 % (5), 5-<10 % (10), 10-<20 % (20), 20-<30 % (30). No. of plots: number of plots in which species newly emerged (+), survived (~), or vanished (–) in the interval 1988/89 - 1996/97. Response: classified from experiments at Pree (PR) and Negrentino (NE), positive (+), negative (–), indifferent (~), or not classified (.). *Asphodelus albus* values reduced due to pulled up plant individuals are marked with an asterisk (*).

Species Response Groups	**Attributes**			**Cover**		**No. of Plots**			**Response**	
	L	**R**	**H**	**1989**	**1996/7**	**+**	**~**	**–**	**PR**	**NE**
Increasing										
Hypericum perforatum	3 b	vs	.	.	1	3	.	.	+	.
Arrhenatherum elatius	3	s	.	1	5	1	1	.	+	.
Carex montana	*2 b*	.	.	1	2	1	2	.	+	.
Brachypodium pinnatum	*3* b	v	a	5	30	.	4	.	+	+
Cruciata glabra	*2*	*v*	.	2	10	.	4	.	+	.
Bromus erectus	*3* b	s	.	1	5	.	4	.	+	–
Festuca ovina	2 b	vs	a	5	10	.	2	.	+	.
Potentilla erecta	*3*	vs	.	1	5	.	2	.	+	+
Teucrium chamaedrys	*2*	.	.	1	5	.	1	.	+	.
Decreasing										
Dianthus carthusianorum	*2 b*	*s*	.	1	.	.	.	4	–	~
Prunella grandiflora	*1*	.	.	1	.	.	.	4	–	.
Ranunculus bulbosus	*2* b	s	m	5	.	.	.	4	–	–
Hieracium pilosella	1 b	vs	.	1	.	.	.	3	–	.
Leontodon hispidus	*3* b	vs	m	1	.	.	.	3	–	~
Trifolium pratense	2	s	m	1	.	.	.	3	–	–
Festuca pratensis	*3* b	vs	m	1	.	.	.	2	–	.
Centaurea nigrescens	*3*	.	.	2	1	.	1	3	–	.
Scabiosa columbaria	1 b	s	m	2	1	.	1	3	–	~
Helictotrichon pubescens	*3*	s	.	1	1	.	2	2	–	+
Dactylis glomerata	3 b	s	.	1	1	.	2	2	–	–
Lotus corniculatus	2	s	m	1	1	.	2	2	–	–
Briza media	*2* b	s	.	1	1	.	2	2	–	–
Luzula campestris	*2* b	v	m	5	1	.	3	1	–	–
Helianthemum nummularium	2	.	.	2	1	.	3	1	–	~
Silene nutans	*2 b*	*s*	.	2	1	.	3	1	–	~
Carex caryophyllea	*2* b	v	m	20	5	.	4	.	–	–
Peucedanum oreoselinum	*3 b*	.	.	10	1	.	4	.	–	.
Anthoxanthum odoratum	2	s	m	10	2	.	4	.	–	–

Table 2 continued

No consistent trend										
Festuca tenuifolia	*2 b*	*vs*	m	20	20	.	4	.	~	–
Festuca rubra/heterophylla	2 b	vs	m	5	5	.	4	.	~	~
Galium rubrum	*2*	*vs*	.	5	5	.	4	.	~	.
Betonica officinalis	1 b	v	.	2	5	.	4	.	~	.
Stellaria graminea	2	*vs*	.	2	1	.	4	.	~	.
Trisetum flavescens	*3*	s	m	2	1	.	4	.	~	+
Agrostis capillaris	*3*	vs	.	1	2	.	4	.	~	~
Hippocrepis comosa	*2 b*	.	.	1	1	.	4	.	~	~
Holcus lanatus	3	s	.	1	1	.	4	.	~	–
Phyteuma betonicifolium	*2 b*	*s*	.	1	1	.	4	.	~	–
Rumex acetosa	2 b	s	.	1	1	.	4	.	~	–
Thymus pulegioides	*2*	*vs*	m	1	1	.	4	.	~	~
Viola hirta	*2* b	vs	.	1	1	.	4	.	~	+
Danthonia decumbens	2 b	s	.	1	1	.	3	.	~	–
Galium verum	2	v	.	1	1	.	3	.	~	.
Asphodelus albus	*3 b*	.	.	5	5*	.	3	1*	~	.
Narcissus verbanensis	*2 b*	.	.	1	1	.	3	1	~	.
Salvia pratensis	*2 b*	.	.	1	1	.	2	1	~	~
Silene vulgaris	*2*	s	.	1	1	.	1	.	~	.
Cerastium strictum	*1*			1	1		1		~	
Euphorbia verrucosa	*1*	.	.	1	2	.	1	.	~	.
Sanguisorba minor	2 b	s	m	1	1	.	1	.	~	~
Anthyllis vulneraria	2 b	*s*	m	1	1	.	1	1	~	–

Abandoning at Negrentino resulted in reversed shifts of the abundant grasses *Brachypodium pinnatum* and *Bromus erectus* over 10 years (Figure 2). During this process the more resolutely clonal *B. pinnatum* became dominant over *B. erectus*, which more relies on seed recruitment. More than half of the 52 common herbs showed successional changes (Table 1). Many more species decreased than increased, eight species completely disappeared from the established vegetation and only one species, *Galium lucidum*, obviously a descendant from a flushing population nearby, newly invaded. Several species increased for a few years and decreased later on, they showed distinct population 'outbreaks' or at least temporarily higher point-frequency means in abandoned plots. Such temporary population 'outbreaks' are due to an increased availability of nutrients and water in abandoned meadows, as indicated by the grasses *Dactylis glomerata* and *Trisetum flavescens* which are commonly found in environments richer in nutrients. Why do these grasses not displace *B. pinnatum* in the long run? The levels of water and nutrients are probably not constantly high enough over time. Furthermore *B. pinnatum* has a well-adapted strategy of producing new shoots under light-limited conditions caused by thick layers of litter: it effectively withdraws nutrients from senescing shoots, translocates them to the rhizome system and remobilizes the reserves in the following spring in support of new shoot growth (de Kroon & Bobbink 1997).

The experiment at Pree shows similar results after 8-10 years of abandonment as far as different methods allow a comparison at all: at Negrentino, the plots were more thoroughly searched for single individuals; at Pree plots were less subjected to trampling and the structure of the litter layer was less affected. At Pree *Brachypodium pinnatum* increased and *Festuca tenuifolia* maintained high abundance, 20 out of 52 locally common species disappeared from at least one and seven species from all 9-m^2 plots (Table 2). Referring to all the species present in 1988 or 1989 26 % disappeared at the study-site scale (4 x 9 m^2) and 32 - 41 % at the single-plot scale.

Most species responded consistently in both experiments (Table 1, 2). Only *Bromus erectus* and *Helictotrichon pubescens* showed weak opposite trends. Inconsistent responses may be a consequence of small-scale heterogeneity within the meadows as in the case of *Scabiosa columbaria:* at Negrentino, this species decreased in three but increased in one abandoned plot with reduced water storage capacity due to a rock near the soil surface. Such conditions are similar to the mown plots, and *S. columbaria* was able to increase because a luxuriant growth of *Brachypodium pinnatum* and the subsequent formation of a thick layer of litter was prevented due to relatively dry conditions. Inconsistent responses between sites may also have originated from other factors such as small mammal activity or pathogens. Moles and voles were not present at Negrentino but showed variable activity between plots and years at Pree. Here, mole hills may have served as favourable microsites for recruitment of seedlings. In 1993 a pathogen caused considerable damage to populations of *Helianthemum nummularium* and *Silene nutans* at Pree, flowering was very limited and cover was strongly reduced.

Although both experimental sites were affected by different stochastic factors a majority of species responded in a similar way to abandonment, allowing the recognition of consistent response groups. Moreover a survey in four valleys of southern Switzerland suggests that the majority of our decreasing species is generally more often found in mown than in abandoned meadows (Häfelfinger et al. 1995, Table 1, 2). This supports the idea that genetically determined attributes of species affect succession in meadows after abandonment. What attributes do ‘meadow species’, as opposed to ‘species of abandoned grasslands’, have in common with one another and why do some herbs disappear more quickly from abandoned meadows than others?

The persistence of species after abandonment can be expected to largely depend on characteristics which enhance their ability to cope with a limiting light resource. In other words, species with low leaf canopies and species which need to regenerate by seeds are handicapped. An examination of life history characteristics revealed that no annual or biennial plants are among the common species of our abandoned plots, and apart from *Anthyllis vulneraria* and *Arabis ciliata* all the species at both sites are long-lived perennials. But despite this monotony in life history types, there are relevant morphological and regenerative characteristics which are not equally distributed between our two groups of species showing either positive (n = 15) or negative (n = 35), but not opposite, responses (Table 1, 2): A leaf-canopy height lower than 10 cm cannot be found among positively responding species, but nine times among negatively responding ones. Leaf-canopy heights lower than 10 cm, or lower than 30 cm and leaves in basal rosettes or concentrated towards the base of the plant, are more numerous among negatively (21 out of 35) than among positively responding (3 out of 15) species. On the other hand, regeneration by seeds as an exclusive principal mode of regeneration can more often be found among negatively responding (16 out of 27) than among positively responding species (3 out of 11). In 43 out of 50 cases leaf-canopy height and regeneration mode can explain the affiliation of responding species to their response group, according to the following key: herbaceous species responding negatively to abandonment are characterized by (a) a low leaf-canopy height (< 10 cm or < 30 cm if leaves are basal), or (b) regeneration by seeds as an exclusive principal mode of regeneration, or (c) basal leaves if two principal modes of regeneration (by seeds and vegetative means) exist. Accordingly, species responding positively to abandonment are characterized by (a) vegetative regeneration as an exclusive principal mode of regeneration and a leaf canopy > 10 cm, or > 30 cm if leaves are basal, or (b) a leaf canopy > 30 cm and leaves not basal if two principal modes of regeneration (by seeds and vegetative means) exist.

Some species adapt to a changing light resource by reducing their reproductive effort (Stampfli & Häfelfinger 1995). However, in the long run their populations may not survive under a dense layer of litter, either because they are too short-lived, too small or unable to hold their leaves in a favourable position (Poschlod et al. 1995), or because they rely more on a reproductive strategy by seeds and their capacity to regenerate vegetatively is not optimized due to a trade-off between regenerative means.

5. Succession after resumed mowing in abandoned grasslands

Can the decline in the number of plant species due to abandonment be reversed by mowing? The dynamic equilibrium theory (Huston 1994) predicts an increase in species diversity if an 'intermediate disturbance' regime like mowing is resumed in a relatively species-poor abandoned grassland. If resource levels are similar to resource levels in meadows, more species should be able to coexist. Competitive dominants should more or less strongly decrease as a consequence of the mortality of modules, depending on their vulnerability to physical damage caused by mowing. Species diversity, however, can only increase if species are available in the surrounding landscape, or in terms of the species-pool hypothesis, at the next larger spatial scale (Taylor et al. 1990; Eriksson 1993; Zobel 1997). If 'new' species appear after mowing in an abandoned grassland they are either expected to have immigrated or recruited from the seed bank or from vegetative parts in the soil.

Since 1988 we have observed the consequences of mowing in an abandoned, species-poor, *Brachypodium pinnatum*-dominated meadow at Poma which clearly differed from the nearby mown meadow at Pree in species-abundance proportions and small-scale species diversity. We assume that species pools of both sites had been similar before 1968 when Poma was abandoned and *Brachypodium pinnatum* and woody species started to competitively exclude many 'meadow species'. As mechanisms of change in species composition could not satisfactorily be understood from monitoring species compositions in our large plots at yearly intervals we carried out an additional experiment on a smaller scale and sampled at shorter time intervals from 1995-1997 (Stampfli & Zeiter 1999).

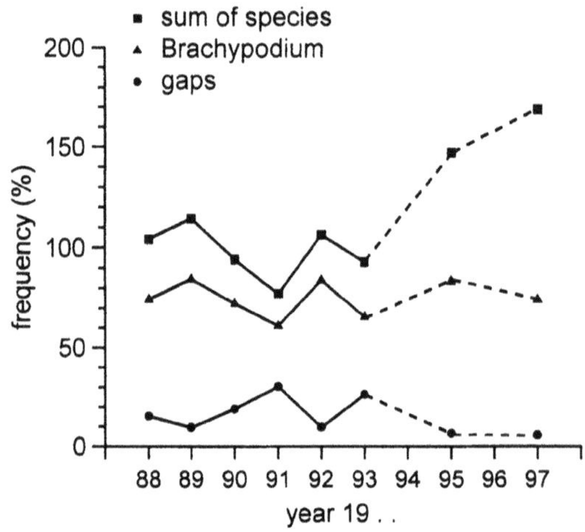

Figure 3. Sum of frequencies (percentages of hits at 456 points) for all species, and frequency of *Brachypodium pinnatum* and gaps after resumed mowing in an abandoned grassland at Poma (unmanured plot, aftermath).

Our experiment on the scale of 200 m^2 showed a slow response of species to resumed mowing in *Brachypodium*-dominated grassland. During 10 years of regular mowing *Brachypodium pinnatum* and *Potentilla erecta* remained the two most abundant species and did not change ranks. *B. pinnatum* maintained a dense cover of shoots: it was present in more than 75 % of the vegetation cover throughout the 10-year period and fluctuated consistently with the humidity of the growth period. The summed up point frequencies of other herb species remained comparably low during 6 years and increased strongly during the last 4 years of the decade (Figure 3), 15 low abundant species showed positive trends in one or both plots, and the number of established species in both plots increased from 46 to 68 in the interval 1988/1989 - 1996/1997. After manuring in November 1994 more 'new' species appeared in the manured than in the unmanured plot (Stampfli & Zeiter 1999).

Although *Brachypodium pinnatum* is considered to be vulnerable to mowing (Briemle & Ellenberg 1994) and this view is supported by field experiments in north-western Europe (i. e. Dierschke 1984; Bobbink & Willems 1991) mowing did not cause increased shoot mortality of the competitive dominant in our experiment. *B. pinnatum* preserved a dense vital network of rhizomes and shoot complexes for 10 years. Six years of regular mowing and a series of dry summers which created gaps in the vegetation cover passed before other species finally started to increase. The most successful ones *(Ajuga reptans, Carex montana, Lathyrus montanus, Veronica officinalis)* increased by clonal growth expanding from populations which had persisted during 20 years of abandonment, either as established or as seed population in the soil

(Stampfli & Zeiter 1999). Although reproducing populations of species relying on seed dispersal *(Centaurea nigrescens, Betonica officinalis, Luzula campestris, Rumex acetosa)* were present before mowing was resumed these species did not increase during the first 6 years. Thus recruitment from seeds did not play an important role after mowing was resumed. As newly invading species were mostly found at short distances from the experimental plots and several other species present in high abundance at a distance of *c.* 100 m in the nearby meadow at Pree *(Arrhenaterum elatius, Helictotrichon pubescens, Festuca ovina, Holcus lanatus, Hypochaeris radicata, Leontodon hispidus, Trisetum flavescens)* did not appear before manuring in 1994, we could assume that seeds were transported by manure, and limitation of seeds from 'meadow species' was probably a major reason for the slow shifts in species composition observed during 10 years.

From monitoring large plots at yearly intervals it was not clear whether seedlings did not appear because seeds were not available or whether they could not establish because of unfavourable weather and/or unsuccessful competition with *B. pinnatum* clones. Therefore we carried out a seed-addition experiment in the abandoned meadow to test whether the establishment of 'meadow species' is limited by a lack of diaspores or favourable microsites for germination and recruitment from the seed bank. We sowed individually *c.* 1.2×10^4 seeds in total of 12 species originating from the nearby meadow at Pree in 30 experimental plots of 60 cm x 60 cm. In this way we did not surpass the natural seed rain density in meadows (Poschlod & Jackel 1993). We monitored the fate of sown and spontaneously emerged individuals for 24 months (Stampfli & Zeiter 1999).

Recruitment limitation was proven by 11 out of 12 species which germinated when added but did not germinate from the seed bank when not added. Once seed limitation was overcome by introducing seeds, individuals of most species established successfully and survived for at least 24 months, some of them even reproduced within this period. Dispersal limitation was indicated by the fact that virtually no species emerged which was not present in the established vegetation of the experimental plots or within a few metres' distance.

6. Conclusions

Even if a meadow is regularly mown its species composition is not 'constant' with time. Apart from considerable year-to-year fluctuations many herb species may also show trends over longer time periods. Variation of humidity is suspected to have a strong effect on the species composition in dry meadows. For ecological field studies this has important consequences: (1) short-term studies are likely to miss the relevant processes, (2) year-to-year variation can only be accounted for if field experiments are replicated with time.

Predicting successional changes of species composition due to any change in management is difficult because various site-specific stochastic factors and a largely unknown small-scale site heterogeneity are involved. After a change in management species may show a temporary response which is opposite to the response in the long run. Wise planning of conservation management cannot simply be guided by large-scale models alone, but has to include genuine studies of the relevant processes on the local scale.

In a landscape of small fragmented meadows of high species richness, as it exists in many valleys in the Swiss Alps today, restoration of meadows by mowing after two or more decades of abandonment is a very slow process because many 'meadow species' do not have persistent seed banks (Thompson et al. 1997) and dispersal over distances of more than 25 m followed by successful establishment is very unlikely. Our results are in accordance with recent findings in other parts of the world (Burke & Grime 1996; Tilman 1997) and we confirm the importance of seed limitation in the dynamics of grasslands (Primack & Miao 1992; Poschlod et al. 1996). From a conservation perspective this implies that maintenance of still existing meadows of high species richness should be given high priority. The same was recently concluded from a soil seed bank study in European grasslands (Bekker et al. 1997).

A promising approach for predicting effects of management changes is based on functional attributes of the species (Grime et al. 1988). However, at the moment, for many species which are common or locally abundant in dry grasslands of the southern Alps and calcareous Prealps basic information on regenerative attributes is not available yet. Important aspects of community invasibility such as germination rates, mortality rates of seedlings after germination and their environmental causes are not well understood. We suggest that regenerative processes and reproductive traits of species, especially dispersal and seed bank characteristics, are important topics for further study.

Acknowledgements

Thanks to support by the Swiss National Science Foundation and the Office for Nature Protection of the Canton of Ticino and to animated collaboration of many colleagues at the former Institute of Geobotany and in Ticino we were able to maintain treatments and records over a period of 10 years. Comments by O. Graf and D.M. Newbery contributed to improve this chapter.

References

Antognoli, C., Guggisberg, F., Lörtscher, M., Häfelfinger, S. & Stampfli, A. 1995. Prati magri ticinesi tra passato e futuro. Porconi, Losone.

Bakker, J.P. 1989. Nature management by grazing and cutting. Kluwer Academic Publishers, Dordrecht.

Bekker, R.M., Verweij, G.L., Smith, R.E.N., Reine, R., Bakker, J.P. & Schneider, S. 1997. Soil seed banks in European grasslands: does land use affect regeneration perspectives? J. Appl. Ecol. 34: 1293-1310.

Bobbink, R. & Willems, J.H. 1991. Impact of different cutting regimes on the performance of *Brachypodium pinnatum* in Dutch chalk grassland. Biol. Conserv. 56: 1-21.

Briemle, G. & Ellenberg, H. 1994. The mowing compatibility of grassland plants. Natur Landsch. 69: 139-147.

Burke, M.J.W. & Grime, J.P. 1996. An experimental study of plant community invasibility. Ecology 77: 776-790.

Chesson, P.L. 1986. Environmental variation and the coexistence of species. pp. 240-256. In: Diamond, J.A. & Case, T.J. (eds), Community ecology. Harper & Row, New York.

Connell, J.H. & Sousa, W.P. 1983. On the evidence needed to judge ecological stability or persistence. Am. Nat. 121: 789-824.

de Kroon, H. & Bobbink, R. 1997. Clonal plant dominance under elevated nitrogen deposition, with special reference to *Brachypodium pinnatum* in chalk grassland. pp. 359-379. In: de Kroon, H. & van Groenendael, J. (eds), The ecology and evolution of clonal plants. Backhuys Publishers, Leiden.

Dierschke, H. 1984. Experimentelle Untersuchungen zur Bestandesdynamik von Kalkmagerrasen (Mesobromion) in Südniedersachsen - I. Vegetationsentwicklung auf Dauerflächen 1972-1984. pp. 9-24. In: Schreiber, K.F. (ed), Sukzession auf Grünlandbrachen. Münstersche Geographische Arbeiten 20. Schöningh, Paderborn.

Dodd, M., Silvertown, J., McConway, K., Potts, J. & Crawley, M. 1995. Community stability: a 60-year record of trends and outbreaks in the occurrence of species in the Park Grass Experiment. J. Ecol. 83: 277-285.

Dunnett, N.P., Willis, A.J., Hunt, R. & Grime, J.P. 1998. Climate and Vegetation: A thirty-nine year study of road verges near Bibury, Gloucestershire. J. Ecol. 86:610-623.

Eriksson, O. 1993. The species-pool hypothesis and plant community diversity. Oikos 68: 371-374.

Gigon, A. 1997. Fluktuationen des Deckungsgrades und die Koexistenz von Pflanzenarten in Trespen-Halbtrockenrasen (Mesobromion). Phytocoenologia 27: 275-287.

Grant, S.A., Hunter, R.F. & Cross, C. 1963. The effects of muirburning *Molinia*-dominant communities. J. British Grassl. Soc. 18: 249-257.

Grime, J.P. 1979. Plant strategies and vegetation processes. Wiley, Chichester.

Grime, J.P., Hodgson, J.G. & Hunt, R. 1988. Comparative plant ecology. Unwin Hyman, London.

Grubb, J.P., Kelly, D. & Mitchley, J. 1982. The control of relative abundance in communities of herbaceous plants. pp. 79-97. In: Newman, E.I. (ed), The plant community as a working mechanism. Blackwell, Oxford.

Häfelfinger, S., Lörtscher, M., Guggisberg, F. & Studer-Ehrensberger, K. 1995. Prati magri e abbandonati della fascia montana del Ticino, una panoramica geobotanica e zoologica. pp. 27-55. In: Antognoli, C., Guggisberg, F., Lörtscher, M., Häfelfinger, S. & Stampfli, A. (eds), Prati magri ticinesi tra passato e futuro. Porconi, Losone.

Herben, T., Krahulec, F., Hadincová, V. & Skálová, H. 1993. Small-scale variability as a mechanism for large-scale stability in mountain grasslands. J. Veg. Sci. 4: 163-170.

Herben, T., Krahulec, F., Hadincová, V. & Pechácková, S. 1995. Climatic variability and grassland community composition over 10 years: separating effects on module biomass and number of modules. Functional Ecol. 9: 767-773.

Huston, M.A. 1994. Biological diversity. University Press, Cambridge.

Keddy, P.A. 1991. Biological monitoring and ecological prediction: from nature reserve management to national state of the environment indicators. pp. 249-267. In: Goldsmith, F.B. (ed), Monitoring for conservation and ecology. Chapman & Hall, London.

Kienzle, U. 1979. Sukzession in brachliegenden Magerwiesen des Jura und des Napfgebietes. Doctoral thesis Univ. Basel.

Krüsi, B.O. 1981. Phenological methods in permanent plot research. Veröff. Geobot. Inst. Stiftung Rübel ETH Zürich 75: 1-115.

Kuhn, U. 1984. Bedeutung des Pflanzenwasserhaushaltes für Koexistenz und Artenreichtum von Trespen-Halbtrockenrasen (*Mesobromion*). Veröff. Geobot. Inst. Stiftung Rübel ETH Zürich 83: 1-118.

Leutert, A. 1983. Einfluss der Feldmaus, *Microtus arvalis* (Pall.), auf die floristische Zusammensetzung von Wiesen-Ökosystemen. Veröff. Geobot. Inst. Stiftung Rübel ETH Zürich 79: 1-126.

Marti, R. 1994. Einfluss der Wurzelkonkurrenz auf die Koexistenz von seltenen mit häufigen Pflanzenarten in Trespen-Halbtrockenrasen. Veröff. Geobot. Inst. Stiftung Rübel ETH Zürich 123: 1-147.

Poschlod, P. & Jackel, A.-K. 1993. Untersuchungen zur Dynamik von generativen Diasporenbanken von Samenpflanzen in Kalkmagerrasen. I. Jahreszeitliche Dynamik des Diasporenregens und der Diasporenbank auf zwei Kalkmagerrasenstandorten der Schwäbischen Alb. Flora 188: 49-71.

Poschlod, P., Kiefer, S. & Fischer, S. 1995. Die potentielle Gefährdung von Pflanzenpopulationen in Kalkmagerrasen auf der Mittleren Schwäbischen Alb durch Sukzession (Brache) und Aufforstung - ein Beispiel für eine Gefährdungsanalyse von Pflanzenpopulationen. Beih. Veröff. Naturschutz Landschaftspflege Bad.-Württ. 83: 199-277.

Poschlod, P., Bakker, J., Bonn, S. & Fischer, S. 1996. Dispersal of plants in fragmented landscapes. pp. 123-127. In: Settele, J., Margules, C.R., Poschlod, P. & Henle, K. (eds), Species survival in fragmented landscapes. Kluwer Academic Publishers, Dordrecht.

Pickett, S.T.A. 1982. Population patterns through twenty years of oldfield succession. Vegetatio 49: 45-59.

Primack, R.B. & Miao, S.L. 1992. Dispersal can limit local plant distribution. Conserv. Biology 6: 513-519.

Rabotnov, T.A. 1974. Differences between fluctuations and successions. pp. 21-24. In: Knapp, R. (ed), Vegetation Dynamics, Handbook of Vegetation Science 8. Junk, The Hague.

Rosén, E. 1985. Succession and fluctuations in species composition in the limestone grasslands of South Öland. pp. 25-33. In: Schreiber, K.F. (ed), Sukzession auf Grünlandbrachen. Münstersche Geographische Arbeiten 20. Schöningh, Paderborn.

Schiefer, J. 1981. Bracheversuche in Baden-Württemberg. Beih. Veröff. Natursch. Landschaftspfl. 22: 1-325.

Schlaepfer, F. 1997. Influence of management on cover and seed production of *Brachypodium pinnatum* (L.) Beauv. in a calcareous grassland. Bull. Geobot. Inst. ETH 63: 3-10.

Silvertown, J., Dodd, M., McConway, K., Potts, J. & Crawley, M. 1994. Rainfall, biomass variation, and community composition in the Park Grass Experiment. Ecology 75: 2430-2437.

Stampfli, A. 1991. Accurate determination of vegetational change in meadows by successive point quadrat analysis. Vegetatio 96: 185-194.

Stampfli, A. 1992a. Effects of mowing and removing litter on reproductive shoot modules of some plant species in abandoned meadows of Monte San Giorgio. Bot. Helv. 102: 85-92.

Stampfli, A. 1992b. Year-to-year changes in unfertilized meadows of great species richness detected by point quadrat analysis. Vegetatio 103: 125-132.

Stampfli, A. 1993. Veränderungen in Tessiner Magerwiesen: Experimentelle Untersuchungen auf Dauerflächen. Doctoral thesis Univ. Bern.

Stampfli, A. 1995. Species composition and standing crop variation in an unfertilized meadow and its relationship to climatic variability during six years. Folia Geobot. Phytotax. 30: 117-130.

Stampfli, A. 1997. Changes in unimproved meadows of canton Ticino. Quaderni Parco Monte Barro 4: 103-108.

Stampfli, A. & Häfelfinger, S. 1995. Cambiamenti della vegetazione dopo la cessazione della gestione agricola. pp. 71-83. In: Antognoli, C., Guggisberg, F., Lörtscher, M., Häfelfinger, S. & Stampfli, A. (eds), Prati magri ticinesi tra passato e futuro. Porconi, Losone.

Stampfli, A. & Zeiter, M. 1999. Plant species decline due to abandonment of meadows cannot easily be reversed by mowing. A case study from the southern Alps. J. Veg. Sci. 10: 151-164.

Stöcklin, J. & Gisi, U. 1989. Auswirkungen der Brachlegung von Mähwiesen auf die Produktion pflanzlicher Biomasse und die Menge und Struktur der Streudecke. Acta Oecologica Oecol. Applic. 10: 259-270.

Taylor, D.R., Aarssen, L.W. & Loehle, C. 1990. On the relationship between r/K selection and environmental carrying capacity: a new habitat templet for plant life history strategies. Oikos 58: 239-250.

Thompson, K., Bakker, J.P. & Bekker, R.M. 1997. The soil seed banks of North West Europe: methodology, density and longevity. University Press, Cambridge.

Tilman, D. 1997. Community invasibility, recruitment limitation, and grassland biodiversity. Ecology 78: 81-92.

Tilman, D. & Pacala, S. 1993. The maintenance of species richness in plant communities. pp. 13-25. In: Ricklefs, R.E. & Schluter, D. (eds), Species diversity in ecological communities. University Chicago Press, Chicago.

van den Berg, J.P. 1979. Changes in the composition of mixed populations of grassland species. pp. 57-80. In: Werger, M.J.A. (ed), The study of vegetation. Junk, The Hague.

Willems, J.H. 1983. Species composition and above ground phytomass in chalk grassland with different management. Vegetatio 52: 171-180.

Willems, J.H. 1990. Calcareous grasslands in continental Europe. pp. 3-10. In: Hillier, S.H., Walton, D.W.H. & Wells, D.A. (eds), Calcareous grasslands - ecology and management. Bluntisham, Huntingdon.

Willis, A.J., Dunnet, N.P., Hunt, R. & Grime, J.P. 1995. Does Gulf stream position affect vegetation dynamics in Western Europe? Oikos 73: 408-410.

Zimmermann, R. 1979. Der Einfluss des kontrollierten Brennens auf Esparsetten-Halbtrockenrasen und Folgegesellschaften im Kaiserstuhl. Phytocoenologia 5: 447-524.

Zobel, M. 1997. The relative role of species pools in determining plant species richness: an alternative explanation of species coexistence? Trends Ecol. Evol. 12: 266-269.

LITTLE FLOWERS IN A MILD WINTER

JÜRG RÖTHLISBERGER

Cantonal School of Zug, Department of Biology, Lüssiweg 24, CH-6300 Zug

Keywords: Climatic change, flowering season, phenology, Switzerland, winter

Abstract
On the basis of a great number of single plant-phenological observations the present study tries to reveal the weather fluctuations during the winters of 1992/93 to 1997/98 in the region of Zug (central Switzerland). A comparison with older documents clearly shows that many plant species in the study area have shifted their time of flowering and development into the winter season. The effects of this shift are ascertainable on three levels: change in species diversity, wider dispersal of heat-preferring ecotypes, adaptive capability of genetically not mutated species. It would be very desirable to increase the volume of the observations also over larger areas and longer time periods and to store the available phenological data in a central place.

1. Precomment

The author thanks Prof. Dr. Conradin A. Burga for the excellent cooperation and for the opportunity to present this contribution in the anthology "Tasks for vegetation science". It is mainly the summary of an earlier publication: Röthlisberger Jürg, "Blümlein im lauwarmen Winter", a phenological study to document the extreme winters of the nineties in the area of Zug; in collaboration with Csilla Nikischer and Jürg Johner; edition of the Cantonal School of Zug, publication No. 11, March 1997 (Röthlisberger 1997). It seems that this publication was read by the respective weather makers around St. Peter and was found to be correct. As if for confirmation, in the winter season of 1997/98, the weather was as in previous years and the development of the vegetation exactly followed the described tendencies.
As far as it was possible for me besides other duties, I again took up the observation work. Especially the strong Christmas thaw and - in comparison to the previous year - a sensationally sunny start of 1998 made it possible to complement the lists. This complementary task was continued beyond the editorial work until the end of the winter season, and the documentation mentioned in the following paragraph was updated with 1983 individual entries from 1 November 1997 to 3 March 1998.

Burga & Kratochwil (eds.), BIOMONITORING, 125-142

2. Introduction and methods

The plant as an open organism must be able to adapt its genetically determined life cycle to the environment. On the one hand the weather fluctuates yearly, on the other hand there are long-term climatic changes and probably also sudden short-lived irreversible climatic changes, which very much challenge the survival strategy of the plant. The authors suggest that the phenological observations count at least as much as the recording of the weather conditions with the required instruments: The observation of the plants creates a net of data in a density which can never be attained with instruments. In addition, the phenological method automatically leads to an appropriate weighting of the meteorological extreme events and the much less spectacular, but mostly more important changes in the average values (Fries 1861-78; Defila 1992).
The various occurring phenomena must be clearly defined in order to be able to examine the plant-phenological observations. It must be tried to define exactly the vegetative stages of development, estimate the percentages of flowering occurrences and then examine if more or less vital fruits develop. The work gets even more difficult when one tries to trace back the developmental phases. Apart from various historical citations which amazingly often contain many correct climatic descriptions, the herbaria are especially interesting. With careful and correct analysis they can at least supplement the current phenological observations.
The present study is based on over 15'000 single observations, in which flowering, development of fruit and vegetative growth of a plant on a certain spot at a certain time during the winter seasons were recorded. Most of the data were collected in field work by the three authors during the winters of 1992/93 until 1997/98. A smaller part of the work, including studies of herbaria, dates from previous years and was mostly done by other people. These recordings in Switzerland cover partly the past 150 years. All the flowering events have been collected in a Microsoft Excel data base. The outcome of that work are two diskettes with 10'597 recordings of 437 plant species, 289 of which flowered in the region of Zug (238.7 km^2) at least once from 1 December to 28/29 February (= winter in the sense of the present work). These data are meant to be the basis for further considerations and conclusions. The two diskettes are available at the price of SFR 20.- at the Kantonsschule Zug, Lüssiweg 24, CH-6300 Zug. For an extract from the database see Table 1.
Not only the contributors of single pieces of information to this data base need to be acknowledged, also the weather was very cooperative: During the six winter seasons of investigation, it was so kind as to be variable and not wintry at all, so that the main task for the authors was to be at the right place at the right time and to record the numerous interesting phenomena there. Certain study areas were regularly visited to obtain uniform and comparable observation material. But single phenomena at conspicuous spots were also of much interest and we could not ignore them: Nature is life and certainly different from a concept artificially created in the laboratory!

Table 1. Extract from the database.

1	2	3	4	5	6	7	8	9
Eranthis hiemalis	6318	681710	217614	535	110	1998	70	birr
Eranthis hiemalis	6415	685780	212010	550	111	1998	80	röth
Eranthis hiemalis	6415	683410	212040	455	111	1998	100	röth
Eranthis hiemalis	6300	681100	222230	425	111	1998	30	röth
Eranthis hiemalis	8933	674610	232220	407	113	1998	10	röth
Eranthis hiemalis	8000	684670	245880	425	116	1998	100	röth
Eranthis hiemalis	8000	682900	246250	410	116	1998	101	röth
Eranthis hiemalis	8000	682950	246420	410	116	1998	101	röth
Eranthis hiemalis	8000	683180	248250	408	116	1998	100	röth
Eranthis hiemalis	6300	681120	222240	423	117	1998	80	birr
Eranthis hiemalis	6300	681760	223525	475	117	1998	50	birr
Eranthis hiemalis	6300	681550	224175	425	117	1998	50	birr
Eranthis hiemalis	6300	681860	224225	455	117	1998	80	birr
Eranthis hiemalis	6300	680340	226820	425	117	1998	80	birr
Eranthis hiemalis	6312	680080	227420	425	117	1998	60	birr
Eranthis hiemalis	6300	681550	224170	420	119	1998	100	röth
Eranthis hiemalis	6318	681710	217614	535	119	1998	100	birr
Eranthis hiemalis	6300	681050	226230	422	122	1998	100	röth
Hepatica nobilis	3257	594000	212900	510	116	1990	5	röcr
Hepatica nobilis	3257	594000	212900	510	131	1991	5	röcr
Hepatica nobilis	3257	594000	212900	510	137	1992	5	röcr
Hepatica nobilis	3257	594000	212900	510	139	1993	5	röcr
Hepatica nobilis	3257	594000	212900	510	120	1994	5	röcr
Hepatica nobilis	8352	702270	261900	480	97	1995	100	röth
Hepatica nobilis	3860	657500	175750	700	99	1995	20	röth
Hepatica nobilis	3860	657600	175800	750	99	1995	5	röth
Hepatica nobilis	3257	594000	212900	510	112	1995	5	röcr
Hepatica nobilis	3000	600700	198700	525	137	1996	50	röth
Hepatica nobilis	3257	594000	212900	510	143	1996	5	röcr
Hepatica nobilis	6777	698700	150700	970	159	1996	100	röth
Hepatica nobilis	6415	685650	211600	535	132	1997	100	röth
Hepatica nobilis	6415	686250	212200	580	132	1997	100	röth
Hepatica nobilis	6416	687170	212950	795	132	1997	100	röth
Hepatica nobilis	6415	686320	212100	580	111	1998	100	röth
Hepatica nobilis	6416	687090	212750	740	111	1998	1	röth
Hepatica nobilis	6340	684620	226950	602	111	1998	5	birr
Hepatica nobilis	6340	684620	226950	602	118	1998	0	birr
Hepatica nobilis	6300	680540	218850	420	119	1998	10	röth
Ranunculus ficaria	4450	628000	257000	370	143	1862	50	uniz
Ranunculus ficaria	8000	685000	250000	600	146	1867	25	uniz
Ranunculus ficaria	1200	502000	118400	375	150	1873	50	uniz
Ranunculus ficaria	3000	602000	199500	560	151	1879	100	uniz
Ranunculus ficaria	6925	715700	94700	400	142	1903	15	uniz
Ranunculus ficaria	1926	573200	108800	500	140	1905	80	uniz
Ranunculus ficaria	8890	744500	217500	455	135	1907	10	uniz
Ranunculus ficaria	5057	646000	233500	500	148	1910	20	uniz
Ranunculus ficaria	8000	682500	247100	415	146	1912	100	uniz
Ranunculus ficaria	8735	720000	237000	900	137	1913	100	uniz
Ranunculus ficaria	9620	724600	242600	620	148	1913	10	uniz
Ranunculus ficaria	6614	697500	108600	400	136	1919	100	uniz
Ranunculus ficaria	4500	608000	229000	450	148	1921	50	uniz
Ranunculus ficaria	5012	643500	247250	450	151	1930	30	uniz
Ranunculus ficaria	5000	647200	250500	365	143	1946	50	uniz
Ranunculus ficaria	6300	681600	223400	440	155	1952	90	merz

Details of the stored data

1 = Scientific name of the plant.

2 = Postal code of the municipality; in municipality districts with several postal codes no distinctions between postal areas, in the case of municipalities without own post office the postal code of the neighbour municipality was used.

3 = Switzerland's west-east co-ordinates, personally determined, copied from documents or reconstructed as accurately as possible from information on the location.

4 = South-north co-ordinates, as 3.

5 = Altitude above sea-level, as 3.

6 = Day. Numbering consecutively from 1 November onwards, e.g. 25 = 25 November, 50 = 20 December, 71 = 10 January, 112 = 20 February, 135 = 15 March (29 February never taken into consideration).

7 = Year. November and December always counted to the next year, that is e.g. 40 – 1995 = 10 December 1994.

8 = Flowering share in % of the maximal flowering capacity of the size of this stock; always a minimum of 1 % is stated, even if only one flower of a very large stock is open.

9 = Abbreviations of the authors' names, most current authors:

- birr = Field notes Marguerite Birrer, Zug
- john = Field notes Jürg Johner, Cham
- merz = Dr. Wolfgang Merz, Zug, 1901-1968, location notes and herbarium references in the Herbarium Tugiense, Cantonal School of Zug
- nalu = Herbarium of the Museum of Natural History Lucerne, including the herbarium of Hans Wallimann (1897-1990, Alpnach-Dorf) deposited there
- niki = Field notes Csilla Nikischer, Zug
- röcr = Phenological observations Dr. Christian Röthlisberger, Grossaffoltern/BE
- röpa = Private herbarium Paul Röthlisberger, Muri/BE, since 1996 in the Botanical Institute of the University of Bern
- röth = Field notes or herbarium references Jürg Röthlisberger, Cham
- unba = Herbarium references Botanical Institute of the University of Basel
- unib = Herbarium references Botanical Institute of the University of Bern
- uniz = Herbarium references Botanical Institute of the University of Zürich, including the combined references from the herbarium of the ETH.

3. Detailed analysis of various locations

In four succeeding winters the survey concentrated on selected individual areas. For technical reasons they were situated within a radius of not more than 10 kilometres, but even though covered quite different ecosystems:

- Cultivated roof garden Röhrliberg Cham, 435 m, Swiss co-ordinates 676'875/226'470, including the surroundings: partly new building site, one old farmhouse garden, intensively cultivated agricultural area
- Fallow fields near the lake (north shore lake of Zug), exact place changing yearly with the agricultural cultivation, approx. 415 m, 679'250-750/225'800-226'000, however always on eutrophic alluvial land
- 2 railway areas around the town of Zug, railway embankment and gravel area, both 416-422 m, around co-ordinates 680'650/225'430 and 681'000/225'330 (Röthlisberger 1995)
- Biotope near the Cantonal School of Zug existing since 1977, naturally cared without any interruption, 430 m, around 682'280/225'420 (Röthlisberger 1992)
- South-exposed border of the forest near Inwil/Baar, 470-500 m, around 683'330/226'000
- Mainly south-exposed excavation at Steren, municipality of Baar, 690-700 m, around 683'220/224'570
- Border of the forest and pasture land Schindellegi/Zugerberg, 910 m, around 683'750/223'000 (see also "Example of a recording", Figure 1)
- Lake shore south of Zug via Oberwil to Walchwil, east-exposed, 413-550 m, 680'000-682'000, 217'000-224'000, many areas covered with buildings, between agricultural land and smaller forest parts.

Anthropogenically strongly influenced areas were deliberately chosen, because in the various pioneer communities the vegetation can most probably easily adapt to the possible climatic changes. As long as there was no snow, the observations took place in a fortnightly rhythm from the end of November till the beginning of March. Thus the flowering plants and the ones with striking vegetative development could quite completely be registered.

The study of a multitude of sites on a small area proved later on to be the right method: Unexpectedly large was the number of species of flowering herbs in the highest site of observation, what can probably primarily be attributed to the many sunshine hours during the inversion periods. In snow-poor winters this could obviously more than make up for the greater night coldness.

Figure 1. Significant change of life conditions in a restricted area: the plants can adapt themselves.

Surprisingly great were also the differences between the various sites on comparable altitudes. The influence of the "Föhn" (characteristic wind from the south in the northern Alps) on the eastern shore of the lake of Zug (after Weggis/Vitznau at the lake of Lucerne, probably the climatically most favoured extreme site in central Switzerland) could also be seen on the list of the plants and the data showing the earliest flowering time. But even train areas and fallow fields with hardly visible climatic and topographic variations often developed in different ways. Besides known and unknown microclimatic factors, the work has shown once more that plants are individuals, whose occurrence and state of development are not only due to temperature, or another ecological factor (Figure 1).

4. Winter flowerers and flowering behaviour

Switzerland lies in a Central European transition climate, which cannot definitely be assigned to the continental or to the maritime type. All Swiss flora works are based on a winter resting time, which is interrupted only in exceptional cases. Thus the very often used Flora by Binz/Heitz (Binz 1986) only mentions - among 2,800 species - one single real winter flowerer (*Helleborus niger*) and three year-round flowerers (*Poa annua, Stellaria media, Senecio vulgaris*). However, alone in the canton of Zug 289 species did flower. Under these conditions, this can be regarded as sensational.

A more detailed analysis of this material shows many individual flowering events outside the normal season. These certainly are of biological importance, as they show the flexibility with which plants respond to changed environmental conditions. In addition to this, there is such a big number of frequently to regularly shifting flowering times that it cannot be dismissed as a lack of exactitude of the former floras. Apart from the 4 named species the spontaneous flora of Zug contains, even by a very cautious interpretation, at least a further five permanent winter flowerers: *Cerastium glomeratum, Cardamine hirsuta, Lamium purpureum, Veronica persica* and *Bellis perennis*. In addition, *Capsella bursapastoris, Euphorbia peplus, Lamium maculatum* and *Calendula officinalis* were recorded in the extremely mild Zug winter of 1997/98. At least 32 other wild species showed in several winters a more or less clear tendency to a permanent winter flowering (Figure 2).

Figure 2. Grazing sheep in the middle of winter: also a symptom of a climatic change.

Even though most authors (e.g. Houghton 1990; Chen 1991; Balling 1992; Veroustraete 1994; Magny 1995; Schönwiese 1995; Broecker 1996; Hänni 1992-97) have lately postulated a general heating up, especially in autumn, while they are more cautious for the winter season, there are many dozens of early flowerers clearly waking up ahead of time (see also Figure 3). More than half of the complete list of winter flowerers is composed of plants, which as late autumn flowerers have extended their flowering time into December and in 1997/98 occasionally even into January, among them many herbs, the flowering of which usually ends in August, July or even in June, as per the relevant Swiss floras (Binz 1986; Lauber & Wagner 1996).

5. Fourteen selected plant species

Fourteen plant species - mostly known as traditional early flowerers - were analysed in detail, namely *Galanthus nivalis*, *Leucojum vernum*, *Alnus incana*, *Alnus glutinosa*, *Corylus avellana*, *Eranthis hiemalis*, *Anemone nemorosa*, *Ranunculus ficaria*, *Cornus mas*, *Primula elatior*, *Lamium purpureum*, *Veronica persica*, *Veronica hederaefolia* and *Tussilago farfara*. Due to their distinct appearance, they can easily be registered by phenological observation records. For these species observation records from third persons have been relatively well documented, partly going back far into the 19th century (Beiche 1872; Fries 1861-78; Merz 1913-68; Merz 1966).

But even with these species it was difficult to find indications of the very mild winters of earlier times in the herbaria. This was only in part successful for the two extreme years 1915/16 and 1920/21. A clearer picture could be gained for the also extraordinarily mild winters of 1982/83, 1988/89 and 1989/90. An advanced flowering time was registered in all observation periods for the present work (since 1992), with the exception of 1995/96. In 1997 all the above-mentioned 14 early flowerers were observed before 1 March, each at least 5 times in the region of Zug. In 1995 and 1994, 12 early flowerers fulfilled this criterion, in 1996, however, only 7. But also in 1997/98, until the end of February, 13 out of these 14 plants had flowered at least 5 times, i.e. *Veronica persica* (flowered 5 x till December 25), *Corylus avellana* (January 6), *Alnus incana* (January 9), *Galanthus nivalis* and *Eranthis hiemalis* (both February 12), *Leucojum vernum* (February 13), *Lamium purpureum* (February 15), *Alnus glutinosa* (February 19), *Anemone nemorosa*, *Ranunculus ficaria* and *Tussilago farfara* (all February 20), *Primula elatior* (February 25) as well as *Cornus mas* (February 27).

With increasing altitude above sea level, the differences between the various delayed flowerings were even more pronounced. This could be established more often for the woody plants, because when the soil is frozen they can hardly take advantage of the short time of heating up due to the sun reflection above the wintry fog cover.

In this respect, herb plants seem to be much more flexible. They can better, at least with regard to flowering, take advantage of local and temporary favourable conditions. In consequence, the chronological sequence of flowering, which is usually graded according to the altitude above sea level, is again and again changed. This can especially well be seen at *Tussilago farfara*. It blooms under extreme conditions at high altitude nearly all the time, almost independent of the season of the year. The further steps for reproduction, however, seem to be much more difficult in the cold season. Already the transfer of the pollen presents a problem, and only relatively seldom vital fruit could be observed in the middle of winter. It was, by the way, also difficult to obtain useful data on the maturing time through flora works, observation records and herbarium references.

6. Complete list of the winter flowerers

The following list shows all plants which in the canton of Zug, in the remaining part of Switzerland north of the Alps* or in the Ticino** (south of Switzerland) were seen flowering by the authors at least once from 1 December to 28/29 February. The nomenclature follows the indications given in the pocket atlas of Thommen (Thommen

1945ff), Hess et al. (Hess et al. 1967), the Flora europaea (Flora europaea 1964ff) and Hay & Singe (Hay & Singe 1979). The names with "+" are complements from the observation season of 1997/98 and not included in the integral version (Röthlisberger 1997). The names of the ornamental plants have been adapted by analogy.

6.1. GENUINE PERMANENT WINTER FLOWERERS

Poa annua, Stellaria media, Cerastium glomeratum, Helleborus niger, Cardamine hirsuta, Chimonanthus praecox, Viola hortensis, +Viola arvensis x hortensis, Erica carnea (cult. form), Erica gracilis, Primula sibthorpii, Jasminum nudiflorum, Lamium purpureum, Veronica persica, Viburnum farreri, +Viburnum tinus, Bellis perennis, Bellis perennis (cult. form), Senecio vulgaris, + Iris unguicularis, +** Sarcococca confusa, +** Camellia japonica*

6.2. PLANT SPECIES WITH A TENDENCY TO PERMANENT WINTER FLOWERING

*Dianthus caryophyllus, Cerastium caespitosum, Arenaria serpyllifolia, Caltha palustris, Ranunculus frieseanus, Mahonia bealei, Fumaria officinalis, Iberis sempervirens, Iberis sempervirens f. nana, Thlaspi arvense, Sinapis arvensis, Nasturtium officinale, Capsella bursapastoris, Arabis albida, Arabidopsis thaliana, Aubrieta deltoidea, Alyssum saxatile, Bergenia crassifolia, Chaenomeles japonica, Fragaria vesca, Potentilla sterilis, +Kerria japonica f. pleniflora, +Pelargonium zonale, Euphorbia helioscopia, Euphorbia peplus, Malva spec. prope silvestris, Fuchsia magellanica, Anthriscus silvestris, Primula elatior, Forsythia intermedia, +Forsythia intermedia f. brachyflora, Gentiana verna, Gentiana clusii (cult. form), Vinca minor, Vinca major, Symphytum tuberosum, Myosotis arvensis, Ajuga reptans, Rosmarinus officinalis, Lamium album, Solanum nigrum, Veronica arvensis, Veronica hederaefolia, Veronica filiformis, Campanula portenschlagiana, Taraxacum officinale, Sonchus oleraceus, *Sesleria caerulea, *Thlaspi silvestre, *Draba aizoides, *Polygala chamaebuxus, +*Mercurialis annua, *Veronica polita, **Parietaria diffusa, **Conyza canadensis x bonariensis*

6.3. PREFERENTIALLY EARLY FLOWERERS

Luzula campestris, Galanthus nivalis, Galanthus ikariae (cult. form),+Galanthus elwesii, Leucojum vernum, Crocus aureus, +Crocus sieberi, Crocus tommasinianus, +Crocus speciosus, Iris reticulata, Alnus incana, Alnus glutinosa, Corylus avellana, Corylus avellana s.l. *f. "Bluthasel", Corylus colurna, Helleborus viridis, Eranthis hiemalis, Hepatica nobilis, Ranunculus ficaria, Erophila praecox, Hamamelis mollis, +Prunus subhirtella, +Acer saccharinum, +Daphne mezereum (cult. form), Cornus mas, Lonicera purpusii, Tussilago farfara, +*Crocus ancyrensis, +*Hamamelis intermedia, +* Rhododendron praecox*

6.4. LESS PREFERENTIALLY EARLY FLOWERERS

*+Carex montana, Narcissus pseudonarcissus (cult. form), +Narcissus minicycla, Crocus chrysanthus, Salix melanostachys, +Salix caprea, +Salix caprea x cinerea, Salix cinerea, Corylus avellana f. contorta, Cerastium semidecandrum, Helleborus foetidus, Anemone nemorosa, +Anemone blanda, Brassica napus, Cardamine pratensis, Erophila verna, Erophila verna x praecox, Spiraea thunbergii, Mercurialis perennis, Viola alba, +Pieris japonica, Primula sibthorpii x elatior, Petasites albus, *Poa annua x supina, *Allium ursinum, *Crocus albiflorus, *Alnus incana x glutinosa, *Corylus maxima*

Figure 3. In the Zug plain rarer and rarer: winter with a real snow cover.

6.5. SIGNIFICANTLY LATE AUTUMN FLOWERERS

Sagina procumbens, Ranunculus bulbosus, Lepidium virginicum, Cardamine hirsuta x flexuosa, Lobularia maritima (cult. form), Cheiranthus cheiri, Rosa spec. (cult. form), Medicago lupulina, Trifolium pratense, Trifolium sativum, Trifolium repens, Vicia sepium, Daucus carota, Cornus stolonifera, +Verbena peruviana, Teucrium scorodonia, Thymus pulegioides, Cymbalaria muralis, Antirrhinum majus, Veronica chamaedrys, Galium album, Knautia silvatica, Knautia arvensis, Conyza canadensis, Erigeron strigosus, Erigeron annuus, Galinsoga quadriradiata, Anthemis tinctoria, Achillea millefolium, Matricaria chamomilla, Matricaria suaveolens, Chrysanthemum ircutianum, +Chrysanthemum ircutianum x pseudadustum, +Chrysanthemum *pseudadustum, Senecio viscosus, Calendula officinalis, Sonchus asper, Crepis capillaris, **Dianthus carthusianorum*

6.6. LESS SIGNIFICANTLY LATE AUTUMN FLOWERERS

Digitaria sanguinalis, Phleum pratense, Holcus lanatus, Arrhenatherum elatius, Trisetum flavescens, Dactylis glomerata, +Poa compressa, Poa pratensis, Poa nemoralis, Festuca gigantea, Bromus erectus, Brachypodium silvaticum, Lolium multiflorum, Colchicum autumnale, Urtica dioica, Rumex obtusifolius, Fallopia convolvulus, Polygonum aviculare s.str., *Polygonum rurivagum, Polygonum persicaria, Chenopodium polyspermum, Chenopodium album, +Attriplex patula, Silene cucubalus, Silene dioica, Silene alba, +Myosoton aquaticum, Cerastium biebersteinii, Ranunculus lanuginosus, Ranunculus repens, Iberis umbellata, Sisymbrium officinale, Brassica oleracea, +Hesperis matronalis, Sedum spurium, Spiraea japonica, Rubus idaeus (cult. form), Rubus caesius, Rubus thyrsanthus, Duchesnea indica, +Fragaria vesca (+/- cult. form), Potentilla fruticosa, Potentilla recta, Potentilla aurea, Geum urbanum, Geum rivale, Filipendula ulmaria, Sanguisorba minor, +Rosa multiflora, Prunus laurocerasus, Ononis arvensis, Geranium robertianum, Geranium columbinum, Geranium pyrenaicum, Euphorbia lathyris, Euphorbia maculata, Euphorbia platyphyllos, Euphorbia stricta, Impatiens parviflora, Althaea rosea, Hypericum patulum, Hypericum perforatum, +Begonia semperflorens, Oenothera muricata* s.l., *Hedera helix, Pimpinella major, Aethusa cynapium, Heracleum sphondylium, Cornus sanguinea, +Borago officinalis, Phacelia tanacetifolia, +Lavandula angustifolia, Galeopsis tetrahit, Lamium maculatum, +Majorana hortensis, Chaenorrhinum minus, Veronica peregrina, Galium aparine, Weigela florida, Symphoricarpos chenaultii, Kentranthus ruber, +Scabiosa columbaria, +Eupatorium cannabinum, +Campanula cochleariifolia, Aster novi-belgii x dumosus, +Aster tradescanti, +Aster versicolor, Callistephus sinensis, Solidago canadensis, Solidago gigantea, Helianthus annuus, +Helianthus tuberosus, Rudbeckia hirta, Tagetes patulus, Chrysanthemum parthenium, Chrysanthemum coronarium* s.str., *+Chrysanthemum coronarium* s.l., *Chrysanthemum frutescens, Chrysanthemum indicum, Senecio jacobaea, Echinops ritro (cult. form), Cirsium oleraceum, Centaurea scabiosa, Centaurea jacea, Centaurea jacea x bracteata, Lapsana communis, Hypochaeris radicata, Leontodon autumnalis, Leontodon hispidus, Mycelis muralis, Hieracium pilosella, Hieracium auricula, Hieracium murorum* s.l., *+*Agrostis stolonifera, *Sinapis alba, +*Raphanus sativus, +*Sedum telephium* s.str., *+*Echium vulgare, +*Tanacetum vulgare, +**Yucca filamentosa, **Dianthus silvester, +**Malva neglecta, +**Peucedanum oreoselinum, +**Stachys officinalis, **Phyteuma betonicifolia, **Tagetes spec., +**Galinsoga parviflora, +**Centaurea dubia, +**Picris hieracioides*

6.7. WITHOUT A CLEAR TENDENCY

Allium schoenoprasum, Anemone coronaria, Ranunculus montanus s.str., *Berberis verrucandi, +Chelidonium majus, Erucastrum gallicum, Potentilla verna, +Alchemilla glabra, Oxalis deppei, Viola silvestris, Viola arvensis, +Gentiana acaulis* s.l., *Phlox* *amoena, +Cyclamen neapolitanum, Omphalodes verna, Glechoma hederacea, Veronica serpyllifolia, +Viburnum plicatum, Lonicera caprifolium x etrusca, +Scabiosa columbaria (cult. form), +Arctotis breviscapa, +Anthemis sect. Maruta spec., Senecio sect. Jacobaea, +*Urtica urens, +*Aubrieta spec., *Anthyllis vulneraria* s.l., **Cyclamen coum, +*Gentiana dinarica, +**Rumex scutatus, +**Dianthus seguieri, **Oxalis corniculata, +**Oxalis stricta x corniculata, **Lamium flavidum, **Lonicera spec., **Erigeron karvinskianus*

Due to the additions made in the winter of 1997/98, the list was extended by a total of 70 names. Moreover, different species could have been moved from the moderately to the significantly late autumn flowerers or interpreted as genuine winter flowerers. To render the comparison with the main publication (Röthlisberger 1997) possible, no changes were made.

7. Discussion and conclusions

To a general climatic heating up the vegetation can basically adapt in three forms:

- Different spectrum of species: What has been varyingly well documented can also be confirmed for the canton of Zug: of 33 plants that have considerably spread lately, 29 belong to the highest temperature number classes 5 and 4, as defined by Landolt (Landolt 1977); only 3 have the moderate temperature number 3, not one belongs to the arctic-alpine temperature numbers 2 or 1. Further two (*Panicum dichotomiflorum* and *Paulownia tomentosa*) are missing in Landolt's list; but according to their height spreading and their favoured substrates they would also have to be classified under 5 or 4 (Merz 1966; Welten & Sutter 1982; Röthlisberger 1992, 1995).
- Other ecotypes of the same species: Since ecotypes cannot be separated morphologically, a discussion of this aspect is very difficult without a genome analysis. Changes of vernalisation mechanisms or even extravagant jumps in the flowering period or another striking phase of development could be interpreted as changed ecotypes.
- Adaptation of genetically unchanged species: Also with unchanged genetic material, all plants can adapt more or less well and regularly their development and their appearance; without that ability they would be helpless in the face of the yearly climatic fluctuations. However, it has probably to be assessed separately for each species if plants can use these mechanisms to cope with climatic changes in the long run.

On 21 December 1992 - i.e. on the shortest day of the year - I observed on a stretch of about 1.2 km between Zug and Baar 23 flowering plants. The climate in central Switzerland is currently no more "as it was in earlier times" and the plants have noticed it as well.
Nevertheless it has to be warned against drawing rash conclusions: It turned out to be very difficult to make records also for the past decades or even centuries. Apart from individual pieces of information dating back to prehistoric times, there are few

connected phenological observations. Much of the documentation material is not accessible, because it is filed in private or public archives, its owners or administrators not being aware of its importance. Furthermore the current active scientific generation hardly has the time required to tackle the backlog of work. This is a deplorable situation, in view of the fact that there are many unemployed university graduates who would be able to perform this work.

Everyone consulting the complete version and the diskette of the present work will have the same objections concerning the fragmentary approach and punctual procedure. The survey for this study has always been effected in the spare time, parallel to many other engagements. Its publication is primarily justified because in the area of central Switzerland, no comparable work is available and, as much as I know, not in the process of being drawn up.

There is a definite need for action. Spot-checking excursions to the Ticino and to the "Berner Oberland" ("Bernese highland") revealed that there are exactly the same differences between the flowering data in the floras and the actually observed flowering events. This also applies to other parts of the country. Alone in the canton of Ticino, the list of the winter flowerers would certainly have been much longer if intensive observations had been made. During two short trips to Prague, Czech Republic, I noticed that even under continental climatic conditions a tendency to winter flowering is also perceptible.

The incentive to a winter flowering is not only given by the rising temperature in the coldest months. As important as the temperature are the conditions of life during the rest of the year. Earlier spring and later autumn lead to a longer shading in the deciduous forest, which does not permit most woody soil plants to take much advantage of the higher temperatures in the summer months. They have no choice but to flower in the winter months, the conditions of life being more favourable then.

Even better represented in the available lists are the vegetation of disturbed habitats and synanthropic plants. When growth from spring to autumn is always disturbed by digging over, mowing, grazing, harvesting, use of herbicide and mechanical pull out, then the winter flowering somehow indirectly gains in importance. It has also to be taken into consideration that many cultivated areas are - independently of whether they are cultivated by farmers or by hobby gardeners - partly or permanently enriched due to fertilisers and pollutants, which certainly intensify the effects of the wintry heating up. It seems therefore logical, where the fewest plants with changed flowering and growing behaviour are found: in the marsh areas - independent of the degree of acidity - and in coniferous forests, where the lighting does not improve in winter.

The short-term concrete need for action mainly exists on two levels:

- Many people in the past and in the present have not only made observations of their environment; they sometimes also had the necessary knowledge to note their observations. For modern information technology it is easy to gather such information and to store on smallest space whatever from a scientific point of view is of importance. Only one thing is required: The correct information must be available at the right time to the right persons. We emphasise once more: In a time where in Switzerland there are far more than 100'000 unemployed people, including a number of qualified scientists, it should be possible to give interested groups access to a data file covering the wide knowledge on the spreading areas of plants and animals, data which for the greater part lie around unread and are so often destroyed without having been recorded. In this case no objection can be made that detailed files of the winter flowerers would encourage their extermination. Only with extensive data banks we can objectively evaluate whether in a given time period indeed a shift has taken place, for example due to a climatic change.

- The second request is along the same lines: In the "screen" age we unfortunately cannot naturally expect that men actually see what happens around them. My neighbour's soup plate shall not be my topic; but it might interest more people to know, whether and what flowers have at what time flowered in front of the residential building, near the access to the motorway and behind the panel of the fitness course. It is clear that not many problems are solved by plant names, and the knowledge of the biological diversity - who does still have it? - has almost become a symbol of superfluous educational subjects for many so-called cultural politicians. But only the one who has, in one form or the other, experienced the biological diversity and its wealth, is in a position to realize in time when something is changing. It would only be a small step then to write this observation down. All those who do so contribute in a small or big way to the understanding of our environment and, after all, also of ourselves.

Example of a recording

Edge of the forest and pasture land up to a cowshed, slightly S-exposed, NE of Schindellegi, municipality of Zug, approx. 905-912 m above sea level, around co-ordinates 683'750/223'000, 26 January 1996, 13.30 h, weather sunny, +/- upper limits of fog, temperature in the sun 13°C, approx. 10 % of the surface covered with white frost.

Agrostis stolonifera	rv rv rv rs rs cm cm	
Holcus lanatus	rv rv rv rv	
Sesleria caerulea	rv rv	
Dactylis glomerata	rv rv rv fm	
Poa annua	cv cv ks km	
Cynosurus cristatus	fm	
Festuca rubra	rv rv rv rs rs	
Lolium multiflorum	rv rv	
Carex verna	rv rv	
Carex silvatica	rv rv	
Juncus inflexus	rv rs	
Rumex obtusifolius	rv	
Stellaria graminea	cv cv	
Stellaria media	cv cv cv kv kv bv bv fs fs fm fm	10 % in flower
Cerastium caespitosum	rv rv cv cv kv kv	
Ranunculus frieseanus	rv rv	
Ranunculus repens	rv rv	
Potentilla sterilis	rv rv rv kv bv bv	10 % in flower
Potentilla verna	rv rv rs	
Geum urbanum	rv rv	

Fragaria vesca	rv rv	
Sanguisorba minor	rv rv	
Ononis repens	rv rv	
Medicago lupulina	rv rv rv cv	
Trifolium pratense	cv cv	
Ajuga reptans	rv rv rv cv	
Prunella vulgaris	rv rv cv cv fm	
Thymus pulegioides	cv cv cv fm fm	
Veronica chamaedrys	cv cv cv cs	
Veronica serpyllifolia	cv cv	
Veronica officinalis	cv cv	
Veronica filiformis	cv cv kv	
Plantago media	rv rv	
Plantago major	fm fm	
Plantago lanceolata	rv rv rv	
Galium album	cv cv	
Sambucus nigra	cv cs 1_{max} = 5 cm (leaves)	
Solidago virgaurea	rv	
Bellis perennis	rv rv bv bv fs	12 % in flower
Achillea millefolium	rv rv	
Taraxacum officinale	rv rv k_{bas} $bv_{l=5\ cm}$	1 % in flower
Hieracium pilosella	rv rv	

Figure 4. Site of the "example of a recording" at Schindellegi.

Explanations

Characterization of the stage of development

"K" = sprout, either only sprout leaves exist or their surface area is at least bigger than all initial stages of first or secondary leaves

"r" = leaf rosette without distinct stem, plant however clearly over the germination stage, i.e. often a hemicryptophyte

"c" = initial in relation to the chamaephytes of a Raunkiaer form of life, plant with leafy, clearly developed stem, independent of the length and of the question whether wooden (can also be a phanerophyte!)

"k" = plant with flower buds, the buds visible also without removal of leaves, sometimes already parts of the crown to be seen from the outside

"b" = flowering plant, at least part of the inner side of the perianth visible, where the style and the stamens are found or at least should be found

"f" = bearing fruit, the bloom clearly in the process of fruit development and production of seeds

"juv"= additional distinction for freshly developed organs, i.e.

"c_{juv}" = freshly sprouted leafy shoot,

"f_{juv}" = still poorly developed fruit etc.

Stage of conservation

"v" = "vital", survives when environmental conditions are not getting worse

"s" = "subvital", clearly reduced capacity, uncertain whether the concerned part will survive

"m" = "mortuus", dead part

Interpretation

The interpretation of fruit was difficult. "fv" only means that the fruit looks healthy, not that the seeds are effectively able to germinate (no germination tests were made). "fm" can mean that the fruit has already fulfilled its biological function of seed production, and only the empty hull can be seen. With "fm" I also describe seeds, which were possibly vital, but at the time of observation were found in a latent life phase.

Frequency

"r" = one leaf rosette that can clearly be identified

"rr" = various leaf rosettes, either rare plants, or only a small part of the rosettes is developed

"rrr" = quite many rosettes, either a well-developed stock of rosettes of a moderately frequent plant, or a respectable minority of a massive stock

"rrrr" = massive stock, at least 50 % of the population in the rosette stadium

Extended symbols

Two examples:

- *Sambucus nigra*, black elder: "cv, cs, l_{max} = 5 cm (leaves)"
  = plant with vital and subvital leaves on well-visible shoots, each sporadic, maximum leaf length (subvital and vital) 5 cm
- *Taraxacum officinale*, dandelion: "rv rv k_{bas} $bv_{l=5cm}$"
  = few vital leaf rosettes, 1 sporadic bud still on the basis of the leaf rosette, 1 vital flowering capitulum on approx. 5 cm-long scape

Acknowledgements

English translation from the original German version: Marguerite Birrer with the collaboration of Janos Turi Nagy and Gregor Achleitner.

References

Balling, R.C. 1992. The heated debate, Greenhouse predictions versus climate reality. Research Institute for Public Policy. San Francisco, USA.

Beiche, E. 1872. Vollständiger Blütenkalender der deutschen Phanerogamen-Flora. 2 vol. Hahn'sche Hofbuchhandlung, Hannover.

Binz, A. 1986. Schul- und Exkursionsflora für die Schweiz. Several editions. Revised by Becherer, A. 1957 and Heitz, C. 1986. Schwabe, Basel.

Birrer, M. Phenological observation records from the area of Zug, 1997-1998. (Unedited.)

Broecker, W. S. 1996. Plötzliche Klimawechsel. Spektrum der Wissenschaft Januar 1996: 86-93.

Chen, J. 1991. Changes of Alpine Climate and Glacier Water Resources. Zürcher Geographische Schriften.

Defila, C. 1992. Pflanzenphänologischer Kalender ausgewählter Stationen in der Schweiz 1951-1990. Schweizerische Meteorologische Anstalt. Zürich.

Flora europaea 1964ff. 5 vol. Cambridge University Press, Cambridge.

Fries, F. E. Notizen zur Flora Raurica, Sissach, 1861-1878. (Unedited manuscript.)

Hänni, W. Unedited temperature measurements, Cham, 1992-1997.

Hay R. & Singe P.M. 1979. Das grosse Blumenbuch. 4th ed. Ulmer, Stuttgart.

Hess, H.E., Landolt, E. & Hirzel, R. 1967. Flora der Schweiz. 3 vol. Birkhäuser, Basel.

Houghton, J.T. (eds) 1990. Climate Change. The IPCC Scientific Assessment. Cambridge University Press.

Johner, J. Unedited phenological observation records from the area of Zug, 1994-1996.

Landolt, E. 1977. Ökologische Zeigerwerte zur Schweizerflora. Veröff. Geobot. Inst. Rübel, Zürich 64.

Lauber, K. & Wagner, G. 1996. Flora Helvetica. Haupt, Bern.

Magny, M. 1995. Une histoire du climat. Errana, Paris.

Merz, W. Unedited field notes, file cards and herbar references on the flora of Zug, 1913-1968.

Merz, W. 1966. Flora des Kantons Zug. Mitteilungen der Naturforschenden Gesellschaft Luzern XX. vol.

Nikischer, C. Unedited phenological observation records from the area of Zug, 1995-1997.

Röthlisberger, C. Unedited notes on the weather and on the phenology of selected plants in the region of Grossaffoltern, canton of Bern, Switzerland, 1976-1997.

Röthlisberger, J. 1992. Wandel der Zuger Flora - Wandel eines Ökosystems. Veröffentlichungen der Kantonsschule Zug 6.

Röthlisberger, J. 1995. Der Güterbahnhof als floristisches Raritätenkabinett. Mitteilungen der Naturforschenden Gesellschaft Luzern 34: 31-83.

Röthlisberger, J. 1997. Blümlein im lauwarmen Winter. Veröffentlichungen der Kantonsschule Zug 11.

Schönwiese, C. 1995. Klimaänderungen. Springer, Berlin.

Thommen, E. 1945ff. Taschenatlas der Schweizer Flora. Several editions. Birkhäuser Verlag, Basel.
Veroustraete, F.E. (ed) 1994. Vegetation, modelling and climate change effects. SPB Academic Publishing, Den Haag.
Welten, M. & Sutter, R. 1982. Verbreitungsatlas der Farn- und Blütenpflanzen der Schweiz. 2 vol. Birkhäuser Verlag, Basel.

VEGETATION MONITORING ON A SMALL-SCALE RESTORATION SITE IN THE ALPINE BELT: PILATUS KULM, SWITZERLAND

ENGELBERT RUOSS[1], CONRADIN A. BURGA[2] & JAKOB ESCHMANN[3]

[1]*Museum of Natural History Luzern, Kasernenplatz 6, CH-6003 Luzern;* [2]*Geographical Institute, University of Zürich, Winterthurerstrasse 190, CH-8057 Zürich;* [3]*Alpine Plant Nursery, Waltwil, CH-6032 Emmen*

Keywords: Vegetation restoration, herbaceous vegetation, species fluctuation, plant cover, Pilatus Kulm, Swiss alpine belt

Abstract
The vegetation restoration of a site (ca. 300 square metres) on Pilatus Kulm (2060 metres a.s.l., Obwalden, Switzerland) located in the alpine belt was observed between 1990 and 1998. Severe damage to the vegetation cover had occurred during construction of a sewer pipeline from Pilatus Kulm down to the valley and establishment of military installations. In a first stage, plant recolonization was induced by planting different seedlings of herbs. In a second stage, seeds of alpine grass species were used to achieve a denser pioneer vegetation cover. A vegetation study made at different plots of the investigation area revealed that over 80 % of the transplants survived the four-year period (1990-1994). On the whole, the plant cover increased to up to 40 % on the slope and up to 80 % on tilted borders within the four-year period. The species diversity increased faster in the random area due to natural immigration. With the introduction of seedlings in 1994, along with favourable circumstances, the establishment of the vegetation has been accelerated, resulting in an area which is now covered by dense herbaceous vegetation. In 1998, the control plots showed a plant cover of 60 % on the slope and up to 95 % on tilted borders. During the eight-year monitoring period, considerable fluctuations of species could be observed.

Kurzfassung
Die Entwicklung einer naturnah wiederbegrünten Fläche (ca. 300 m²) in der alpinen Stufe auf Pilatus Kulm (2060 m ü.M., Obwalden, Schweiz) wurde von 1990 bis 1998 beobachtet. Die Vegetation war durch Bauarbeiten an der Kanalisation Pilatus Kulm - Alpnach (1988/1989) sowie durch Installationen für Bauten des Bundesamts für Militärflugplätze 1992/1993 weitgehend zerstört worden. Mit der Pflanzung von Setzlingen wurde zuerst die Wiederbesiedlung vorbereitet und die Pioniervegetation mit der Ansaat von standortgerechten Grassamen anschliessend verdichtet (1990-1994). An Kontrollstandorten (Dauerbeobachtungsflächen) hat sich die Bedeckung in vier Jahren auf 40 % am Hang und auf 80 % an den randlichen Kanten erhöht. Die Artenvielfalt hat

Burga & Kratochwil (eds.), BIOMONITORING, 143-152

in randlichen Bereichen natürlicherweise stärker zugenommen. Rund 80 % der eingesetzten Pflanzen überlebten die Vierjahresperiode. Mit der zusätzlichen Ansaat von 1994 konnte die Begrünung beschleunigt werden, so dass die Fläche nach acht Jahren, dank geeigneter Rahmenbedingungen, eine weitgehend geschlossene Vegetationsdecke aufweist. Die Deckung auf den Kontrollflächen betrug zwischen 60 % am Hang und 95 % an den randlichen Kanten, wobei eine hohe Fluktuation der Arten feststellbar war.

1. Introduction

The restoration site (ca. 300 square metres) is located at ca. 2060 metres a.s.l. in the summit area of Pilatus Mountain (Central Swiss Alps, Canton of Obwalden, Figure 1). The substratum consists of limestone and base-rich schist. The site was destroyed by construction of a sewer pipeline in 1988/1989 and by military installations in 1992/1993 (Figure 2).

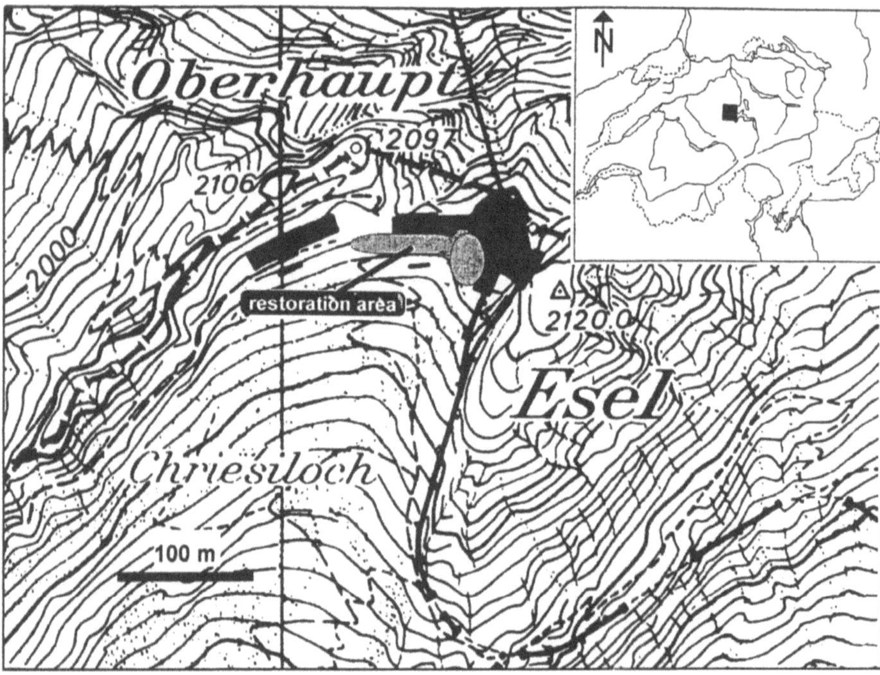

Figure 1. Location of the restoration area of Pilatus Kulm: southern-exposed slope of the hotel and station.

Pilatus Kulm is easily accessible by a cogwheel railway and attracts many tourists; it is a nature reserve well-known for its rich flora and fauna. Vegetation restoration was commissioned by the Pilatus Train Company and the Federal Airforce Office. The goals were to provide a suitable place for visitors and a starting point for hang-gliders, an attractive plant cover, slope stabilization, and visual improvement of the summit area

landscape. A plant cover similar to the surrounding vegetation (Wallimann 1971) should be established within five years. The problems of vegetation dynamics and restoration of areas in the Swiss alpine belt after human impact have been described by different authors (e.g. Stolz 1984; Delarze 1994 or Urbanska & Fattorini 1998).
The restoration work was initiated in 1990, interrupted for the period 1992-1993, and continued in 1994. The assessment was made in 1994 (Ruoss et al. 1995). Control analyses were realized in 1996 and 1998.

Figure 2. General view of the restoration site in 1990.

2. Methods

The methods of planting (Figure 3) are based mainly on the studies and the methods published by Urbanska et al. (1986, 1988).
The topsoil was amended with a 0.5-metre layer of rendzina mixed with limestone gravel; a 10-centimetre humus layer was added later. Alpine plants from the same area were used for seed and clonal transplant production at a regional nursery. Transplants were established in 1990 and 1991; in 1994, an alpine grass mixture was hand-seeded (300 g seeds/square metre). The transplanted grasses included: *Poa alpina, Briza media, Sesleria caerulea* and *Deschampsia caespitosa* (3000 young plants). The propagation of *Festuca* sp., *Luzula* sp. and *Carex* sp. was not successful. Also planted were *Achillea atrata, Antennaria dioica, Arabis alpina, Astragalus alpinus, Biscutella laevigata, Campanula cochleariifolia, C. scheuchzeri, Draba aizoides, Erinus alpinus, Eryngium alpinum, Globularia cordifolia, Gypsophila repens, Helianthemum alpestre, Hutchinsia alpina, Linaria alpina, Lotus alpinus, Phyteuma betonicifolium, Scabiosa lucida, Silene*

acaulis, Solidago virgaurea ssp. *minuta, Thymus serpyllum, Trifolium badium, T. montanum, T. repens* (4400 young plants) and dwarf shrubs, such as *Arctostaphylos uva-ursi, Dryas octopetala, Rhododendron hirsutum, Salix retusa* and *S. reticulata* (70 young plants).

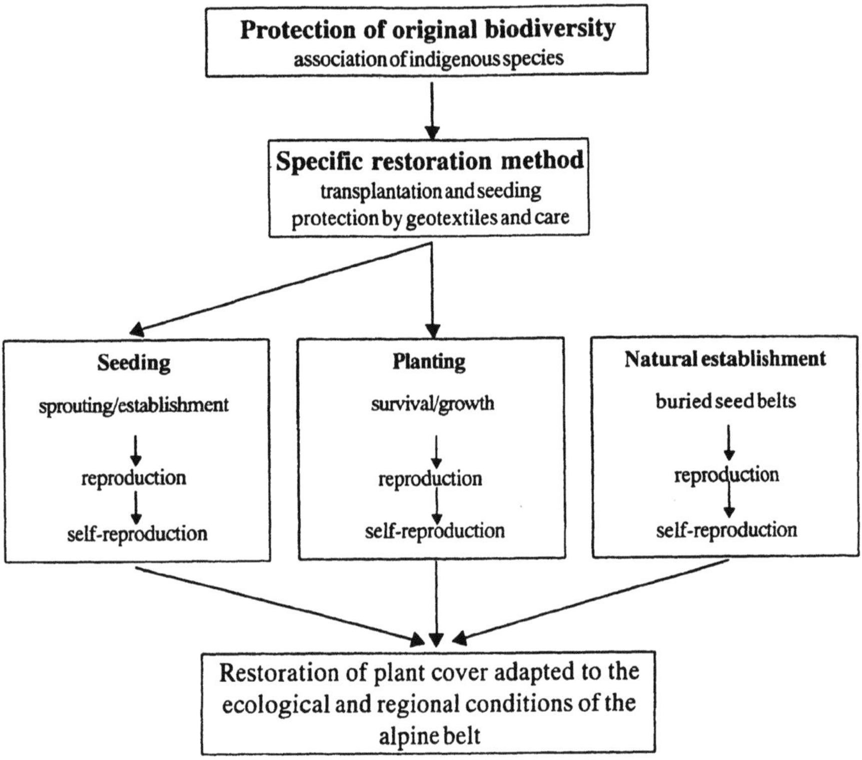

Figure 3. Concept of near-natural vegetation restoration in the alpine belt.

The grasses *Festuca violacea* and *Poa alpina* (15 kg of seeds) were seeded in 1994 after application of 10 centimetres of humus. Additional species were introduced together with the humus. The site was covered with jute and watered during summer (Figure 4).
The progress of the vegetation restoration was studied by the method of Braun-Blanquet (1951) at three different plots (slope, tilted border) in 1994, 1996 and 1998 and documented with photos. The nomenclature of the plants follows the Flora Helvetica (Lauber & Wagner 1996).

Figure 4. General view of the restoration site after seeding in 1994. The seeded slope was partly covered with jute.

3. Results

Over 80 % of the transplants survived the four-year period (1992-1996). The vegetation development after restoration led to a species-rich situation (Table 1). On the whole, the plant cover increased to up to 40 % on the NW slope and up to 80 % on the tilted borders until 1994 (Figure 5a). The seeded grasses produced a loose plant cover within five weeks.

During the eight-year period, considerable fluctuations of species could be observed. Among the most successful plant species were *Alchemilla monticola, Cerastium holosteoides, Deschampsia caespitosa, Doronicum grandiflorum, Lotus alpinus, Myosotis alpestris, Poa alpina, Trifolium alpestre, T. badium* and *T. repens.* Some species of the pioneer vegetation and other species introduced together with the humus layer disappeared completely, such as *Arabis alpina, Barbarea intermedia, Erucastrum nasturtiifolium, Hutchinsia alpina, Phleum pratense* and *Phyteuma scheuchzeri.* In 1998, a lot of new plant species, immigrated from the surrounding area, were observed in the control plots: *Adenostyles alliariae, Agrostis rupestris, Bellis perennis, Campanula cochleariifolia, C. scheuchzeri, Chrysanthemum leucanthemum, Cirsium spinosissimum, Festuca halleri, Plantago major, Rumex scutatus* or *Sagina procumbens.*

In 1998, the control plots showed a plant cover of 60 % on the NW slope and up to 95 % on tilted borders (Figure 5a). The number of plant species increased at the NW slope from 21 to 26 (Figure 5b). At the inclined borders it increased in the first period from 22 to 41 and decreased to 35 plant species until 1998. On a control plane site without planting (only hay seeds), the plant cover increased after 1994 from 25 to 60 %, mainly due to natural plant immigration from the surrounding areas. In this area, the species number decreased at the beginning of the monitoring from 16 to 11 and increased to 26 species until 1998.

Table 1. Vegetation of three permanent plots in 1991, 1994 and 1998, analysed by the method of Braun-Blanquet (1951).

		NW slope			**plane site**			**tilted border**		
	area	3 x 2 metres			4 x 3 metres			5 x 3 metres		
Plant species	relevé	1991	1994	1998	1991	1994	1998	1991	1994	1998
	method	plantation			hay seeds			plantation		
	plant cover	20 %	40 %	75 %	25 %	25 %	60 %	30 %	80 %	95 %
Achillea atrata		+	1	1			+			
Adenostyles alliariae							r		+	+
Agrostis rupestris							+			
Agrostis cf. *tenuis*										+
Alchemilla conjuncta agg.									1	r
Alchemilla monticola		r	+	+	r	+	+	r	1	+
Alopecurus pratensis								r	1	+
Antennaria dioica								+		
Anthyllis vulneraria s.l.									+	
Arabis alpina		1	1		+			2	1	
Astragalus alpinus								r	1	
Avenula versicolor		r		r						
Barbarea intermedia		2	+		+			1	+	
Bellis perennis				r			+			
Biscutella laevigata									+	
Briza media								r		
Campanula cochleariifolia										r
Campanula scheuchzeri										+
Carduus defloratus		r	+	+		+	+		1	+
Cerastium holosteoides		+	+		+			r	1	+
Chrysanthemum leucanthemum				+			+			r
Cirsium spinosissimum				+						+
Cynosurus cristatus										r
Dactylis glomerata									+	+
Deschampsia caespitosa		+	1	+			+		2	2
Doronicum grandiflorum		+	+	+				r	1	+
Epilobium montanum				+					+	+
Erinus alpinus								r		
Erucastrum nasturtiifolium		2	+		+			1	+	
Festuca halleri							+			+
Festuca rubra s.latissimo							+	r	1	1
Festuca violacea									+	
Galium anisophyllum				+			+		+	+

Table 1 continued

Gentianella campestris								+	
Hieracium sp.			r						
Hutchinsia alpina	+	+		r	+		2	2	
Myosotis alpestris		+	+		r			+	
Leontodon helveticus		+	+	r	+	+		1	
Leontodon hispidus									r
Lolium perenne				r				1	
Lotus alpinus	1	2	1	r	+			2	1
Pedicularis verticillata			+					r	r
Phleum alpinum								1	
Phleum pratense s.l.							r	1	
Phyteuma betonicifolium	r								r
Phyteuma scheuchzeri	r	r		r					
Plantago alpina						r			
Plantago lanceolata			r	r	r	+		+	r
Plantago major			+			+		+	1
Poa alpina	+	2	2	2	2	2	1	2	1
Poa annua	+	1	+			+		+	
Ranunculus alpestris	r	+							
Ranunculus montanus agg.		+	+			+			
Ranunculus repens									r
Rumex scutatus			+		+	r			
Sagina procumbens				r		1			+
Saxifraga paniculata							r	+	
Scabiosa lucida								+	+
Sedum album							r		
Silene vulgaris ssp. *glareosa*								1	
Solidago virgaurea ssp. *minuta*	r						r		
Taraxacum officinale			+			+		1	+
Thymus serpyllum agg.				r	+	+	+	1	+
Trifolium alpestre			+			+	+	2	3
Trifolium badium	2	2	1	r		+	+	3	2
Trifolium montanum	+	1							
Trifolium repens	r	1	+	r	+	+	r	+	+
Tussilago farfara								2	2
Veronica serpyllifolia						1			
number of plant species	21	21	26	16	11	26	21	41	35

In 1998, pioneer mosses were collected at the tilted border: *Bryum argenteum* Hedw., *Barbula unguiculata* Hedw., *Brachythecium rutabulum* (Hedw.) B. & S. & G. and *Hypnum revolutum* (Mitt.) Lindb., on the plane site *Brachythecium rutabulum* (Hedw.) B. & S. & G., *Ceratodon purpureus, Cratoneuron filicinum* (Hedw.) Brick and *Didymodon fallax* (Hedw.) Zander.

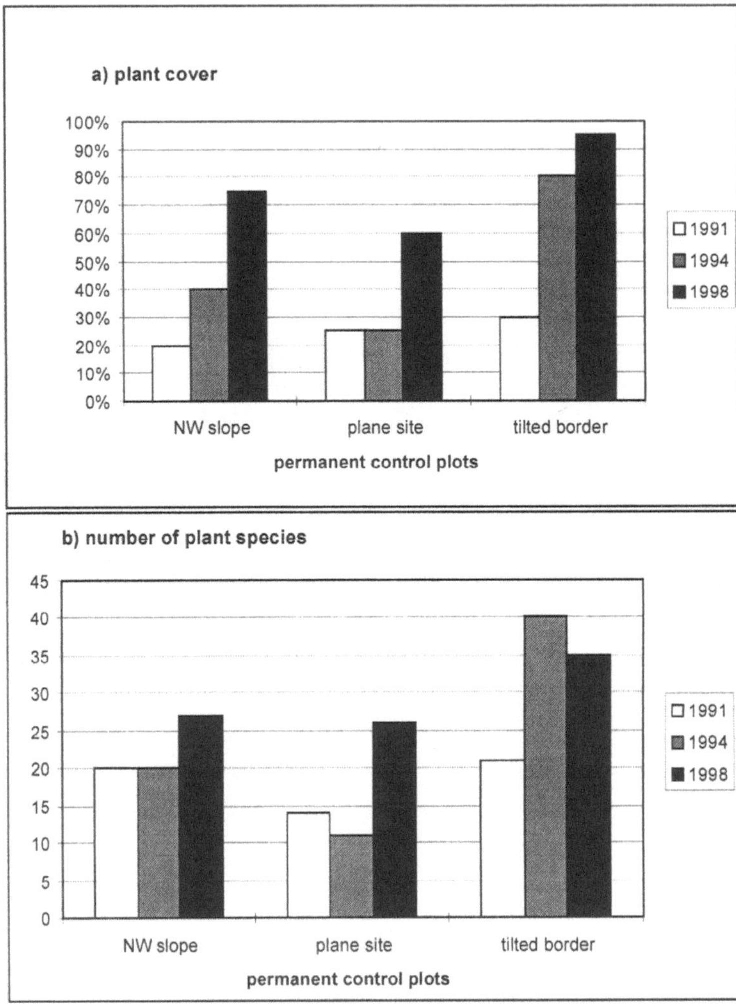

Figure 5. Comparison of vegetation cover (a, top) and number of plant species (b, bottom) of permanent plots in 1991, 1994 and 1998.

4. Discussion

In 1998, the experimental area had fully restored itself, except in two areas. The area reached its natural state and showed a plant cover between 60 and 95 %. The pioneer plants introduced at the beginning of the project had mostly disappeared. Several species from the surrounding vegetation established themselves only after the eight-year period.

Two small areas had a poorer plant cover. An area which had not been covered with jute showed signs of serious erosion (Figure 6). The other area was constantly exposed to human impact, mainly visitors and hang-gliders.

Figure 6. General view of the restoration site in 1998.

During the monitoring period many problems arose: visitors and hang-gliders desired free access to the site despite the ongoing restoration, which had to be interrupted for two years on account of the military construction. Clients and tourists expected full recovery already within one year. The employees of the train company viewed restoration with mistrust because it involved neither high technology nor extensive use of machines. In addition, wildlife (*Ibex ibex*) damaged the site by grazing and trampling.

Restoration in the alpine belt at around 2000 metres a.s.l. is laborious and thus costly, but it is possible to obtain good results. The soil character seems to be the major limiting factor for the used high mountain plants. A restoration plan worked out before damaging saves much trouble and money. Minimizing human impact must be considered already at the beginning of each new intervention in an area.

Acknowledgements

We would like to thank the Pilatus Train Company and the Federal Airforce Office for financial support, Elizabeth Gosselin, Luzern, for correcting the English and Eva Maier, Bernex (GE), for determining moss samples.

References

Braun-Blanquet, J. 1951. Pflanzensoziologie. 2nd ed. Springer, Wien.

Delarze, R. 1994. Dynamique de la végétation sur les pistes ensemencées de Crans-Montana (Valais, Suisse). Effets de l'altitude. Botanica Helvetica 104: 3-16.

Lauber, K. & Wagner, G. 1996. Flora Helvetica. Haupt, Bern, Stuttgart, Wien.

Ruoss, E., Burga, C.A. & Eschmann, J. 1995. Naturnahe Wiederbegrünung auf Pilatus Kulm (2060 m ü.M.). Mitteilungen der Naturforschenden Gesellschaft Luzern 34: 85-96.

Stolz, G. 1984. Entwicklung von Begrünungen oberhalb der Waldgrenze aus der Sicht der Botanik. Zeitschrift für Vegetationstechnik 7: 29-34.

Urbanska, K.M. & Fattorini, M. 1998. Seed bank studies in the Swiss Alps. I. Un-restored ski run and the adjacent intact grassland at high elevation. Botanica Helvetica 108: 93-104.

Urbanska, K.M. & Schütz, M. 1986. Reproduction by seed in alpine plants and revegetation research above timberline. Botanica Helvetica 96: 43-60.

Urbanska, K.M., Schütz, M. & Gasser, M. 1988. Revegetation trials above the timberline - an exercise in experimental population ecology. Ber. Geobot. Inst. ETH, Stiftung Rübel, Zürich 54: 85-109.

Wallimann, H. 1971. Flora des Kantons Obwalden. Mitteilungen der Naturforschenden Gesellschaft Luzern 7: 47-309.

CHAPTER C

ASPECTS OF GLOBAL CHANGE IN THE ALPS AND IN THE HIGH ARCTIC REGION

LONG-TERM MONITORING OF MOUNTAIN PEAKS IN THE ALPS

GEORG GRABHERR, MICHAEL GOTTFRIED & HARALD PAULI

Department of Vegetation Ecology and Conservation Biology, Institute of Ecology and Conservation Biology, University of Vienna, Althanstrasse 14, A-1090 Vienna

Keywords: Alpine, climate change, migration, nival, permanent plots, vegetation

Abstract

Historical records of nival summit floras from the Alps were compared with the recent species composition at the same summits revisited in the years 1992/1993. A general trend to increased species richness could be detected. For one particular summit, Piz Linard, quantitative data were provided by the historical author; our revisitation showed clearly that species abundance has increased substantially at this mountain during the last 50 years. The increase in species richness as well as the increase in abundance suggest that the observed warming of 1 - 2°C since the turn of the century has affected the biota at these low-temperature limits of plant life. This is in agreement with unexpected observations, such as the record of the uppermost population of *Carex curvula* at 3468 m. *Carex curvula* is the dominating species of the late successional grassland vegetation which occupies an altitudinal belt between 2400 - 2800 m. Furthermore, the uppermost storage of Swiss stone pine seeds (3103 m; seeds were germinating) was recorded. The seeds are distributed by the bird *Nucifraga caryocatactes* that stores them in clusters at places which are less snow-covered in winter time. In this case the bird brought the seeds from the timber line more than 700 meters below up to the discovered storage-place. These observations prove an upward migration of alpine biota.

Many plant species in alpine/nival environments are long-lived species (e.g. *Carex curvula*). Unexpected observations, like uppermost records (or higher frequencies of germination events), might indicate that the current climate change effects are different from those in the past centuries.

Burga & Kratochwil (eds.), BIOMONITORING, 153-177

Most of the indications of biotic, but also physical, effects of the ongoing climate change have been derived from observations in mountain areas. Therefore long-term monitoring in mountains, alpine/nival environments in particular, is generally considered as useful. Establishing networks of monitoring sites in high mountain areas has to be considered as urgent.

Kurzfassung

Historische Aufnahmen zur Gefäßpflanzenflora hochalpiner Alpengipfel wurden mit der aktuellen Florenzusammensetzung verglichen (rezente Aufnahmen in den Sommern 1992/93). Ein klarer Trend zu höheren Artenzahlen konnte nachgewiesen werden. Für den Gipfel des Piz Linard waren in den historischen Daten auch quantitative Angaben zur Häufigkeit vorhanden, aus denen klar hervorging, daß die Häufigkeit der meisten Arten in den letzten 50 Jahren stark zugenommen hat. Sowohl die Zunahme der Artenvielfalt als auch der Häufigkeiten sind ein klarer Beweis, daß die Erwärmung seit der Jahrhundertwende um 1 bis 2°C die Lebensgemeinschaften an den Kältegrenzen des Pflanzenlebens beeinflußt hat. Dies ist auch in Übereinstimmung mit unerwarteten Beobachtungen wie dem Nachweis der höchsten Population von *Carex curvula* auf 3468 m. *Carex curvula* ist die dominante Art der zonalen alpinen Rasenvegetation, welche typischerweise eine Zone zwischen 2400 - 2800 m einnimmt. Weiter gelang auch der Höchstfund eines Reservoirs von Zirbensamen (3103 m; Samen keimten zum Zeitpunkt der Beobachtung), welches von einem Zirbenhäher von der Waldgrenze, die hier ca. 700 m tiefer liegt, heraufgebracht wurde. Aufgrund all dieser Beobachtungen kann als gesichert gelten, daß die alpine Lebewelt nach oben wandert.

Viele Arten der alpin/nivalen Lebensräume sind sehr langlebig (z.B. *Carex curvula*). Unerwartete Beobachtungen wie Höchstfunde, Häufung von Keimereignissen, lassen vermuten, daß die aktuellen Effekte des Klimawandels neuartig sind.

Die meisten Beobachtungen, welche auf bereits einsetzende biotische, aber auch physikalische Effekte des Klimawandels hindeuten, stammen von Gebirgsregionen. Langzeitbeobachtungen in Gebirgen, Hochgebirgen im speziellen, werden daher allgemein empfohlen. Die Etablierung eines Netzwerks von Beobachtungsstationen in Hochgebirgen erscheint dringlich.

1. Introduction

Plant life under the harsh climatic conditions of alpine environments, especially at their limits, has been a stimulating object of study since the early days of ecological botany (Heer 1885; Schibler 1898; Rübel 1912; Braun 1913; Klebelsberg 1913). It has been common practice both for professional scientists and for amateur field botanists to note down the plant species, or collect specimens, when having climbed high alpine peaks. A search for such data sets from the Alps revealed more than 300 species lists older than 40 years (most of them for vascular plants only), all from mountains reaching to or higher than 3000 m a.s.l., thus representing the so-called nival environment.

As already recognised by Klebelsberg (1913), Braun-Blanquet (1955, 1957), these summits with old records of their flora could be considered as permanent plots, and might be excellent indicators of climate change. The recent debate on climate change effects on vegetation (Solomon & Shugart 1993; Markham et al. 1993; Watson et al. 1996), in mountain environments in particular (Nilsson & Pitt 1991; Ozenda & Borel 1991; Holten 1993; Beniston 1994; Beniston & Fox 1996; Guisan et al. 1995; Price & Barry 1997) has been the incentive to reconsider this "summit approach", for the following reasons:

1. Climate change effects on vegetation, effects of warming in particular, may be most obvious at these temperature-limited habitats;
2. Other effects of the changing environment, i.e. increased carbon dioxide, nitrogen deposition, changes in precipitation, might be less relevant at these extreme altitudes;
3. Biotic interference has a less selective effect within the scattered plant assemblages of the nival summits;
4. Human influence can be excluded, except on those summits which are visited very frequently.

Based on the findings of the previous authors (Klebelsberg 1913; Braun-Blanquet 1955, 1957), in combination with the expected upward movement of the altitudinal vegetation belts (e.g. Ozenda & Borel 1991; Markham et al. 1993), and the observed climate warming since the turn of the century (for the Alps see Böhm 1986; Auer et al. 1996), the hypothesis was proposed that the species diversity must have increased at the Alpine summits, indicating a general trend towards upward migration of alpine plants. This hypothesis was the starting point of a now seven-year Austrian research initiative

1. which provided evidence for the expected migration (based on the revisitation approach, see Grabherr et al. 1994; Grabherr et al. 1995; Gottfried et al. 1994; Pauli et al. 1996);
2. during which more than 1000 exactly surveyed and documented permanent plots at the alpine/nival ecotone were established at Mt. Schrankogel (Tyrol, Austria), which was selected as a model mountain (Pauli et al. 1999);
3. during which computer models have been developed to explore the migration dynamics at high altitudes (Gottfried et al. 1998).

In this paper we present the original data set of the revisitation approach, summarise the most important aspects as well as mention, and comment on, additional "exciting" observations, and discuss briefly the opportunities for future permanent plot research in alpine/nival environments.

2. Proofing the migration hypothesis

2.1. STUDY SITES AND HISTORICAL RECORDS

As already mentioned, more than 300 old records of summit floras from various parts of the European Alps have been published so far. However, only comparatively few locations appeared to be appropriate for revisitation as most of the authors did not indicate exactly the sector within which the plants were recorded. Finally we selected 30 records by professional botanists well known for their excellent knowledge of alpine plants, i.e. Heer (1866), Schibler (1898), Rübel (1912), Klebelsberg (1913), Braun (1913), Braun-Blanquet (1957, 1958), Reisigl & Pitschmann (1958). The mountains are located in the Central Alps of Switzerland, Austria, and Italy (Figure 1). The data sets date from 1895 to 1953. They provide complete presence/absence lists of vascular plants at the uppermost 15 to 30 m (with some exceptions, see Appendix 1). At 12 of the 30 peaks the uppermost position of each species within the summit area was recorded with the accuracy of one-meter isolines (Braun 1913; Braun-Blanquet 1957, 1958). For the mountain Piz Linard (Switzerland) which has been visited several times, a time series (1835, 1864, 1895, 1911, 1937, 1947, 1992) was available, as well as a map which shows the abundance of the registered species for 1947.

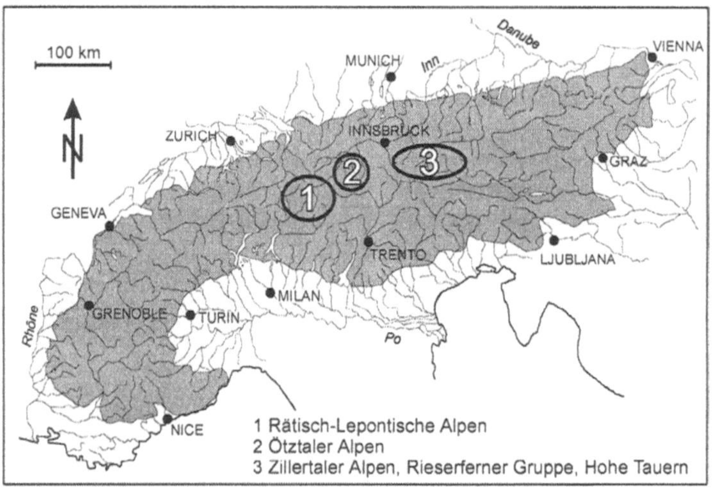

Figure 1. Geographical position of the areas where the investigated summits are located: 1. Rätisch-Lepontische Alpen (the Grisons; Switzerland; 22 summits); 2. Ötztaler Alpen (Tyrol, Austria/Italy; 7 summits); 3. Zillertaler Alpen, Rieserferner Gruppe, Hohe Tauern (Tyrol, Carinthia; Austria/Italy; 6 summits).

2.2. METHODS

The selected summits were revisited during the summers of 1992 and 1993, and the historical investigation was repeated, i.e. listing the occurring species in the summit area considered. In addition to the historical approach the abundance of each species was recorded, as well as the exact position (in meters a.s.l.) of the uppermost individual of each species population. Altitudinal positions were measured using an altimeter and a clinometer. One summit record took 2 to 7 hours (about 4 hours in most cases). In general, field investigations were conducted by two persons. For further details see Pauli et al. (1998).

The historical data set was checked for obvious misidentifications. The recent data set was then compared with the historical one. Qualitative indicators for changes were 1. the invaders, 2. species which were not found again, and 3. constant species. As quantitative measure a weighted species increase was calculated according to the following formula:

wSID = (wnS - wmS)/(eS + wmS) *100/[(pY - hY)/10]

wSID = weighted species increase per decade, wnS = weighted number of newly found species, wmS = weighted number of species not found again (missing species), eS = number of species found in both records, pY = year of the present investigation, hY = year of the historical investigation.

The newly found species (wnS) were weighted for abundance, which was 1 for species with high abundance, 0.5 for species with low abundance, and 0.25 for species with very low abundance, respectively (for details see Appendix 1). This weighting procedure takes into account that rare species might not have been recognised by the historical author. Species which were not found again were down-weighted by 0.7. Species occurring in the historical as well as in the recent record were not weighted.

2.3. RESULTS

The historical and the recent data sets for the 30 mountains are given in Appendix 1 (data sets for 5 recently investigated summits are additionally presented). The entire original data sets (35 recent, 30 historical ones) have been added here to provide the basis for future revisitations.
As a matter of fact, those summits, the historical records from which are older than 80 years, can be considered as belonging to the oldest permanent plots in the world; Piz Linard is probably the oldest anyway.

Species richness of the considered summits
The species richness of the investigated summit areas varies remarkably, from 10 to more than 60 species, and decreases with increasing altitude in a non-linear way (Figure 2). Summits with siliceous bedrock material seem to be richer compared to those with carbonates (limestone, dolomite). In total 116 species were recorded at the reinvestigated summits. Almost all of these peaks exceed 3000 m. Altogether 98 species were recorded for the first time for at least one summit, 8 species were recorded more than four times as new (Table 1a). Among the unexpected observations the following two are especially noteworthy: 1. For *Carex curvula* All. the uppermost population ever recorded in the Alps was found at Hohe Wilde at an altitude of 3468 m, where it was also recorded as new species; 2. The uppermost seed storage of Swiss stone pine (*Pinus cembra* L.) ever recorded was found at 3103 m (just germinating when found). *Pinus cembra* seeds are stored by the bird *Nucifraga caryocatactes*. In this case the bird transported the seeds from the tree line more than 700 m below up to this exceptionally high altitude.
At the summits a core group of species occur frequently. These species could be considered as "nival species" in a strict sense, although they sometimes also occur far below the nival zone (Table 1b). They were often recorded in the historical data sets, i.e. *Ranunculus glacialis* L., *Poa laxa* Haenke, *Saxifraga bryoides* L., and, to a lesser extent, *Cerastium uniflorum* Clairv. and *Androsace alpina* (L.)Lam. As migrating

species, four groups could be identified (Table 1a): 1. species which can be considered as alpine ruderals, such as *Potentilla frigida* Vill., *Cardamine resedifolia* L., *Draba fladnizensis* Wulf., *Draba dubia* Suter, *Agrostis rupestris* Scop. and *Poa alpina* L.; 2. species from open alpine/subnival early successional grasslands, such as *Minuartia sedoides* (L.)Hiern and *Silene exscapa* All.; 3. species from late successional alpine grasslands, such as *Elyna myosuroides* (Vill.)Fritsch, *Carex curvula* All., *Juncus jaquinii* L. and *Senecio incanus* ssp. *carniolicus* (Willd.)Br.-Bl.; and 4. snow bed species, such as *Gnaphalium supinum* L. and *Saxifraga androsacea* L. Some of the invaders, however, cannot be assigned to any of these groups, e.g. *Saxifraga oppositifolia* L. which inhabits now many more summits than some decades ago.

Table 1. The most frequently newly found species (1a), and the most common species (1b), at the 25 reinvestigated siliceous mountains. "Weighted values" = down-weighting if species abundance is low (for explanation see Table 2).

a) most frequently newly found species			b) most common species	
species	**no. of summits where species was new**	**weighted value**	**species**	**no. of summits where species was present**
Saxifraga oppositifolia	6	5.5	*Ranunculus glacialis*	23
Potentilla frigida	8	4	*Poa laxa*	22
Draba fladnizensis	6	3.75	*Saxifraga bryoides*	22
Cardamine resedifolia	5	3.75	*Cerastium uniflorum*	20
Agrostis alpina	5	3.2	*Androsace alpina*	19
Agrostis rupestris	5	3.2	*Saxifraga oppositifolia*	19
Geum reptans	4	3	*Tanacetum alpinum*	19
Minuartia sedoides	4	2.95	*Draba fladnizensis*	18
Draba dubia	6	2.75	*Gentiana bavarica*	17
Gentiana brachyphylla	4	2.5	*Saxifraga exarata*	17
Poa alpina	4	2.5	*Luzula spicata*	16
Cerastium uniflorum	3	2.5	*Minuartia sedoides*	16
Juncus jacquinii	3	2.5	*Erigeron uniflorus*	15
Gnaphalium supinum	5	2.45	*Potentilla frigida*	14
Elyna myosuroides	4	2.25	*Carex curvula*	13
Senecio incanus ssp.*carniolicus*	4	2.25	*Festuca halleri agg.*	13
Euphrasia minima	3	2.25	*Oreochloa disticha*	12
Saxifraga seguieri	3	2.25	*Silene exscapa*	10
Silene exscapa	3	2.25	*Geum reptans*	10
Erigeron uniflorus	4	2	*Phyteuma globulariif.*ssp.*pedemontanum*	10
Carex curvula	2	2	*Doronicum clusii*	10
Eritrichum nanum	2	2	*Senecio incanus* ssp.*carniolicus*	10
Saxifraga androsacea	2	2		

Quantitative analysis of change

The results of calculating the weighted species increase standardised on increase per decade are given in Table 2. The majority of the summits show a clear increase of more than 1 % per decade. Without weighting the increase would be even higher. The weighted values represent a kind of "at least" increase, taking into account errors made by the historical author, and down-weighting those species which occur with low abundance. A comparison of the records covering a long time span (about 80 years; see Table 2) with those covering a short one (about 40 years) showed no acceleration of the speed of migration (see values for the weighted species increase per decade in Table 2). No significant difference between carbonate summits and siliceous ones was found, either. Most summits are only occasionally visited by tourists, and no correlation was ascertained between tourist frequency (estimated as high, low, very low) and species increase. Furthermore, tourists keep usually very close to the trails or climbing routes. In our experience, the impact of visitors can be neglected at least for this study.

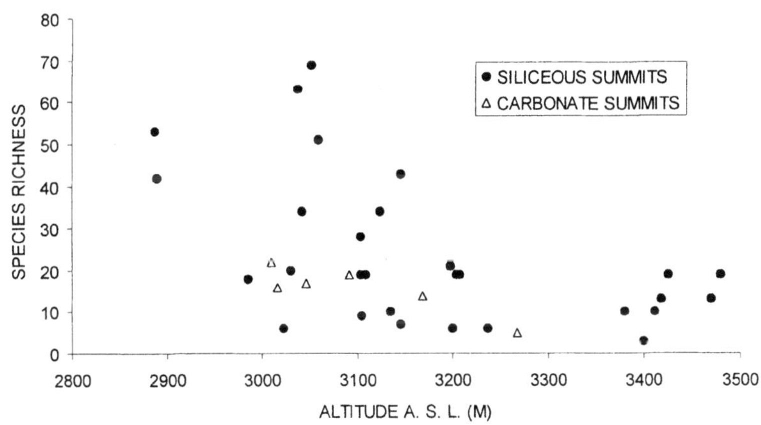

Figure 2. Species richness (= number of vascular plants) of 35 high alpine summits (for original data and details on the investigated sites see Appendix 1).

Three groups of summits can be distinguished according to the rates of increase: a group where species richness has increased dramatically (> 4 % per decade), a group of obvious, but moderate increase (> 1 % per decade), and, finally, a group of summits where species richness has increased either at a low rate or even has decreased. To the first group belong rocky mountains, where the numerous rock outcrops offer many migration pathways for the flora. Along them seeds can be trapped in small crevices which, at these high altitudes, provide safe, warmer sites free of severe cryoturbation. Most of the summits are connected with the lower alpine grassland zone by such migration pathways, and the invaders belong to the alpine ruderal group as well as to the "climax" species group of the alpine belt. The latter are almost absent in the second and the third group where the surface of the summits consists of block fields or scree without rock outcrops and connections to the alpine belt.

Table 2. Change of species richness at 30 subnival and nival summits of the Eastern Alps; car. = dolomite or limestone as bedrock material, sil. = siliceous bedrock; Br.Bl. = Braun-Blanquet, Kleb. = Klebelsberg, Rei.&Pit. = Reisigl & Pitschmann; weighted number of species not found again (wmS) - a value of 0.7 was taken for these species as an average derived from the abundance weightings of the recent floras; weighted number of newly found species (wnS) - newly found species with low abundance were down-weighted by 0.5, those of very low abundance by 0.25; wSI (weighted species increase) = weighted number of newly found species minus missing species standardised on the sum of species in both records plus the weighted number of newly found species; wSID (weighted species increase per decade) = wSI standardised over time (given in increase per decade).

summit	**altitude [m]**	**bedrock**	**author of the historical record**	**year of historical record**	**year of present record**	**interval [m]**	**historical number of species**	**present number of species**	**no. of species in both records**	**weighted no. of species not found again**	**weighted no. of newly found species**	**wSI [%]**	**wSID [%]**
Piz dals Lejs	3042	sil.	Rübel	1907	1992	30	11	34	10	0,7	16	143,0	**16,8**
Piz Laschadurella	3046	car.	Br.-Bl.	1919	1992	30	7	17	6	0,7	7,25	97,8	**13,4**
Hohe Wilde	3480	sil.	Rei.&Pit.	1953	1992	20	11	19	11	0	4,75	43,2	**11,1**
Piz Tavrü	3168	car.	Br.-Bl.	1927	1993	15	5	14	5	0	3	60,0	**9,1**
Stockkogel	3109	sil.	Rei.&Pit.	1953	1992	9	12	19	12	0	4,25	35,4	**9,1**
Piz Foraz	3029	car.	Br.-Bl.	1921	1993	30	9	19	9	0	5	55,6	**7,7**
Napfspitze	2888	sil.	Kleb.	1911	1992	3	31	53	30	0,7	18,3	57,3	**7,1**
Monte Vago	3059	sil.	Br.-Bl.	1905	1992	30	28	51	27	0,7	14,75	50,7	**5,8**
Piz Nuna	3124	sil.	Br.-Bl.	1919	1992	30	19	34	19	0	7,25	38,2	**5,2**
Piz Uertsch	3268	car.	Br.-Bl.	1911	1993	30	2	5	2	0	0,75	37,5	**4,6**
Großer Lenkstein	3236	sil.	Kleb.	1912	1992	2	4	6	3	0,7	2	35,1	**4,4**
Wilde Kreuzspitze	3135	sil.	Kleb.	1907	1992	3	8	10	6	1,4	3,25	25,0	**2,9**
Piz Kesch	3418	sil.	Br.-Bl.	1902	1992	30	8	13	8	0	2	25,0	**2,8**
Munt Pers	3207	sil.	Rübel	1906	1992	11	15	19	14	0,7	4	22,4	**2,6**
Hint. Spiegelkogel	3426	sil.	Rei.&Pit.	1953	1992	15	15	19	15	0	1,45	9,7	**2,5**
Piz Forun	3052	sil.	Br.-Bl.	1903	1992	30	48	69	47	0,7	11	21,6	**2,4**
Gorihorn	2986	sil.	Schibler	1895	1992	7	14	18	12	1,4	3,75	17,5	**1,8**
Festkogel	3038	sil.	Rei.&Pit.	1953	1992	35	58	63	54	2,8	6,5	6,5	**1,7**
Hint. Seelenkogel	3470	sil.	Rei.&Pit.	1953	1992	12	11	13	11	0	0,75	6,8	**1,7**
Flüela Schwarzhorn	3146	sil.	Br.-Bl.	1902	1992	50	34	43	32	1,4	6	13,8	**1,5**
Piz Stretta	3104	sil.	Br.-Bl.	1903	1992	30	20	28	20	0	2,5	12,5	**1,4**
Liebenerspitze	3399	sil.	Rei.&Pit.	1953	1992	20	2	3	1	0,7	0,75	2,9	**0,8**
Piz Plazer	3104	sil.	Br.-Bl.	1918	1992	30	16	19	14	1,4	2,25	5,5	**0,7**
Piz Sesvenna	3204	sil.	Br.-Bl.	1918	1992	30	17	19	17	0	0,5	2,9	**0,4**
Piz Blaisun	3200	sil.	Br.-Bl.	1911	1993	20	5	6	4	0,7	0,75	1,1	**0,1**
Piz Julier	3380	sil.	Br.-Bl.	1900	1992	30	9	10	8	0,7	0,75	0,6	**0,1**
Piz Nair	3010	car.	Br.-Bl.	1918	1993	10	22	22	20	1,4	0,5	-4,2	**-0,6**
Piz Linard	3411	sil.	Br.-Bl.	1947	1992	30	10	10	8	0,75	0,5	-2,9	**-0,6**
Piz Trovat	3146	sil.	Rübel	1907	1992	24	8	7	7	0,7	0	-9,1	**-1,1**
Radüner Rothorn	3022	sil.	Schibler	1897	1992	5	8	6	5	2,1	1	-15,5	**-1,6**

An excellent example for a mountain with no appropriate pathways from the alpine belt to the far top is Piz Linard, from the summit of which the oldest record is available (from 1835; Heer 1866). Piz Linard has been revisited several times after 1835 (Table 3). Each visit contributed some new species to the list. Surprisingly, there was no increase between the last two visits, i.e. 1947 and 1992. Obviously, the invasion of the "nival species group" ended about 50 years ago. New invaders would have been the alpine grassland species which could not migrate through the block fields of which the summit of Piz Linard consists. However, a comparison of a small distribution map of the plants found in 1947 with today's situation provided evidence that the abundance of the nival plants remarkably increased.

Table 3. Time series of species composition and abundance for the uppermost 30 altitudinal meters of Piz Linard (Switzerland, 3411 m a.s.l.); p = present (for those data no figures for abundance are given in the historical data), c = common, i = intermediate abundance (relative to the common species), r = rare, ia = abundance increased since last observation.

Species				**Year**			
	1835	**1864**	**1895**	**1911**	**1937**	**1947**	**1992**
Androsace alpina	p	p	p	p	p	c	c, ia
Ranunculus glacialis		p	p	p	p	c	c, ia
Saxifraga bryoides			p	p	p	c	c, ia
Saxifraga oppositifolia			p	p	p	c	c, ia
Poa laxa				p	p	c	c, ia
Draba fladnizensis				p	p	c	c
Gentiana bavarica				p	p	i	i
Cerastium uniflorum				p	p	i	r
Tanacetum alpinum		p			p	i	
Saxifraga exarata					p	r	
Cardamine resedifolia							r
Luzula spicata							r

For a couple of species migration rates could be calculated on the basis of the altitude of the uppermost individual in the historical record (provided for 12 summits by Braun-Blanquet 1957, 1958) and the recent position of the uppermost individuals. Maximum migration rates were about 4 m/decade, the majority of values range between zero and 1.5 m/decade. These rates can now be compared with expected rates derived from the observed warming of 1 to 2°C (annual mean temperature) since the turn of the century (Auer et al. 1996; about 10 decades) and the average decrease of 0.5°C per 100 m increase in altitude, i.e. = 20 - 40 m/decade. Although this rate should not be taken in absolute terms, it indicates that the true migration rates are much lower.

However, the estimated migration rates of species should be considered as minimum values, because: 1. some species have already been observed at the very summit in the historical investigation, and therefore their moving rates amount to zero, 2. the historical upper limit of a species was unknown, if below the investigation area (in these cases we had to take the lower boundary of the investigation area as reference). Nevertheless, these results provide the first empirical approximations of migration rates in alpine/nival environments.

2.4. DISCUSSION

The high number of newly found species at the investigated summits as well as the overall increase in species richness must be taken as proof that the migration hypothesis has to be accepted. The comparably slight warming of 1 to 2°C since the turn of the century has induced the migration of many plant species to higher altitudes. At least for the ecological systems at the low-temperature limits of plant life the registered warming has been ecologically relevant.

That this conclusion may also be valid for alpine environments at lower altitudes is supported by the following observations: Reproduction by seeds of the alpine grassland species *Carex curvula* is only possible in a sequence of warm years, as it became evident by an extraordinary fruiting year in 1992. Warm years in the 40's and 80's of this century also caused an exceptional regeneration success of this species. Scattered populations of *Carex curvula* from snow beds in the alpine environments of the mountain Hohe Muth (Tyrol) were found to consist of even-aged cohorts of 45 - 50-year-old individuals, or 10 - 20-year-old individuals, respectively (Grabherr 1997). Similarly, the Scandinavian mountain birch (*Betula tortuosa* Ledeb.) expanded beyond the actual timber line in the Scandinavian mountains during the warming in the 30's and 40's. In contrast to *Carex curvula* most individuals of these cohorts died during the relatively cold period between 1950 and 1970 (Holtmeier 1994). The latter example shows clearly that the alpine vegetation types may react very differently. Warming might even cause damage. Individuals of the evergreen cushion plant *Diapensia lapponica* L. were severely damaged during last years' blizzards at Latnjajaure station (Northern Sweden) as a result of too early snow melting which also induced early breaking of winter dormancy. Even 500-year-old individuals had probably never experienced such a combination of an early warm season and strong winds (Molau 1995, Molau pers. comm.).

Regardless of positive or negative effects biological signs of climate change are evident in alpine or nival biota. A general trend towards higher altitudes is obvious. Exceptional observations of very long-lived species, i.e. establishment of *Carex curvula* at the highest altitude ever recorded, damage of old *Diapensia* individuals, even indicate that the recent change might be more effective than anything experienced during the last centuries. This is in agreement with other observations, such as the spread of vector-borne diseases to higher altitudes in tropical mountains (summarised by Beniston & Fox 1995; Epstein et al. 1998).

For the summits investigated in our study no extinction events for the near future are to be expected when taking the upward moving of vegetation belts as a model (see Ozenda & Borel 1991; Markham et al. 1993; Halpin 1994). Only the endemic species *Draba ladina* Br.-Bl. which was found exclusively at the uppermost areas of the visited summits might be endangered. However, the simple model of isohypsal upward migration of vegetation belts is far from reality. A computer model of distribution patterns of species and species assemblages along the alpine/nival ecotone at the mountain Schrankogel indicates a very complex situation (Gottfried et al. 1998). The present vegetation structure at Schrankogel with a pronounced alpine grassland belt, an alpine/nival ecotone, and the belt of nival herb fields will disappear in a climate warming scenario of more than 1.0°C. Many plants are "pushed to the wall" when migrating upwards, as favourable sites are becoming increasingly scarce. Thus extinction events might occur before plants have reached the uppermost habitats.

3. Outlook

As Epstein et al. (1998) concluded recently, the most serious signals for climate change effects are coming from mountain environments. In consequence, they consider a monitoring of mountain areas to be particularly useful. The demand for monitoring sites in mountain regions is also stated in a report of the International Geosphere-Biosphere Program (Becker & Bugmann 1997) asking for an initiative which may end up as an intercore project of the IGBP (GCTE/BAHC), with the creation of a mountain-based global monitoring network on climate change, its effect on the cryosphere and alpine biota. In our experience such monitoring of alpine/nival biota must be simple, and strictly standardised. Establishing monitoring sites now will help future generations to get a more precise picture of how ecological systems react in the time span corresponding to the time horizons of political decision making. Neither predictive modelling nor studies of vegetation history as proxy for the future can provide such hard facts; these can only be obtained by permanent observations.

Acknowledgements
This Austrian research initiative was financed by the Austrian Academy of Sciences, and the Austrian Federal Ministry of Science and Transport. The authors would express their gratitude to H. Reisigl who commented on his historical data, to the Scientific Committee of the Swiss National Park for the permission to visit mountains in the park closed to the public, as well as to the wildlife wardens of the park for their help. Thanks to our colleagues and friends who assisted in the field, and in the computer lab, K. Reiter in particular.

References

Auer, I., Böhm, R. & Mohnl, H. 1996. Übersicht über aktuelle Arbeiten der Wiener Arbeitsgruppe für klimatologische Zeitreihenanalyse. ÖGM - Österreichische Gesellschaft für Meteorologie, Bulletin 96(1).

Becker, A. & Bugmann, H. (eds) 1997. Predicting global change impacts on mountain hydrology and ecology: integrated catchment hydrology/altitudinal gradient studies. Workshop Report - Documentation resulting from an International Workshop Kathmandu, Nepal, 30 March - 2 April 1996. IGBP Report 43. IGBP Secretariat, The Royal Swedish Academy of Sciences, Stockholm.

Beniston, M. & Fox, D.G. 1996. Impacts of climate change on mountain regions. pp. 191-213. In: Watson, R.T., Zinyowera, M.C. & Moss, R.H. (eds), Climate change 1995. Impacts, adaptations and mitigation of climate change: scientific-technical analysis. Cambridge University Press, Cambridge.

Beniston, M. (ed) 1994. Mountain environments in changing climates. Routledge, London.

Böhm, R. 1986. Der Sonnblick. Die 100jährige Geschichte des Observatoriums und seiner Forschungstätigkeit. Österreichischer Bundesverlag, Wien.

Braun, J. 1913. Die Vegetationsverhältnisse der Schneestufe in den Rätisch-Lepontischen Alpen. Neue Denkschr. Schweiz. Naturforsch. Ges. 48: 1-348.

Braun-Blanquet, J. 1955. Die Vegetation des Piz Languard, ein Maßstab für Klimaänderungen. Svensk Botanisk Tidskrift 49(1-2): 1-9.

Braun-Blanquet, J. 1957. Ein Jahrhundert Florenwandel am Piz Linard (3414 m). Bull. Jard. Botan. Bruxelles, Vol. Jubil. W. Robyns: 221-232.

Braun-Blanquet, J. 1958. Über die obersten Grenzen pflanzlichen Lebens im Gipfelbereich des Schweizerischen Nationalparks. Kommission der Schweiz. Naturforsch. Ges. zur Wiss. Erforschung des Nationalparks 6: 119-142.

Ehrendorfer, F. (ed) 1973. Liste der Gefäßpflanzen Mitteleuropas. 2nd ed. Fischer, Stuttgart.

Epstein, P.R., Diaz, H.F., Elias, S., Grabherr, G., Graham, N.E., Martens, W.J.M., Mosley-Thompson, E. & Susskind, J. 1998. Biological and physical signs of climate change: Focus on mosquito-borne diseases. Bulletin of the American Meteorological Society 79: 409-417.

Gottfried, M., Pauli, H. & Grabherr, G. 1994. Die Alpen im "Treibhaus": Nachweise für das erwärmungsbedingte Höhersteigen der alpinen und nivalen Vegetation. Jahrbuch des Vereins zum Schutz der Bergwelt 59: 13-27.

Gottfried, M., Pauli, H. & Grabherr, G. 1998. Prediction of vegetation patterns at the limits of plant life: a new view of the alpine-nival ecotone. Arctic and Alpine Research 30(3): 207-221.

Grabherr, G. 1997. The high-mountain ecosystems of the Alps. pp. 97-121. In: Wielgolaski, F. (ed), Polar and Alpine Tundra. Ecosystems of the World 3. Elsevier, Amsterdam.

Grabherr, G., Gottfried, M. & Pauli, H. 1994. Climate effects on mountain plants. Nature 369: 1-448.

Grabherr, G., Gottfried, M., Gruber, A. & Pauli, H. 1995. Patterns and current changes in alpine plant diversity. pp. 167-181. In: Chapin III, F. S. & Körner, C. (eds), Arctic and Alpine Biodiversity: Patterns, Causes and Ecosystem Consequences. Ecological Studies 113.

Guisan, A., Holten, J.I., Spichiger, R. & Tessier, L. (eds) 1995. Potential Ecological Impacts of Climate Change in the Alps and Fennoscandian Mountains. Ed. Conserv. Jard. Bot. Genève: 1-194.

Halpin, P.N. 1994. Latitudinal variation in the potential response of mountain ecosystems to climatic change. pp. 180-203. In: Beniston, M. (ed), Mountain Environments in Changing Climates. Routledge, London.

Heer, O. 1866. Der Piz Linard. Jahrbuch des Schweiz. Alpin Club III: 457-471.

Heer, O. 1885. Ueber die nivale Flora der Schweiz. Neue Denkschr. d. Allgemeinen Schweizerischen Ges. f.d. gesamten Naturwissenschaften Bd. XXIX: 1-114.

Holten, J.I. 1993. Potential effects of climatic change on distribution of plant species, with emphasis on Norway. pp. 84-104. In: Holten, J.I., Paulsen, G. & Oechel, W.C. (eds), Impacts of climatic change on natural ecosystems with emphasis on boreal and arctic/alpine areas. NINA, Trondheim.

Holtmeier, F.-K. 1994. Ecological aspects of climatically-caused timberline fluctuations: review and outlook. pp. 220-233. In: Beniston, M. (ed), Mountain Environments in Changing Climates. Routledge, London.

Klebelsberg, R. 1913. Das Vordringen der Hochgebirgsvegetation in den Tiroler Alpen. Österr. Bot. Zeitschr., Jahrg. 1913: 177-186, 241-254.

Markham, A., Dudley, N. & Stolton, S. 1993. Some like it hot: climate change, biodiversity and the survival of species. WWF, Gland.

Molau, U. 1995. Climate change, plant reproductive ecology, and populations dynamics. pp. 67-71. In: Guisan, A., Holten, J. I., Spichiger, R. & Tessier, L. (eds), Potential ecological impacts of climate change in the Alps and Fennoscandian mountains. Ed. Conserv. Jard. Bot. Genève.

Nilsson, S. & Pitt, D. 1991. Mountain world in danger - Climate change in the forests and mountains of Europe. Earthscan Publications, London.

Ozenda, P. & Borel, J.-L. 1991. Mögliche ökologische Auswirkungen von Klimaveränderungen in den Alpen. Internationale Alpenschutz-Kommission CIPRA, Kleine Schriften 8/91: 1-71.

Pauli, H., Gottfried, M. & Grabherr, G. 1996. Effects of climate change on mountain ecosystems - upward shifting of alpine plants. World Resource Review 8(3): 382-390.

Pauli, H., Gottfried, M., Reiter, K. & Grabherr, G. 1998. Monitoring der floristischen Zusammensetzung hochalpin/nivaler Pflanzengesellschaften. pp. 320-343. In: Traxler, A. (ed), Handbuch des vegetationsökologischen Monitorings, Teil A: Methoden. Umweltbundesamt, Wien.

Pauli, H., Gottfried, M. & Grabherr, G. 1999. Vascular plant distribution patterns at the low-temperature limits of plant life - the alpine-nival ecotone of Mount Schrankogel (Tyrol, Austria). Phytocoenologia 29(3):297-325.

Price, M.F. & Barry, R.G. 1997. Climate change. pp. 409-445. In: Messerli, B. & Ives, J.D. (eds), Mountains of the World. The Parthenon Publishing Group, New York.
Reisigl, H. & Pitschmann, H. 1958. Obere Grenzen von Flora und Vegetation in der Nivalstufe der zentralen Ötztaler Alpen (Tirol). Vegetatio 8: 93-129.
Rübel, E. 1912. Pflanzengeographische Monographie des Berninagebietes. Engelmann, Leipzig.
Schibler, W. 1898. Über die nivale Flora der Landschaft Davos. Jahrbuch des Schweiz. Alpenclub 33: 262-291.
Solomon, A.M. & Shugart, H.H. (eds) 1993. Vegetation Dynamics and Global Change. Chapman & Hall, New York.
Watson, R.T., Zinyowera, M. C. & Moss, R. H. (eds) 1996. Climate change 1995. Impacts, adaptations and mitigation of climate change: scientific-technical analysis. Cambridge University Press, Cambridge.

Appendix 1:

Data set for the comparison of historical records of vascular plants on 30 high alpine summits and their reinvestigation during 1992 and 1993, and additionally for 5 summits recently investigated for the first time. Part 1: Rätisch-Lepontische Alpen; Part 2: Ötztaler Alpen; Part 3: Zillertaler Alpen, Rieserferner Gruppe; Part 4: summits recently investigated for the first time (1 summit in the Rätisch-Lepontische Alpen; 1 summit in the Ötztaler Alpen, 3 summits in the Hohe Tauern); see also Figure 1.
Summit data are arranged from west to east (for Part 1 - 3, and for Part 4 separately) and indicated as: Geographical name (Altitude a.s.l.): Geographical coordinates / Geographical area / Date: date of reinvestigation / Interval: altitudinal investigation and comparison interval from summit downwards / Hist. date: date of historical investigation / Name of historical investigator / Ref: reference of historical investigation. The latter three indications are not given for Part 4.
The species data for Part 1 - 3 have been divided into three parts: eS: equal species, found in historical and recent investigation; mS: missing species, found in historical but not in recent investigation; nS: new species, found in recent but not in historical investigation. Indications for eS: Species code / Altitude of uppermost specimen in recent investigation / Exposition (in capitals); and/or ev: verbal indication of exposition; and/or po: verbal indication of position / Abundance in recent investigation / [Altitude of uppermost specimen in historical investigation]. Indications for mS: Species code / [Altitude of uppermost specimen in historical investigation]. Indications for nS (and for species data (S) in Part 4): Species code / Altitude of uppermost specimen in recent investigation / Exposition (in capitals); and/or ev: verbal indication of exposition; and/or po: verbal indication of position / Abundance in recent investigation. In po: distances which are indicated as m (meters) are to be interpreted as horizontal distance to the uppermost summit point, and distances which are indicated as am (altitudinal meters) are to be interpreted as vertical distance to the summit. For abbreviations of species names (i.e., species codes) see Appendix 2.
Abbreviations of abundance values: d..dominant; c-d..common to dominant; c,Ld..common, locally dominant; c..common; s-c,Lc..scattered to common, locally common; s-c..scattered to common; s,Lc-d..scattered, locally common to dominant; s,Lc..scattered, locally common; s,Ls-c..scattered, locally scattered to common; s..scattered; r-s,Ls..rare to scattered, locally scattered; r-s..rare to scattered; r,Lc..rare, locally common; r,Ls..rare, locally scattered; r..rare; r!..very rare; Lc-d..locally common

to dominant; Lc..locally common; Ls-c..locally scattered to common; Ls..locally scattered; Lr-s..locally rare to scattered; no data..no abundance data available.
Definition of abundance values: dominant: very abundant, making up a large part of the vascular plant phytomass, but not necessarily forming a dense vegetation layer; common: frequent and wide-spread within the investigation area; scattered: wide-spread within the investigation area; rare: some specimens at several locations, but hardly to be overlooked; very rare: one or a few specimens; locally: only at one or a few locations within the investigation area (prefix in addition to the abundance values).
Species abundance weighting for the calculation of wSID (see Table 2): eS (equal species): weight 1; mS (missing species): weight 0.7 (calculated as mean of all weighted abundance values recorded in the recent investigation); nS (new species): weight 1: d; c-d; c,Ld; c; s-c,Lc; s-c; s,Lc-d; s,Lc; s,Ls-c; s; weight 0.5: r-s,Ls; r-s; r,Lc; r,Ls; r; Lc-d; Lc; Ls-c; weight 0.25: r!; Ls; Lr-s; weight 0.7: no data.
For the calculation of wSID (see Table 2) and the definition as eS, mS, or nS, every species was taken into account which reaches the lower boundary of the altitudinal investigation interval, or beyond. Some altitudinal values of the uppermost specimen of a species are indicated as interval (in the recent investigation: cases where the distinct altitude was not precisely definable; in the historical investigation: cases where altitudes are indicated as interval of the investigation). In such cases the uppermost altitude of that species was defined as mean of the interval boundaries, rounded to the next higher altitudinal meter.

Part 1: Rätisch-Lepontische Alpen

Piz Julier (3380m): 46°29'32"N 9°45'40"E / southern Err-Gruppe / Date: 13 08 1992 / Interval: 30am / Hist. date: 21 08 1900 / Braun-Blanquet / Ref.: Braun (1913);
eS: ANDRALPI/3374/SE;po:10-20m, SE-ridge/Ls/[3373]; **CERAUNIF**/3345-55/W;po:100-150m N, W-face of N-ridge/r/[3380]; **DRABFLAD**/3376/po:rocky outcrops SE of summit/s/[3365]; **POA_LAXA**/3376/SE;po:10m, SE-ridge/s-c/[3370]; **RANUGLAC**/3376/SE;po:10m, SE-ridge/s-c/[3375]; **SAXIBRYO**/3372/SE;po:10-20m, SE-ridge/Ls/[3373]; **SAXIOPPO**/3375/W;po:20-30m NNE, N-ridge/r,Ls/[3365]; **TANAALPI**/3355/S;po:S-face/s/[3355];
mS: PHYTGL_P/[3355];
nS: AGRORUPE/3350/SSW;po:ESE-summit, S-face/r; **SAXIEXAR**/3355/S;po:15-20m SE of ESE-notch/r!;

Piz Üertsch (3268m): 46°35'51"N 9°50'16"E / Albulapass, Graubünden / Date: 05 08 1993 / Interval: 30am / Hist. date: 13 09 1911 / Braun-Blanquet / Ref.: Braun (1913);
eS: DRABTOME/3238-68//s,Lc/[3245]; **SAXIOPPO**/3268/po:6m W of a stone-pile at the summit area/s,Lc/[3255];
nS: CERALATI/3238-68/po:E-ridge/r!; **HUTCALPI**(cf)/3238-68//r!; **MINUGERA**/3238-68//r!;

Piz Blaisun (3200m): 46°36'15"N 9°51'52"E / Albulapass, Graubünden / Date: 04 08 1993 / Interval: 20am / Hist. date: 13 09 1911 / Braun-Blanquet / Ref.: Braun (1913);
eS: ANDRALPI/3196/SW;po:W-ridge/Lc/[3196]; **CAMPCENI**/3192-93/SE,S;po:SE-face/s/[3186]; **GEUMREPT**/3192-93/SE,S;po:SE-face/s,Lc/[3186]; **SAXIOPPO**/3200/N,SW;ev:summit;po:5m NW/c/[3200];
mS: ARTEGENI/[3186];
nS: LINAALPI/3187/SE,S;po:SE-face/r; **RANUGLAC**/3181-83/SE,S;po:SE-face/r!;

Piz Forun (3052m): 46°39'23"N 9°51'59"E / Kesch-Gruppe / Date: 22 08 1992 / Interval: 30am / Hist. date: 27 09 1903 / Braun-Blanquet / Ref.: Braun (1913);
eS: AGRORUPE/3044-46/NE,SE;po:NW-ridge + SE-face/c-d/[3051]; **ANDRALPI**/3051-52/po:summit-ridge/s/[3052]; **ANTECARP**/3035/NE,SE;po:SE- + NE-face of NW-ridge/s/[3041]; **ANTHALPI**/3042/SE;po:SE-face/s/[3046]; **AVENVERS**/3044-46/NE;po:NW-ridge/s/[3051]; **CARECURV**/3051-52/W;ev:ridge;po:summit-ridge/d/[3051]; **CERAUNIF**/3050/E;po:summit-

ridge/s/[3051]; **DOROCLUS**/3048-49/NE;po:NW-ridge/s-c/[3052]; **EMPEHERM**/3040/E;po:SE-face, at rocky outcrops with sward-fragments/r/[3041]; **ERIGUNIF**/3048/NE,SE;po:NW-ridge + SE-face/s,Ls-c/[3051]; **ERITNANU**/3044-46/NE;po:NW-ridge/Ls/[3041]; **EUPHMINI**/3042-47/SE;po:SE-face/s/[3051]; **FESTINTE**/3051/E;po:summit-ridge/s-c/[3041]; **GENTBAVA**/3051-52/W;po:summit-ridge/s/[3052]; **GENTPUNC**/3046/SE;po:SE-face/c/[3051]; **GEUMMONT**/3044/SE;po:SE-face/Ls-c/[3051]; **GEUMREPT**/3051-52/E;po:summit-ridge/c/[3041]; **GNAPSUPI**/3050/SE;po:summit-ridge, ca.30m S of the S-stone-pile/s/[3051]; **HOMOALPI**/3045/SE;po:SE-face/s/[3051]; **JUNCTRIF**/3045-46/SE;po:SE-face/s-c/[3041]; **LEONHELV**/3047/SE;po:SE-face/s-c/[3051]; **LINAALPI**/3042/SE;po:SE-face, ca.50m S of the stone-pile/Ls/[3041]; **LUZUALPI**/3043/SE;po:SE-face/r,Ls/[3041]; **LUZULUTE**/3047/SE;po:SE-face of S-summit-ridge/c/[3051]; **LUZUSPIC**/3047/NE;po:NW-ridge/s/[3041]; **MINUSEDO**/3051-52/po:summit-ridge/c/[3052]; **OREODIST**/3051-52/W;po:summit-ridge/c-d/[3052]; **PEDIKERN**/3037/SE;po:SE-face/r/[3041]; **PHYTGL_P**/3051-52/po:summit-ridge/c/[3052]; **PHYTHEMI**/3046/SE;po:SE-face/r-s/[3041]; **POA_LAXA**/3051-52/po:summit-ridge/c/[3051]; **PRIMHIRS**/3044-46/NE;po:NW-ridge/s/[3041]; **PRIMINTE**(cf)/3040/SE;po:SE-face/r/[3046]; **PULSVERN**/3038/SE;po:SE-face, ca.50m S of the S stone-pile/Ls/[3041]; **RANUGLAC**/3051-52/po:summit-ridge/s/[3052]; **SAXIBRYO**/3051-52/po:summit-ridge/c/[3052]; **SAXIEXAR**/3051/S;po:summit-ridge/s/[3052]; **SAXISEGU**/3047/NE;po:NW-ridge/Ls/[3041]; **SEDUALPE**/3051-52/W;po:summit-ridge/r-s/[3046]; **SEMPMONT**/3042/SE;po:ridge + SE-face ca. 50m S of a stone-pile/Ls/[3041]; **SIBBPROC**/3051-52/po:summit-ridge/s/[3051]; **SILEEXSC**/3044-46/NE;po:NW-ridge/s/[3046]; **TANAALPI**/3051-52/po:summit-ridge/s/[3052]; **TRISSPIC**/3047/NE;po:NW-ridge/r-s/[3051]; **VACCVITI**/3038/SE;po:SE-face, ca.50m S of the stone-pile/r/[3041]; **VEROALPI**/3032/SE;po:SE-face/r/[3046]; **VEROBELI**/3041/SE;po:SE-face/s/[3046];
mS: GENTBRAC/[3046];
nS: ACHIMOSC/3040/SE;po:SE-face, ca.50m S of the stone-pile/r; **ARNIMONT**/3037/SE;po:SE-face/r,Lc; **AVENFLEX**/3043/SE;po:SE-face/r; **CARDRESE**/3050/E;po:S-ridge/r-s; **CERAPEDU**/3051/NW;po:summit-ridge, ca.20m N/r-s; **DRABFLAD**/3047/N;po:NW-ridge/r; **ELYNMYOS**/3035/NE;po:NE-face of NW-ridge/Lc; **HIERALPI**(cf)/3040/SE;po:SE-face, close to S-summit-ridge/r!; **HUPESELA**/3044-46/NE;po:NW-ridge/Ls; **JUNCJACQ**/3042/SE;po:SE-face/s,Lc; **JUNICO_A**/3042/SE;po:SE-face, ca.50m S of the stone-pile/r; **LLOYSERO**/3037/SE;po:SE-face/Ls; **NARDSTRI**/3041/SE;po:SE-face/r; **POA_ALPI**/3043-47/NE;po:NE-face of NW-ridge/r; **POLYVULG**/3040/NE;po:NE-face of NW-ridge/r!; **POTEAURE**/3041/SE;po:SE-face/Lc; **POTEFRIG**/3044-46/NE;po:NW-ridge/r; **SALIHERB**/3044-46/NE;po:NW-ridge/Ls; **SAXIOPPO**/3050/W;po:summit-ridge/s; **SENEIN_C**/3051-52/E;po:summit-ridge/s-c; **TARAALPI**/3042/SE;po:SE-face, ca.50m S of the stone-pile/Lr-s; **VACCGAUL**/3033-37/SE;po:SE-face/r;

Piz Kesch (3418m): 46°37'20"N 9°52'27"E / Kesch-Gruppe / Date: 23 08 1992 / Interval: 30am / Hist. date: 27 07 1902 / Braun-Blanquet / Ref.: Braun (1913);
eS: ANDRALPI/3407/S;po:SW-face/s/[3398]; **CERAUNIF**/3405/S;po:SW-face/s/[3398]; **DRABFLAD**/3417/SE;po:ca.4m from the cross on the summit/c/[3388]; **ERITNANU**/3403-5/SE;po:S of main-summit, E of the SW rock-tower in the ridge/s/[3388]; **POA_LAXA**/3413/S;po:SW-face/c/[3398]; **RANUGLAC**/3408/E;po:E-face/r-s/[3388]; **SAXIBRYO**/3408/S;po:SW-ridge of NW-summit/c/[3398]; **SAXIOPPO**/3416/ev:ridge;po:NW-ridge/c-d/[3413];
nS: ARTEGENI(cf)/3407/S;po:SW-face/r!; **DRABDUBI**(cf)/3388-92/S;po:SW-face, below S-rock-face of SW-ridge of NW-summit/Lr-s; **POA_ALPI**/3403-5/SE;po:S, E of the SW-rock-tower in the ridge/r-s; **POTEFRIG**/3400-5/SE;po:S-rock-face of SW-ridge of NW-summit/r; **SAXIEXAR**/3407/W;po:SW-face/r,Ls;

Flüela Schwarzhorn (3146m): 46°44'13"N 9°56'34"E / northern Vadret-Gruppe / Date: 08 08 1992 / Interval: 50am / Hist. date: 09 08 1902 / Braun-Blanquet / Ref.: Braun (1913);
eS: ANDRALPI/3136/ev:ridge;po:ca.20m, NE-pre-summit/s/[3141]; **CARDRESE**/3135/SE;po:ca.20m, NE-pre-summit/s/[3136]; **CARECURV**/3128/S;po:E-ridge of the NE-pre-summit/Lc-d/[3126]; **CERAUNIF**/3145/ev:ridge;po:5m/c/[3136]; **DOROCLUS**/3121/ESE;po:E-ridge of the NE-pre-summit/s/[3126]; **DRABFLAD**/3130/SW,W;po:W-face of S-ridge/r-s/[3096]; **ERIGUNIF**/3141-42/SE;po:SE-face/s/[3139]; **FESTINTE**(cf)/3143/SE;po:SE-face/c/[3136]; **GENTBAVA**/3141-42/SE;po:SE-face/s/[3141]; **GENTBRAC**/3100-10/ev:ridge;po:E-ridge of the NE-pre-summit/r/[3126]; **GEUMREPT**/3131/S;po:S-face of NE-pre-summit/s,Lc/[3131]; **GNAPSUPI**/3121/S;po:E-ridge of the NE-pre-summit/r-s,Ls/[3126]; **LINAALPI**/3115-20/SE;po:SSW of NE-pre-summit, scree-path/s/[3136]; **LUZUSPIC**/3141-42/SE;po:SE-face/c/[3141]; **MINUSEDO**/3141-42/SE;po:SE-face/s/[3136]; **OREODIST**/3126/S;po:E-ridge of the NE-pre-summit/Lc-d/[3136]; **PEDIKERN**/3115/SE;po:S-face (SE)/Ls/[3096]; **PHYTGL_P**/3135/SE;po:ca.20m, NE-pre-summit/s-c/[3136];

POA_LAXA/3145/ev:ridge;po:5m/c/[3141]; **POTEFRIG**/3140/SE;po:SE-face/r-s/[3126]; **PRIMHIRS**/3128/S;po:E-ridge of the NE-pre-summit/s/[3126]; **RANUGLAC**/3144/S;po:5m/s,Lc/[3146]; **SAXIBRYO**/3145/ev:ridge;po:5m/c/[3136]; **SAXIEXAR**/3144/SE;po:SE-face/s,Lc/[3131]; **SAXIOPPO**/3141-42/SE;po:SE-face/r-s/[3126]; **SEDUALPE**/3129/S;po:S-face of NE-pre-summit/s,Lc/[3136]; **SENEIN_C**/3135/SE;po:ca.20m, NE-pre-summit/s-c/[3136]; **SIBBPROC**/3100-10/SSE;po:E-ridge of the NE-pre-summit/Ls/[3126]; **SILEEXSC**/3132/S;po:S-face of NE-pre-summit/s/[3136]; **TANAALPI**/3136/ev:ridge;po:ca.20m, NE-pre-summit/s/[3136]; **TRISSPIC**/3115/SE;po:S-face (SE)/Ls/[3136]; **VEROALPI**/3127/S;po:S-face of NE-pre-summit/s,Lc/[3136];
mS: LEONHELV/[3126];**POA_ALPI**/[3126];
nS: AGROALPI/3121/S;po:E-ridge of the NE-pre-summit/r-s; **AGRORUPE**/3127/S;po:S-face of NE-pre-summit/c; **EMPEHERM**/3100-10/SSE;po:SE-face of NE-pre-summit/r!; **EUPHMINI**/3141-42/SE;po:SE-face/s; **LIGUMUT1**(cf)/3100-10/SSE;po:E-ridge of the NE-pre-summit/r!; **LLOYSERO**/3100-10/SSE;po:SE-face of NE-pre-summit/Ls; **MINUGERA**/3115/SE;po:S-face (SE)/r; **OXYRDIGY**/3125/SE;po:SSW of NE-pre-summit, scree-path/r!; **PHYTHEMI**/3131/S;po:S-face of NE-pre-summit/r; **SAXISEGU**/3130/SW;po:W-face of S-ridge/s; **VACCGAUL**/3122/S;po:E-ridge of NE-pre-summit, in rock-crevices/r;

<u>Radüner Rothorn (3022m):</u> 46°43'31"N 9°56'51"E / northern Vadret-Gruppe / Date: 09 08 1992 / Interval: 5am / Hist. date: 03 10 1897 / Schibler / Ref.: Braun (1913);
eS: POA_LAXA/3022/po:10-15m ESE, summit-ridge/s/[3022]; **RANUGLAC**/3022/po:10-15m ESE, summit-ridge/c/[3022]; **SAXIBRYO**/3018/po:30m ESE, summit-ridge/s/[3022]; **SAXISEGU**/3017/E;po:NE-face/s/[3022]; **TANAALPI**/3017/NE;po:NE-face/s/[3022];
mS: ANDRALPI/[3022];**CERAPEDU**/[3022];**GENTBAVA**/[3022];
nS: CERAUNIF/3020/NE;po:4m, NE-face/s;

<u>Munt Pers (3207m):</u> 46°25'20"N 9°57'17"E / Bernina / Date: 29 07 1992 / Interval: 11am / Hist. date: 11 08 1906 / Rübel / Ref.: Rübel (1912);
eS: CARECURV/3197/ev:ridge;po:50m S/r/[3196]; **CERAUNIF**/3207/E;po:ca. 15m, summit-ridge/c/[3196-07]; **DOROCLUS**/3204/E;po:NE/s/[3196]; **ERIGUNIF**/3204/E;po:15-20m, E-face/s/[3196-07]; **GENTBAVA**/3206-7/E;po:ca.15m, summit-ridge/s/[3196-07]; **LUZUSPIC**/3204/E;po:ca.15-20m, E-face/r/[3196]; **MINUSEDO**/3203/E;po:ca.15-20m, E-face/c/[3196]; **OREODIST**/3205/E;po:ca.15m, summit-ridge + E-face/c/[3196]; **PHYTGL_P**/3204/E;po:ca.15-20m, E-face/Lc/[3196-07]; **POA_LAXA**/3207/E;po:ca. 15m, summit-ridge/c/[3196-07]; **RANUGLAC**/3207/ev:all exposures;po:ca.15m, summit-ridge/c/[3196-07]; **SAXIBRYO**/3207/po:ca.15m, summit-ridge/c/[3196-07]; **SAXIEXAR**/3201/E,W;po:30m N/r/[3196-07]; **TANAALPI**/3206-7/E;po:ca.15m, summit-ridge/s/[3196-07];
mS: ANDRALPI/[3196];
nS: DRABDUBI/3204/E;po:15-20m, E-face/r; **FESTHAL.**/3205/E;po:ca.15-20m, E-face/c; **GEUMREPT**/3207/E;po:uppermost E-face/s; **POTEFRIG**/3200/E;po:E-face, E of summit/r; **SAXISEGU**/3206/E;po:uppermost E-face/s;

<u>Gorihorn (2986m):</u> 46°46'47"N 9°57'35"E / south-western Silvretta / Date: 30 07 1992 / Interval: 7am / Hist. date: 19 08 1895 / Schibler / Ref.: Braun (1913);
eS: AGRORUPE/2983/S;po:S/s/[2986]; **CARECURV**/2983-84/S;po:W of a small notch/s/[2984]; **GENTBAVA**/2982/S;po:10m SW/r/[2984]; **MINUSEDO**/2983/S;po:S of the notch/s/[2984]; **OREODIST**/2982/S;po:summit-area/s/[2984]; **PHYTGL_P**/2984/S;po:summit-area, 10m/c/[2984]; **POA_LAXA**/2986/S;po:summit-area, 10m/c/[2982]; **RANUGLAC**/2984/N,S;po:close to the S of the small saddle between W- and E-summit/s/[2984]; **SAXIBRYO**/2984/N,S;po:ca. 1m W of a small notch/c/[2986]; **SAXIEXAR**/2984/N;po:8m NW, ridge-area/s/[2984]; **SENEIN_C**/2983/S;po:S of the notch/s/[2984]; **TANAALPI**/2980/S;po:S-face/s/[2979];
mS: LINAALPI/[2984];**LUZUSPIC**/[2984];
nS: CARDRESE/2983/S;po:S of the notch/s; **DRABFLAD**/2983/N;po:uppermost N-face/r; **GNAPSUPI**/2980/S;po:15-20m S/r; **LEONHELV**(cf)/2983/S;po:S of the notch/r!; **PRIMHIRS**/2983/S;po:15-20m S/r; **SAXIOPPO**/2984/N;po:8m NW/s;

<u>Piz Trovat (3146m):</u> 46°24'23"N 9°58'18"E / Bernina / Date: 29 07 1992 / Interval: 24am / Hist. date: 29 08 1907 / Rübel / Ref.: Rübel (1912) & Braun (1913);
eS: CARDRESE/3145/SW;po:5-10m S/Ls/[3142]; **MINUSEDO**/3130/SSE;po:SSE-ridge/Ls/[3142]; **POA_LAXA**/3146/ev:ridge;po:5-10m S/c/[3146]; **RANUGLAC**/3145/ev:ridge;po:5-10m S/s/[3146];

SAXIEXAR/3145/SW;po:5-10m S/r/[3146]; **SENEIN_C**/3120-25/SSE;po:ESE-face/Ls/[3122]; **TANAALPI**/3140/SSE;po:60-80m ESE/Ls/[3146];
mS: PRIMHIRS/[3132];

Piz dals Lejs (3042m): 46°26'58"N 10°2'18"E / Puschlaver Berge / Date: 01 08 1992 / Interval: 30am / Hist. date: 06 10 1907 / Rübel / Ref.: Rübel (1912) & Braun (1913);
eS: ANDRALPI/3042/ev:ridge;po:ca.10m, summit-plateau/c/[3042]; **CARECURV**/3040/ev:ridge;po:ca.40-50m E/Ls/[3037]; **GENTBAVA**/3042/ev:ridge;po:ca.10m, summit-plateau/c/[3042]; **MINUSEDO**/3042/ev:ridge;po:ca.10m, summit-plateau/c/[3042]; **POA_LAXA**/3042/ev:ridge;po:ca.10m, summit-plateau/c/[3042]; **RANUGLAC**/3039/N;po:30m WNW/r-s/[3042]; **SAXIBRYO**/3042/ev:ridge;po:ca.10m, summit-plateau/c/[3042]; **SAXIEXAR**/3042/ev:ridge;po:ca.10m, summit-plateau/c/[3042]; **SAXIOPPO**/3042/ev:ridge;po:ca.10m, summit-plateau/c/[3042]; **TANAALPI**/3042/ev:ridge;po:ca.10m, summit-plateau/c/[3042];
mS: PRIMHIRS/[3042];
nS: AGROALPI/3015-20/ev:ridge;po:ca.200m ESE-SE/r; **ARTEGENI**/3030/S;po:S-face/r; **CERAUNIF**/3041/ev:ridge;po:5-15m NW/c; **DOROCLUS**/3037/ev:ridge;po:ca.30m E/r; **DRABDUBI**/3038-39/S;po:S-SW-face/r; **DRABFLAD**/3042/ev:ridge;po:ca.10m, summit-plateau/s; **ERIGUNIF**/3042/S;po:summit-plateau/s; **ERITNANU**/3042/ev:ridge;po:ca.10m, summit-plateau/c; **EUPHMINI**/3015-20/N;po:ca.200m ESE-SE, ridge-area/Lr-s; **FESTHAL.**/3040/ev:ridge;po:5m W/Lc-d; **GEUMREPT**/3042/S;po:ca.10m E, uppermost SW-face/s; **GNAPSUPI**/3042/S;po:8m E, uppermost S-SW-face/r; **LINAALPI**/3037/S;po:S-SW-face/s; **LUZUSPIC**/3040/S;po:10-50m E/c; **OREODIST**/3015-20/N;po:ca. 200m ESE-SE, ridge-area/Lc; **PHYTGL_P**/3042/ev:ridge;po:ca.10m, summit-plateau/s; **POA_ALPI**/3032-37/S;po:10-50m E/s; **POLYVIVI**/3015-20/N;po:ca.200m ESE-SE, ridge-area/Lr-s; **POTEFRIG**/3015-20/N;po:ca.200m ESE-SE, ridge-area/Ls; **SALIHERB**/3015-20/N;po:ca.200m ESE-SE, ridge-area/Lc; **SENEIN_C**/3041/ev:ridge;po:ca. 10m E/r; **SILEEXSC**/3037/S;po:10-50m E, S-SW-face/c; **TARAALPI**(cf)/3032-37/S;po:10-50m E/r; **TRISSPIC**/3015-20/ev:ridge;po:ca.230-250m ESE-SE/Ls;

Piz Stretta (3104m): 46°28'40"N 10°2'46"E / Puschlaver Berge / Date: 02 08 1992 / Interval: 30am / Hist. date: 23 08 1903 / Braun-Blanquet / Ref.: Braun (1913);
eS: ANDRALPI/3102/E;po:15m ENE, ENE-face/Ls/[3102]; **CARECURV**/3090/SSE;po:SE-slope/Lc/[3086]; **CERAUNIF**/3103/E;po:15m ENE, ENE-face/Lc/[3101]; **DOROCLUS**/3080/NE;po:ESE, ENE-face, NE of a snow-cornice/r/[3091]; **DRABFLAD**/3098/E;po:ca.15-20m ENE/s/[3096]; **EUPHMINI**/3090/SSE;po:SE-slope, upper Carex curv.-patch/Ls/[3091]; **GENTBAVA**/3095/NE;po:ESE, ENE-face, NE of a snow-cornice/Ls/[3096]; **LUZUSPIC**/3090/SSE;po:SE-slope, upper Carex curv.-patch/Lc/[3096]; **MINUSEDO**/3090/SSE;po:SE-slope, upper Carex curv.-patch/Lc/[3091]; **OREODIST**/3080/SSE;po:SE-slope, Carex curv.-patch/Ls/[3086]; **PHYTGL_P**/3090/SSE;po:SE-slope, uppermost Carex curv.-patch/Ls-c/[3096]; **POA_ALPI**/3085/SSE;po:SE-slope, Carex curv.-patch/Ls/[3086]; **POA_LAXA**/3104/po:summit/s/[3096]; **POLYVIVI**/3085/SSE;po:SE-slope, Carex curv.-patch/Ls/[3086]; **RANUGLAC**/3103/S;po:S-slope/r-s/[3096]; **SAXIBRYO**/3103/E;po:15m ENE, ENE-face/Lc/[3102]; **SAXIEXAR**/3100/E;po:20-25m E, ENE-face/Ls/[3102]; **SAXIOPPO**/3102/E;po:15m ENE, ENE-face/Lc/[3102]; **TANAALPI**/3101/S;po:S-slope/r-s/[3096]; **TRISSPIC**/3090/SSE;po:SE-slope, upper Carex curv.-patch/Lc/[3086];
nS: ANTECARP/3075/SSE;po:SE-slope, E-Carex curv.-patch/r; **ARABCOER**/3075-85/S;po:ca.100m SW, S-slope/r!; **ELYNMYOS**/3085/SSE;po:SE-slope, Carex curv.-patch/Ls; **ERIGUNIF**/3080/SSE;po:SE-slope, Carex curv.-patch/Ls; **GEUMREPT**/3075-85/S;po:ca.100m SW, S-slope/r; **LINAALPI**/3070-80/S;po:ca.100m SW, S-slope/Ls; **SALIHERB**/3080/SSE;po:SE-slope, Carex curv.-patch/Ls; **SAXISEGU**/3085/NE;po:ESE, ENE-face, NE of a snow-cornice/Ls;

Piz Linard (3411m): 46°48'0"N 10°4'22"E / southern Silvretta / Date: 17 08 1992 / Interval: 30am / Hist. date: 01 10 1947 / Braun-Blanquet / Ref.: Br.-Bl.1957;
Comment: Two missing species were weighted with respect to abundance values indicated in the historical record (TANAALPI: 0.5; SAXIEXAR: 0.25).
eS: ANDRALPI/3407/E;po:ca.15m N, ridge-area/c/[3409]; **CERAUNIF**/3380-82/ESE;po:E-face, S below the southernmost rock-rib/r/[3397]; **DRABFLAD**/3400/E;po:15-20m S, E-face/s/[3405]; **GENTBAVA**/3390-95/E;po:ca.40m S, 10-15m N of a stone-pile/r/[3387]; **POA_LAXA**/3409-10/E;po:summit-ridge/c/[3405]; **RANUGLAC**/3410/po:5-7m N, summit-ridge/c/[3409]; **SAXIBRYO**/3408/E;po:10m S, summit-ridge/s-c/[3405]; **SAXIOPPO**/3407/E;po:ca.15m N, summit-ridge/s-c/[3407];
mS: SAXIEXAR/[3382];**TANAALPI**/[3402];

nS: CARDRESE/3385-95/E;po:30-40m N, E-face/r!; **LUZUSPIC**/3380-82/ESE;po:E-face, S of the southernmost rock-rib/r!;

<u>**Monte Vago (3059m):**</u> 46°26'30"N 10°4'49"E / south of Livigno / Date: 31 07 1992 / Interval: 30am / Hist. date: 30 07 1905 / Braun-Blanquet / Ref.: Braun (1913);
eS: ACHIMOSC/3030/S;po:SW-S-face/s/[3059]; **ANDRALPI**/3059/SW;po:5m/s/[3059]; **CARDRESE**/3058/S;po:5-10m/s/[3059]; **CARECURV**/3058/S;po:5-10m/c/[3059]; **CERAUNIF**/3058/W;po:5m/s/[3059]; **ERIGUNIF**/3054/SE;po:SE-ridge/s/[3059]; **FESTINTE**/3058/SW;po:5-10m/c/[3059]; **GENTBAVA**/3045-50/SE;po:2m below NE-ridge/r!/[3059]; **GEUMREPT**/3052/E;po:S/r/[3059]; **LEONHELV**/3058/SE;po:5-10m/r-s/[3059]; **LINAALPI**/3052/S;po:10-15m SW, S-SW-face/r/[3059]; **LUZULUTE**/3050/S;po:S-face/r/[3059]; **LUZUSPIC**/3058/SW;po:5-10m/c/[3059]; **OREODIST**/3058/SE;po:5-10m/c/[3059]; **PEDIKERN**/3053/SE;po:10-20m SE/s/[3059]; **PHYTGL_P**/3058/S;po:5-10m/c/[3059]; **POA_LAXA**/3059/ev:all exposures;po:5m/c/[3059]; **POTEFRIG**/3055/SE;po:SE-ridge/s/[3059]; **PRIMHIRS**/3052/SE;po:10-20m/s/[3059]; **SAXIBRYO**/3059/SW;po:5m/c/[3059]; **SAXIEXAR**(cf)/3058/SE;po:5-10m/c/[3059]; **SAXISEGU**/3053/N;po:N-face/s/[3059]; **SEDUALPE**/3047/S;po:ca.20m SW + SE-ridge/r,Ls/[3059]; **SENEIN_C**/3059/NE;po:5m/s/[3059]; **TANAALPI**/3059/W;po:5-10m/s-c/[3059]; **VEROALPI**/3047/S;po:ca.20m SW/r/[3059]; **VEROBELI**/3040-45/S;po:ca.20m SW/r/[3059];
mS: PRIMLATI/[3059];
nS: AGROALPI/3045/S;po:SW-face/Ls-c; **AGRORUPE**/3055/SW;po:S-face/Ls-c; **ANTECARP**/3030-35/SE;po:SE-face/Ls; **DOROCLUS**/3049/E;po:E-face/r-s; **DRABDUBI**/3033-37/S;po:SW-ridge/r; **ELYNMYOS**/3030-35/SE;po:SE-face/Lc; **EMPEHERM**/3053/SE;po:10-20m SE/r; **ERITNANU**/3055/SW;po:S-face/s; **EUPHMINI**/3053/SE;po:10-20m SE/s; **GENTBRAC**/3045-50/S;po:S-face/r; **GNAPSUPI**/3047/S;po:ca.20m SW/r!; **HIERALPI**(cf)/3030-35/SE;po:SE-face/r; **HIERGLAN**(cf)/3053/SE;po:NE-ridge/r!; **JUNCTRIF**/3040-45/S;po:ca.20m SW, below a flat saddle/r; **JUNICO_A**/3052/S;po:S-face/r; **MINUSEDO**/3059/SW;po:5-10m/c; **PULSVERN**/3053/SE;po:10-20m SE/r; **RANUGLAC**/3059/W;po:5m/c; **SALIHERB**/3055/ev:ridge;po:NE-ridge/Lc; **SALISERP**/3045/SE,S,SW;po:SW-SE-face/s-c; **SAXIOPPO**/3045-50/S;po:ca.20m SW/s; **SIBBPROC**/3050/S;po:S-face/r; **SILEEXSC**/3056/SE;po:5-10m/c; **VACCGAUL**/3040-45/S;po:ca.20m SW/r;

<u>**Piz Nuna (3124m):**</u> 46°43'28"N 10°9'13"E / east-north-east of Zernez / Date: 12 08 1992 / Interval: 30am / Hist. date: 01 08 1919 / Braun-Blanquet / Ref.: Braun-Blanquet (1958);
eS: ANDRALPI/3118/SE;po:W-summit/Ls/[3108]; **CARECURV**/3122/ESE;po:5-7m S/c-d/[3118]; **CERAUNIF**/3123/ev:ridge;po:immediate summit-area/c/[3118]; **DRABFLAD**/3119-20/ENE,W;po:summit-ridge/s-c/[3118]; **ERIGUNIF**/3119/E;po:6-7m S/s/[3113]; **FESTINTE**/3122-23/ESE;po:5m S/c-d/[3118]; **GENTBAVA**/3119/SSE;po:W-summit, S-face/s/[3118]; **LUZUSPIC**/3119/SE;po:6-7m S, E-face/s/[3113]; **MINUSEDO**/3119/SE,S;po:W-summit, S-face/s,Lc/[3108]; **OREODIST**/3121/E;po:ENE-face/s-c/[3118]; **PHYTGL_P**/3118/NE;po:SSE-ridge S of main-summit/Lc/[3113]; **POA_LAXA**/3123-24/ev:ridge;po:immediate summit-area/c/[3118]; **RANUGLAC**/3123/ev:ridge;po:immediate summit-area/s-c/[3118]; **SAXIBRYO**/3123-24/ENE;po:immediate summit-area/c/[3118]; **SAXIEXAR**/3123-24/ENE;po:immediate summit-area/c/[3118]; **SAXIOPPO**/3123-24/ev:ridge;po:immediate summit-area/s/[3118]; **SAXISEGU**/3115/W;po:10-12m S, W-face/r/[3108]; **SILEEXSC**/3118/W;po:W-face S of the W-notch/r-s/[3108]; **TANAALPI**/3121/S;po:6-7m S/s/[3118];
nS: ARTEGENI/3114/SE;po:close to the N of the N-notch/r; **AVENVERS**/3110/ENE;po:ENE-face, at flatter sites/r!; **CARDRESE**/3112/ENE;po:ENE-face, at flatter sites/s; **DOROCLUS**/3095-0/ENE;po:SEern ENE-face, flatter site/r; **GENTBRAC**/3112-14/SE;po:N-notch/r; **GEUMREPT**/3116-17/NE;po:SSE-ridge/r; **GNAPSUPI**/3104/ENE;po:SEern ENE-face, flatter site/r; **LEONHELV**/3115/W;po:10-12m S, W-face/r; **LLOYSERO**/3110/ENE;po:SSE-ridge/Ls; **MINURECU**/3108/ENE;po:ENE-face, at flatter sites/r; **POA_ALPI**/3110/ENE;po:SEern ENE-face, flatter site/r; **PRIMHIRS**/3118/ENE;po:1-2m, ENE-face/r; **SEDUALPE**/3104/ENE;po:SEern ENE-face, flatter site/r; **SENEIN_C**/3110-15/SW;po:WSW-face of SSE-ridge/r,Ls; **TARAALPI**(cf)/3112/ENE;po:ENE-face, at flatter sites/r!;

<u>**Piz Laschadurella (3046m):**</u> 46°41'57"N 10°11'50"E / Swiss National Park, Engadiner Dolomiten / Date: 05 08 1992 / Interval: 30am / Hist. date: 29 07 1919 / Braun-Blanquet / Ref.: Braun-Blanquet (1958);
eS: ARABCOER/3035/S;po:S-face/s/[3041]; **CERALATI**/3046/E,S,W;po:summit/c/[3046]; **DRABTOME**/3044/S;po:ca.20m, E-ridge/s/[3046]; **MOERCILI**/3035/S;po:S-face/r-s/[3041]; **SAXIAPHY**/3046/SE;po:summit-area/s-c/[3046]; **SAXIOPPO**/3046/E,S,W;po:summit/c/[3046];

mS: FESTPUMI/[3016];
nS: ACHIATRA(cf)/3025/N;po:E-ridge-S-face, at flatter sites/r!; **ANDRHELV**/3015-20/S;po:E-ridge-S-face/r; **ARABALPI**/3030/SW;po:SW-face/r; **ARENCILI**/3015-20/S;po:E-ridge-S-face/r; **CAMPCOCH**/3025/S;po:E-ridge-S-face/r; **DRABLADI**/3025/S;po:E-ridge-S-face/r; **FESTALPI**/3015-20/S;po:E-ridge-S-face/s; **MINUGERA**/3030/S;po:E-ridge-S-face/s; **POA_MINO**/3043/po:SW-face/s; **TARAALPI**/3025/S;po:E-ridge-S-face, at flatter sites/s; **TRISDIST**/3025/S;po:E-ridge-S-face, at flatter sites/r,Ls;

<u>Piz Nair (3010m):</u> 46°40'5"N 10°16'22"E / Swiss National Park, Engadiner Dolomiten / Date: 06 08 1993 / Interval: 10am / Hist. date: 27 07 1918 / Braun-Blanquet / Ref.: Braun-Blanquet (1958);
eS: ANDRHELV/3009-10/S,W;po:S-rock-tower of the S-ridge/c/[3010]; **ARABPUMI**/3008-9/S;po:S-ridge/r-s/[3009]; **ARENCILI**/3007-8/SW;po:S-rock-tower in a ridge/s/[3007]; **CAMPCOCH**/3009-10/W;po:summit/r-s/[3010]; **CERAUNIF**/3009-10/E,W;po:summit/s/[3010]; **DRABDUBI**(cf)/3008/W;po:S-ridge/s/[3001]; **DRABTOME**/3009-10/S,W;po:summit-area/r-s/[3010]; **DRYAOCTO**/3000-1/S;po:S-ridge/r/[3001]; **ERIGUNIF**/3009/SW;po:S-rock-tower in a ridge/r/[3001]; **FESTALPI**/3009-10/S;po:S-rock-tower in a ridge/s/[3010]; **HUTCALPI**(cf)/3009-10/W;po:uppermost W-face/r/[3010]; **MINUGERA**/3009/SW;po:S-rock-tower in a ridge/r/[3007]; **POA_ALPI**/3009/W;po:uppermost W-face/r/[3009]; **POA_MINO**/3003-5/E,SE,S;po:S-ridge, E-face/s/[3001]; **SALISERP**/3000-1/S;po:S-ridge/r/[3003]; **SAXIAPHY**/3008/W;po:uppermost W-face/r-s/[3009]; **SAXICAES**/3001-3/E;po:rocky outcrop in the E-face/r!/[3010]; **SAXIMOSC**/3008-10/S,W;po:uppermost W-face/s/[3010]; **SAXIOPPO**/3008/S;po:summit-area/s-c/[3010]; **TRISDIST**/3009-10/S;po:SE pre-summit/r-s/[3010];
mS: SILEACAU/[3001];**TARAALPI**/[3001];
nS: DRABLADI/3008/SW;po:S-ridge/r!; **JUNICO_A**/3001-2/SE,S;po:rocky outcrop in the E-face/r!;

<u>Piz Foraz (3092m):</u> 46°41'30"N 10°16'28"E / Swiss National Park, Engadiner Dolomiten / Date: 09 08 1993 / Interval: 30am / Hist. date: 28 08 1921 / Braun-Blanquet / Ref.: Braun-Blanquet (1958);
eS: ANDRHELV/3090/S;po:S-face/s/[3087]; **CERAUNIF**/3091/SW;po:SW-ridge/s-c/[3089]; **DRABLADI**/3092/S;po:summit-ridge + S-face/s/[3082]; **DRABTOME**/3092/S,W;po:summit-ridge + S-face/s-c/[3082]; **FESTALPI**/3083-84/S;po:SW-ridge/s/[3077]; **MINUGERA**/3089/S;po:S-face/s/[3082]; **POA_MINO**/3085-86/S;po:S-face, at the beginning of a flatter scree-area/r/[3082]; **SILEACAU**/3083-84/S;po:S-face/r,Ls/[3082]; **TRISDIST**/3062-67/S;po:S-face, SSE-ridge-area/Ls/[3072];
nS: ARABCOER/3072-74/S;po:S-face/r-s; **ARABPUMI**/3087-89/W;po:SW-ridge/r; **CAMPCOCH**/3067-69/S;po:S of a rocky outcrop in the S-SSE-ridge/Ls; **ERIGUNIF**/3072-74/S;po:S-SSE-ridge, at a prominent rocky outcrop/r; **POA_ALPI**/3083-84/S;po:S-face/r,Ls; **SAXIAPHY**/3062-67/S;po:S-face, close to SSE-ridge/r; **SAXICAES**/3087/S;po:small rock-tower in the S-face/r; **SAXIMOSC**/3083-84/S;po:SW-ridge/r!; **SAXIOPPO**/3092/W;po:summit-ridge/s-c; **TARAALPI**/3062-67/S;po:S-face/r;

<u>Piz Tavrü (3168m):</u> 46°40'49"N 10°17'50"E / Swiss National Park, Engadiner Dolomiten / Date: 07 08 1993 / Interval: 15am / Hist. date: 31 07 1927 / Braun-Blanquet / Ref.: Braun-Blanquet (1958);
eS: CERAUNIF/3162-63/SE,S;po:S-face/s/[3158]; **DRABTOME**(cf)/3162-63/po:S-face, SW-ridge, S-rock-tower in the ridge/s-c/[3158]; **POA_MINO**/3153-54/S;po:S-face, S of the S-rock-tower in the ridge/r,Ls/[3158]; **SAXIAPHY**/3161-62/S;po:ESE-ridge/s/[3158]; **SAXIOPPO**/3166/po:W-ridge/s/[3168];
nS: ARABCOER/3158-59/S;po:S-face/r; **ARABPUMI**/3158/S;po:S-face/r!; **DRABLADI**/3162-63/SSE,S;po:SW-ridge/r!; **HUTCALPI**/3160-61/S;po:ESE-ridge/r; **MINUGERA**/3158/S;po:S-rock-tower of the W-ridge/r!; **MOERCILI**/3156-58/SE;po:S-SW-ridge W of a notch/r!; **SEDUATRA**/3153/S;po:S-face/r!; **TARAALPI**(cf)/3155-57/E;po:S-rock-tower of the W-ridge/r!; **TRISDIST**/3155-56/S;po:S-face, below a rock-tower of the ESE-ridge/r,Ls;

<u>Piz Plazèr (3104m):</u> 46°42'35"N 10°23'22"E / Sesvenna-Gruppe / Date: 26 07 1992 / Interval: 30am / Hist. date: 31 07 1918 / Braun-Blanquet / Ref.: Braun-Blanquet (1958);
eS: ANDRALPI/3101/ev:ridge;po:30m NW, NW-pre-summit/r/[3098]; **CARECURV**/3088-92/SW;po:60m S, SW ridge-branching/Lc/[3093]; **CERAUNIF**/3103-4/ev:ridge;po:1-5m/s/[3103]; **ERIGUNIF**/3080/S,W;po:ca. 80-100m, SSE-ridge/Ls/[3093]; **FESTHAL**/3103/ev:ridge;po:10m S, exposed rock (SW)/r-s/[3098]; **GENTBAVA**/3101/ev:ridge;po:30m NW, NW-pre-summit/s/[3098]; **LUZUSPIC**/3097/ev:ridge;po:ca. 20m NW (notch towards the NW-pre-summit)/s/[3098]; **MINUSEDO**/3088-92/SW;po:60m S, SW ridge-branching/Lc/[3098]; **OREODIST**/3088-92/SW;po:60m S, SW ridge-branching/Lc/[3093]; **POA_LAXA**/3103/ev:ridge;po:10m S, exposed rock (SW)/c/[3098]; **RANUGLAC**/3103-4/ev:ridge;po:1-5m/c/[3103]; **SAXIBRYO**/3103/S,SW;ev:ridge;po:10m S, exposed rock (SW)/c/[3104]; **SAXIEXAR**/3103-4/SW;po:1-5m/c/[3098]; **TANAALPI**/3103/ev:ridge;po:10m S, exposed rock (SW)/s/[3098];

mS: CARDRESE/[3098];**PHYTHEMI**/[3098];
nS: AVENVERS/3102/ev:ridge;po:NW-pre-summit/r!; **DRABFLAD**/3103-4/SW;po:1-5m/r!; **PHYTGL_P**/3088-92/SW;po:60m S, SW ridge-branching/Lc; **POTEFRIG**/3085/po:ca. 80m, SSE-ridge/Ls; **SAXIOPPO**/3103-4/SW;po:1-5m/s-c;

Piz Sesvenna (3204m): 46°42'25"N 10°24'15"E / Sesvenna-Gruppe / Date: 26 07 1992 / Interval: 30am / Hist. date: 31 07 1918 / Braun-Blanquet / Ref.: Braun-Blanquet (1958);
eS: ANDRALPI/3190/N;po:SE-ridge/r,Lc/[3187]; **CERAUNIF**/3194-96/N;po:SE-ridge/s/[3177]; **DOROCLUS**/3172-77/SE;po:SE of SE-pre-summit/Ls/[3177]; **DRABFLAD**/3188/S;po:100-150m SE (pre-summit)/r/[3187]; **ERIGUNIF**/3188/S;po:100-150m SE (pre-summit)/r-s/[3187]; **FESTHAL.**/3188/S;po:100-150m SE (pre-summit)/r/[3177]; **GENTBAVA**/3180-85/E;po:E of SE-pre-summit/Lc/[3187]; **GEUMREPT**/3180-85/E;po:E of SE-pre-summit/r/[3177]; **LUZUSPIC**/3175/SE;po:SE of SE-pre-summit/Ls/[3177]; **PHYTGL_P**/3172-77/SE;po:SE of SE-pre-summit/Ls-c/[3177]; **POA_LAXA**/3200-2/ev:ridge;po:ca. 10m WNW/c/[3197]; **POTEFRIG**/3175/ESE;po:SE of SE-pre-summit/Ls/[3177]; **RANUGLAC**/3203-4/ev:ridge;po:25m WNW (pre-summit)/c/[3204]; **SAXIBRYO**/3203-4/po:at the cross on the summit/c/[3197]; **SAXIEXAR**/3200/ev:ridge;po:ca. 15m WNW/c/[3197]; **SAXIOPPO**/3202/S;po:at the cross on the summit/s/[3197]; **TANAALPI**/3200/ev:ridge;po:between the two main-summits/r-s/[3187];
nS: SENEIN_C/3180/E;po:E of SE-pre-summit/r!; **TARAALPI**(cf)/3188/S;po:SE-pre-summit/r!;

Part 2: Ötztaler Alpen

Hinterer Spiegelkogel (3426m): 46°49'49"N 10°57'33"E / Ramolkamm, Ötztaler Alpen / Date: 21 07 1992 / Interval: 15am / Hist. date: 1953 / Reisigl & Pitschmann / Ref.: Reisigl & Pitschmann (1958);
eS: ANDRALPI/3424-25/SSE;ev:ridge;po:10m E/s/[3411-26]; **ARTEMUTE**/3414/SSE;po:SSE-face/s/[3411-26]; **CERAUNIF**/3424-25/SSE;ev:ridge;po:5m E/s/[3411-26]; **DRABDUBI**/3420/SE,S;po:1am below S-ridge/r/[3411-26]; **DRABFLAD**/3425/SW;po:2m SW/r/[3411-26]; **ERIGUNIF**/3418/SE,S;po:1-2am below S-ridge/s/[3411-26]; **FESTHAL**/3418/SE,S;po:1-2am below S-ridge/r-s/[3411-26]; **GENTBAVA**/3416/SE,S;po:S-ridge/r/[3411-26]; **POA_LAXA**/3424-25/SSE;ev:ridge;po:10m E/s-c/[3411-26]; **POTEFRIG**/3416/SE,S;po:S-ridge/s/[3411-26]; **RANUGLAC**/3424-25/ev:ridge;po:10m E/s-c/[3411-26]; **SAXIBRYO**/3424-25/SSE;ev:ridge;po:10m E/c/[3411-26]; **SAXIEXAR**(cf)/3423/SW;po:ca. 10m W/s/[3411-26]; **SAXIOPPO**/3426/ev:ridge;po:1m S/c/[3411-26]; **TRISSPIC**/3420/SSE;po:SE-face, 15m E/Ls/[3426];
nS: LUZUSPIC/3414/SE,S;po:S-ridge/no data; **MINUSEDO**/3420/SSE;po:SE-face, 15m E/Ls; **SILEEXSC**/3420/SSE;po:SE-face, 15m E/Ls; **TANAALPI**/3420/SSE;po:SE-face, 15m E/Ls;

Stockkogel (3109m): 46°53'46"N 11°0'0"E / Ramolkamm, Ötztaler Alpen / Date: 23 07 1992 / Interval: 9am / Hist. date: 1953 / Reisigl & Pitschmann / Ref.: Reisigl & Pitschmann (1958);
eS: ANDRALPI/3105/ev:ridge;po:5-10m W-SW/s/[3100-9]; **CERAUNIF**/3105/ev:ridge;po:5-10m W-SW/c/[3100-9]; **FESTHAL.**/3101/S;po:5-10m W-SW/r/[3100-9]; **GENTBAVA**/3105/ev:ridge;po:5-10m W-SW/s/[3100-9]; **LUZUSPIC**/3105/S;po:5-10m W-SW/s/[3100-9]; **MINUSEDO**/3105/ev:ridge;po:5-10m W-SW/c/[3100-9]; **POA_LAXA**/3105/ev:ridge;po:5-10m W-SW/s-c/[3100-9]; **RANUGLAC**/3105/ev:ridge;po:5-10m NW/c/[3100-9]; **SAXIBRYO**/3105/N,S;ev:ridge;po:ca. 5-10m W-NW/c/[3100-9]; **SAXIEXAR**(cf)/3101/S;po:5-10m W-SW/s/[3100-9]; **SILEEXSC**/3105/S;po:5-10m W-SW/c/[3100-9]; **TANAALPI**/3105/ev:ridge;po:ca.6-10m W/c/[3100-9];
nS: AGRORUPE/3101/S;po:ca. 10m ESE, below a snow-cornice/r; **CARECURV**/3105/S;ev:ridge;po:5-10m W-SW/c-d; **ERIGUNIF**/3105/S;po:5-10m W-SW/r-s; **OREODIST**/3105/S;ev:ridge;po:5-10m W-SW/c; **PINUCEMB**/3103/S;po:5-10m W-SW/r!; **POTEFRIG**/3101/S;po:5-10m W-SW/r-s; **PRIMGLUT**/3105/S;po:5-10m W-SW/r;

Hohe Wilde-Südgipfel (3480m): 46°45'58"N 11°1'21"E / southern Ötztaler Alpen / Date: 27 08 1992 / Interval: 20am / Hist. date: 1953 / Reisigl & Pitschmann / Ref.: Reisigl & Pitschmann (1958); Comment: Reisigl & Pitschmann indicated the investigation-interval 3460-3475m, which excludes the uppermost 5 am. According to Reisigl (pers. comm.) all species were found below the uppermost 5 am, with the exception of DRABFLAD (compare Reisigl & Pitschmann 1958, Table 16, Table 17).
eS: ANDRALPI/3477/SSE;po:ESE, uppermost SE-face/s/[3460-75]; **CERAUNIF**/3477/SSE;po:ESE, uppermost SE-face/s-c/[3460-75]; **DRABFLAD**/3480/SE,W;po:4m NW/s/[3480]; **GENTBAVA**/3466/SE;po:SE-face/s/[3460]; **LUZUSPIC**/3472/SE;po:uppermost SE-face, E of

summit/c/[3460-75]; **MINUSEDO**/3470-74/SSE;po:ESE, uppermost SE-face/c/[3460-75]; **POA_LAXA**/3480/SE;po:5m E/c-d/[3460-75]; **RANUGLAC**/3479/SSE;po:ESE, uppermost SE-face/s-c/[3460-75]; **SAXIBRYO**/3477/SSE;po:ESE, uppermost SE-face/c-d/[3460-75]; **SILEEXSC**/3472/SE;po:uppermost SE-face, E of summit/c/[3460-75]; **TANAALPI**/3472/SSE;po:ESE, uppermost SE-face/c/[3460];
nS: CARECURV/3468/SE;po:SE in the SE-face/c; **ERIGUNIF**/3460/SSE;po:SSE-face/Ls; **FESTINTE**/3460/SSE;po:SSE-face/Ls; **POTEFRIG**/3466/SE;po:SE-face/s; **PRIMGLUT**/3465/SSE;po:SSE-face/r; **PRIMMINI**/3468/SSE;po:SSE-face/r!; **SAXIEXAR**/3477/SSE;po:ESE, uppermost SE-face/s; **SAXIOPPO**/3472/SSE;po:SE-face/r-s;

Hinterer Seelenkogel (3470m): 46°48'7"N 11°2'40"E / southern Ötztaler Alpen / Date: 16 09 1992 / Interval: 12am / Hist. date: 1953 / Reisigl & Pitschmann / Ref.: Reisigl & Pitschmann (1958);
eS: ANDRALPI/3462-63/ev:ridge;po:ca.30m SW/s/[3458-70]; **DRABDUBI**(cf)/3458/S;po:S-face/r!/[3458-70]; **DRABFLAD**/3467/S;po:25m SW, S-face/s/[3458-70]; **ERIGUNIF**/3467/S;po:S-face/s,Lc-d/[3458-70]; **GENTBAVA**/3466/S;po:ca.15m, S-face/s-c/[3458-70]; **LUZUSPIC**/3460/S;po:S-face/r!/[3458-70]; **POA_LAXA**/3468/S;po:ca. 5m, S-face/c/[3458-70]; **POTEFRIG**/3466/S;po:ca. 15m, S-face/c,Ld/[3458-70]; **RANUGLAC**/3467/S;po:5m SW, S-face/c/[3458-70]; **SAXIBRYO**/3468/SE;po:4-5m SW/c/[3458-70]; **SAXIOPPO**/3469/SE;po:4-5m SW/s/[3458-70];
nS: CERAUNIF/3460/ESE;po:small slope at the SW-ridge/r,Lc; **TANAALPI**/3461/S;po:15m S, S-face/r!;

Festkogel (3038m): 46°50'9"N 11°3'5"E / Obergurgl, Ötztaler Alpen / Date: 24 07 1992 / Interval: 35am / Hist. date: 1953 / Reisigl & Pitschmann / Ref.: Reisigl & Pitschmann (1958);
Comment: Four species additionally mentioned by Reisigl & Pitschmann (1958, Table 17: CAMPCOCH, FESTRUBR s.l., SAXIPANI, TRIFPALL) were excluded here: they were found outside the investigation area in the ridge further south (Reisigl, pers. comm.).
eS: ACHIMOSC/3033-38/po:S-ridge/r,Lc/[3003-38]; **AGRORUPE**/3018-28/SE;po:40m NNE, E-face/s/[3003-38]; **ALCHFLAB**/3018-28/SE;po:50m NNE, E-face/r/[3003-38]; **ANDRALPI**/3033-38/SW,W;po:ridge-area/s/[3003-38]; **ANTECARP**/3018-28/SE;po:50m NNE, E-face/Ls/[3003-38]; **ANTHALPI**/3008-13/SE;po:60m NNE, E-face/r/[3003-38]; **AVENVERS**/3008-13/SE;po:60m NNE, E-face/r/[3003-38]; **CARDRESE**/3018-28/SE;po:50m NNE, E-face/s/[3003-38]; **CARECURV**/3033-38/SW,W;po:ridge-area/d/[3003-38]; **CERAUNIF**/3033-38/SW,W;po:ridge-area/c/[3003-38]; **DOROCLUS**/3028-31/SW;po:W at the summit/s/[3003-38]; **DRABFLAD**/3033-38/SW,W;po:ridge-area/r/[3003-38]; **ELYNMYOS**/3018-28/SE;po:50m NNE, 5m E, E-face/Lc/[3003-38]; **EMPEHERM**/3008-13/NE,E;po:60m NNE/r/[3003-38]; **ERIGUNIF**/3033-38/SW,W;po:ridge-area/s/[3003-38]; **EUPHMINI**/3018-28/SE,S;po:50m NNE, E-face (Ls at 100m)/r,Ls/[3003-38]; **FESTINTE**/3033-38/SW,W;po:ridge-area/c/[3003-38]; **GENTACAU**/3018-28/SE;po:50m NNE, E-face/Ls/[3003-38]; **GENTBAVA**/3033-38/SW,W;po:ridge-area/s/[3003-38]; **GEUMMONT**/3008-13/SE;po:60m NNE, E-face/r/[3003-38]; **GEUMREPT**/3028-33/E;po:E-face, ridge-area/s/[3003-38]; **GNAPSUPI**/3018-28/SE;po:50m NNE, E-face/s/[3003-38]; **HOMOALPI**/3018-28/SE;po:50m NNE, E-face/r/[3003-38]; **JUNCTRIF**/3018-28/SE;po:50m NNE, E-face/Ls/[3003-38]; **JUNICO_A**/3018-28/SE;po:50m NNE, E-face/r/[3003-38]; **LEONHELV**/3018-28/SE;po:50m NNE, E-face/r/[3003-38]; **LUZULUTE**/3018-28/SE;po:50m NNE, E-face/s/[3003-38]; **LUZUSPIC**/3033-38/SW,W;po:ridge-area/c/[3003-38]; **MINUSEDO**/3033-38/SW,W;po:S-ridge + W-face/c/[3003-38]; **OREODIST**/3033-38/SW,W;po:ridge-area/d/[3003-38]; **PHYTHEMI**/3033-38/po:S-ridge/r,Ls/[3003-38]; **POA_ALPI**/3018-28/SE;po:50m NNE, E-face/s/[3003-38]; **POA_LAXA**/3033-38/SW,W;po:ridge-area/c/[3003-38]; **POLYVIVI**/3028-33/SW;po:SW below S-ridge/r,Ls/[3003-38]; **POTEAURE**/3008-13/SE;po:60m NNE, E-face/r/[3003-38]; **POTEFRIG**/3033-38/SW,W;po:ridge-area/s/[3003-38]; **PRIMGLUT**/3033-38/SW,W;po:S-ridge + W-face (+ N)/c/[3003-38]; **PRIMHIRS**/3018-28/SE;po:50m NNE, E-face/Ls/[3003-38]; **PULSVERN**/3013-18/SE;po:50m NNE, E-face/s/[3003-38]; **RANUGLAC**/3033-38/E,W;po:ridge-area/c/[3003-38]; **SALIHERB**/3028-33/S,SW;po:area of S-ridge/c/[3003-38]; **SALISERP**/3018-28/SE;po:E-face/s/[3003-38]; **SAXIBRYO**/3033-38/SW,W;po:ridge-area/c/[3003-38]; **SAXIEXAR**/3033-38/SW,W;po:ridge-area + W-face/c/[3003-38]; **SEDUALPE**/3018-28/SE;po:40m NNE/r/[3003-38]; **SEMPMONT**/3023-33/SW;po:W-face/s/[3003-38]; **SENEIN_C**/3033-38/po:S-ridge/r,Ls/[3003-38]; **SIBBPROC**/3018-28/SE;po:50m NNE, E-face/r/[3003-38]; **SILEEXSC**/3033-38/SW,W;po:ridge-area/c/[3003-38]; **TANAALPI**/3033-38/SW,W;po:S-ridge + W-face/c/[3003-38]; **TRISSPIC**/3033-38/SW,W;po:ridge-area/r/[3003-38]; **VACCGAUL**/3003-8/NE,E;po:60m NNE/r/[3003-38]; **VACCVITI**/3013-18/SE;po:50m NNE, E-face/s/[3003-38]; **VEROBELI**/3018-28/SE;po:50m NNE, E-face/Ls/[3003-38];
mS: HIERALPI/[3003-38];**HUPESELA**/[3003-38];**RHODFERR**/[3003-38];**SOLDPUSI**/[3003-38];

nS: AGROALPI/3028-33/SE;po:2-5m E/s,Lc; **DRABDUBI**(cf)/3028-33/SE,S;po:2-5m E, E-face/r; **GENTBRAC**/3028-33/E,SE;po:2-5m E, E-face/r; **JUNCJACQ**/3028-33/E,SE;po:2-5m E, E-face/s,Lc; **LUZUALPI**/3018-28/SE/s,Lc; **MINUGERA**/3033-38/SW,W;po:ridge-area/s; **OXYRDIGY**/3028-33/W/r; **SALIRETU**/3018-28/SE;po:50m NNE, E-face/r; **VEROALPI**/3018-28/SE;po:5-10m SE, E-face/r;

Liebenerspitze (3399m): 46°49'15"N 11°4'35"E / southern Ötztaler Alpen / Date: 17 09 1992 / Interval: 20am / Hist. date: 1953 / Reisigl & Pitschmann / Ref.: Reisigl & Pitschmann (1958);
Comment: Reisigl & Pitschmann (1958) indicated the interval 3380-3395m, which excludes the uppermost 5 am. This uppermost summit area was bare of species (Reisigl, pers. comm.).
eS: SAXIOPPO/3399/E,S,W;po:2m E, ridge + S-face + NNW of summit/c,Ld/[3379-94];
mS: POA_LAXA/[3379-94];
nS: DRABDUBI/3390/S;po:ca.100-130m W, rock-tower in the crest E of the W-summit/r; **SAXIBRYO**/3383-84/E;po:NNW, at the N-ridge-edge/r!;

Part 3: Zillertaler Alpen, Rieserferner Gruppe

Wilde Kreuzspitze (3135m): 46°54'48"N 11°35'38"E / south-western Zillertaler Alpen / Date: 18 07 1992 / Interval: 3am / Hist. date: 1907 / Klebelsberg / Ref.: Klebelsberg (1913);
eS: ARENCILI/3130-35/S;po:SSE-face/c/[3135]; **CERAUNIF**/3130-35/S;po:SSE-face/s/[3135]; **DRABFLAD**/3130-35/S;po:SSE-face/c/[3135]; **POA_ALPI**(cf)/3130-35/S;po:SSE-face/s/[3135]; **SAXIMOSC**/3130-35/S;po:SSE-face/c/[3135]; **SAXIOPPO**/3130-35/NE,S;po:N + SSE-face/c/[3135];
mS: GENTBAVA/[3135];**HUTCBREV**/[3135];
nS: ANDRALPI/3130-35/SE,S;po:SSE-ridge (3m S), E-ridge (15m E)/s; **SAXIANDR**/3130-35/S;po:E-ridge/s; **SAXIRUDO**/3130-35/SE,S;po:SSE-face (3-5m S), E-ridge/s-c; **TRISSPIC**/3130-35/S;po:SSE-face (3m S)/Ls;

Napfspitze (2888m): 46°56'27"N 11°44'26"E / southern Zillertaler Alpen / Date: 17 07 1992 / Interval: 3am / Hist. date: 1911 / Klebelsberg / Ref.: Klebelsberg (1913);
eS: ANDRALPI/2885/ev:ridge;po:10-30m NE/s/[2888]; **AVENVERS**/2885-88/SE;po:5-10m SW/s/[2888]; **CARECURV**/2885-88/SE;po:5-10m SW,S,ENE/d/[2888]; **CERAUNIF**/2884-86/NW;po:N + NE of the summit-marking/s-c/[2888]; **DOROCLUS**/2885/SE;po:10-30m NE/s/[2888]; **ERIGUNIF**/2888/po:summit-area/s/[2888]; **EUPHMINI**/2885/SE;po:30m NE/r/[2888]; **FESTINTE**(cf)/2885-88/SE;po:1-3m S/r/[2888]; **GENTBAVA**/2885/NW;po:N, NE/s/[2888]; **GEUMMONT**/2885/SE;po:10-30m NE/s/[2888]; **LEONHELV**/2885/SE;po:10-30m NE/s/[2888]; **LIGUMUT1**/2885-88/SE;po:1-3m S/s/[2888]; **LUZUALPI**/2883-88/E,SE;po:1-10m ESE/s/[2888]; **LUZUSPIC**/2885-88/SE;po:1-3m S/s/[2888]; **MINUGERA**/2885-88/SE;po:1-3m S/s/[2888]; **OREODIST**/2885-88/SE;po:5-10m SW,S,ENE/c/[2888]; **PHYTGLOB**/2885-88/SE;po:5-10m SW/s/[2888]; **PHYTHEMI**/2885-88/SE;po:5-10m SW/s/[2888]; **POA_LAXA**/2885-88/SE,NW;po:SE + NW-face/c/[2888]; **POLYVIVI**/2885-88/SE;po:5-10m SW/s/[2888]; **PRIMGLUT**/2885/SE,NW;po:5-10m SE,N,NE (common at NW-face)/c/[2888]; **PRIMMINI**/2885-88/SE;po:5-10m SW/s/[2888]; **RANUGLAC**/2885/NW;po:N, NE/c/[2888]; **SALIHERB**/2885-88/SE;po:5-10m SW/s/[2888]; **SAXIBRYO**/2885/NW;po:N, NE of the summit-marking/d/[2888]; **SAXIMOSC**/2885/NE,SW;po:S, N/s/[2888]; **SENEIN_C**/2885-88/SE;po:5-10m SW/s/[2888]; **SILEEXSC**/2885-88/SE;po:5-10m SW, 3m S/c/[2888]; **TANAALPI**/2885-88/SE;po:5-10m SW/s/[2888]; **VEROALPI**/2883-88/E,SE;po:1-10m ESE/r/[2888];
mS: SAXIPANI/[2888];
nS: AGROALPI/2885-88/SE;po:1-3m S/no data; **AGRORUPE**/2885/SE;po:30m NE/no data; **ARABALPI**/2883-88/E,SE;po:1-10m ESE/no data; **ARENCILI**/2885-88/SE;po:3m S/r; **ARTEMUTE**(cf)/2883-88/E,SE;po:1-10m ESE/s; **CARDRESE**/2888/po:<1m/s; **CAREPARV**/2885/SE;po:10-30m NE/r; **DRABFLAD**/2885/ev:ridge;po:SSW/s; **ELYNMYOS**/2885-88/SE;po:1-3m S/c; **GENTACAU**/2885-88/SE;po:5-10m SW/s; **GENTBRAC**/2885-88/SE;po:5-10m SW/s; **GNAPSUPI**/2885/SE;po:10-30m NE/no data; **JUNCJACQ**/2885/SE;po:10-30m NE/r; **JUNCTRIF**/2885-88/SE;po:1-3m S/s; **MINUBIFL**(cf)/2887/ev:ridge;po:2m SW/r; **MINUSEDO**/2885/NW;po:N, NE of the summit-marking/no data; **PEDIASPL**/2885-88/SE;po:5-10m SW/s; **SALISERP**/2883-88/E,SE;po:1-10m ESE/no data; **SAXIANDR**/2885/NW;po:N, NE/c; **SAXIOPPO**/2885/NW;po:N, NE/c; **SEDUALPE**/2885/SE;po:10-30m NE/no data; **SIBBPROC**/2883-88/E,SE;po:1-10m ESE/no data; **VEROBELI**/2885-88/SE;po:5-10m SW/no data;

Großer Lenkstein (3236m): 46°56'25"N 12°10'1"E / Rieserferner-Gruppe / Date: 16 07 1992 / Interval: 2am / Hist. date: 1912 / Klebelsberg / Ref.: Klebelsberg (1913);
eS: POA_LAXA/3236/S;po:1m S/c/[3236]; **RANUGLAC**/3234/SSW;po:ca. 2-3m SSW of the summit-marking/c/[3236]; **SAXIBRYO**/3236/E;po:2m E/c/[3236];
mS: SILEEXSC/[3236];
nS: DRABFLAD/3235/E;po:3m E/r; **MINUSEDO**/3235/E;po:2m E/s; **POTEFRIG**/3234/E;po:5m E/r;

Part 4: recently investigated summits

Igl Compass (3016m): 46°35'26"N 9°49'28"E / Albulapass, Graubünden, Rätisch-Lepontische Alpen/ Date: 05 08 1993 / Interval: 10am;
S: ANDRALPI/3014/E;po:SE-edge/Ls; **ANDRHELV**/3015-16/SSW;po:SE-edge/s,Lc; **ARABALPI**/3015-16/E;po:SE-edge/s; **ARTEGENI**/3008/SSW;po:SSW-face/r-s; **CAMPCENI**/3011/SSW;po:SSW-face/Ls; **CERALATI**/3015/E;po:SE-edge, ridge-area/c; **DRABDUBI**(cf)/3014-15/SSW;po:SE-edge/s; **FESTALPI**/3006-8/S,SW;po:SSW-face/r-s; **GEUMREPT**/3015/E;po:SE-edge/s-c; **POA_ALPI**/3012-14/E;po:SE-edge/r; **RANUGLAC**/3006-8/NE,E;po:NE-face/r!; **SAXIMOSC**/3006-8/SE,SW;po:SE-edge/r,Ls; **SAXIOPPO**/3015-16/E,S;po:SE-edge/c; **SAXIPANI**/3006-8/SE,SW;po:SE-edge/r; **TARAALPI**(cf)/3008/SSW;po:SSW-face/r!; **TRISSPIC**/3015-16/ev:ridge;po:SE-edge/s;

Gschrappkogel (3197m): 46°57'44"N 10°54'24"E / Geigenkamm, Ötztaler Alpen / Date: 02 08 1993 / Interval: 15am;
S: ACHIMOSC/3182-84/SE;po:SSE-face/Ls; **AGRORUPE**/3182-84/SE;po:SSE-face/r; **ANDRALPI**/3185/NE/r; **CARDRESE**/3182-84/SE;po:SSE-face/r!; **CERAUNIF**/3196-97/ev:ridge;po:ca. 10m N of a stone-pile at the summit area/s; **DOROCLUS**/3182-84/SE;po:SSE-face/r; **DRABFLAD**/3195-96/W/s; **FESTHAL.**(cf)/3189/S;po:S-ridge/Ls; **GENTBAVA**/3196-97/po:summit-ridge/s; **GNAPSUPI**/3186/SE;po:SSE-face/r-s; **LUZUSPIC**/3193/W/s; **MINUSEDO**/3186/SSE;po:SSE-face/Ls; **POA_LAXA**/3196-97/po:summit-ridge/c; **POTEFRIG**/3185/NW/r!; **RANUGLAC**/3196-97/po:summit-ridge/c; **SAXIBRYO**/3196-97/po:summit-ridge/c; **SAXIEXAR**/3191-92/W/r,Lc; **SEDUALPE**/3196/SE;po:SSE-face/Lr-s; **SENEIN_C**/3186/SE;po:SSE-face/Ls; **SILEEXSC**/3182-84/SE;po:SSE-face/r; **TANAALPI**/3196-97/ev:ridge;po:ca. 10m N of a stone-pile at the summit area/s;

Reichenberger Spitze (3030m): 46°59'10"N 12°15'38"E / western Lasörling-Gruppe, Hohe Tauern, East-Tyrol / Date: 30 07 1993 / Interval: 5am;
S: ARENCILI/3025/SE/r-s; **ARTEGENI**/3025/SE;po:SE-face/s; **CERAUNIF**/3028-30/po:S-summit-ridge/s; **DRABDUBI**(cf)/3025/SE/r; **DRABFLAD**/3028-30/W;po:S-summit-ridge/s; **DRABHOPP**/3025/SE;po:SE-face/s; **FESTPUMI**/3025-26/SE;po:SE-face/Lc; **GENTORBI**/3025/SE;po:SE-face/s; **HUTCALPI**/3025/SE;po:SE-face/r-s; **LINAALPI**/3028-30/po:S-summit-ridge/r-s; **MINUGERA**/3028-30/SE;po:SE-face/s; **MINUSEDO**/3028-30/po:SE-face of S-ridge/c; **POA_ALPI**/3028-30/SE;po:SE-face/s; **SAXIMOSC**/3025/SE;po:SE-face/r; **SAXIOPPO**/3028-30/po:S-summit-ridge/c; **SAXIRUDO**/3028/N;po:N-face/Ls; **SEDUATRA**/3025/SE;po:SE-face/s; **SESLOVAT**/3024/SE;po:SE-face, close below investigation area/r; **TRISDIST**/3028-30/SE;po:SE-face/Lc; **TRISSPIC**/3025/po:E-ridge/r;

Glatte Schneid (2890m): 47°2'32"N 12°42'27"E / Glocknergruppe, Hohe Tauern / Date: 21 08 1993 / Interval: 10am;
S: ANDRALPI/2890/N,NE,NW;po:summit-ridge/r-s; **ARABALPI**/2886/N,NE;po:N-face of the E-ridge/r; **ARENCILI**/2886-87/SW,W;po:W-face, ridge-area/r-s; **ARTEGENI**/2890/ev:all exposures;po:summit/c-d; **CAMPCOCH**/2880/SE;po:SE-face/r!; **CARRRUPE**/2877-80/W;po:rocky outcrop of the NNW-ridge, close below investigation area/r; **CERAUNIF**/2890/ev:all exposures;po:summit/c-d; **DOROGLAC**/2885-86/NE;po:NE-face/r; **DRABFLAD**/2890/ev:all exposures;po:summit/c; **DRABHOPP**/2890/E,S,W;po:summit-ridge/s; **ELYNMYOS**/2885/SE;po:SE-face/r; **ERIGUNIF**/2888/SE;po:E-ridge/c; **EUPHMINI**/2886-87/SE;po:SE-face/Lc; **GENTNANA**/2885/SE;po:SE-face/r,Ls; **GENTPROS**/2885-86/SE;po:SE-face/Ls; **HUTCALPI**(cf)/2890/po:summit-ridge/r!; **LEONALPI**/2880/SE;po:SE-face/r; **LINAALPI**/2881-82/SE;po:SE-face/r!; **MINUGERA**/2888/W;po:W-face/s; **MINUSEDO**/2889/E;po:E-ridge/c-d; **OXYTCAMP**/2884-85/SE;po:SE-face/r; **PEDIASPL**/2877-80/W;po:rocky outcrop of the NNW-ridge, close below investigation area/r; **PHYTGLOB**/2882-84/E;po:E-ridge/r; **POA_ALPI**/2890/SE;po:mainly in the uppermost SE-face/c-d; **POA_LAXA**(cf)/2887/SW;po:W-face/Lr-s; **POLYVIVI**/2886/SE;po:SE-face/Ls; **POTEAURE**/2885/SE;po:SE-face/r!; **POTECRAN**/2883/SE;po:SE-face/r!; **PRIMMINI**/2882-84/E;po:E-ridge/r; **RANUGLAC**/2882-83/SE;po:SE-face/r; **SAXIADSC**/2886/N;po:N-face of the E-ridge/Ls; **SAXIANDR**/2889/po:NNW-ridge, E-

face/s-c,Lc; **SAXIBRYO**/2882-83/N,E,SE;po:E-ridge/Ls; **SAXIMOSC**/2890/SE;po:summit/s,Lc; **SAXIOPPO**/2890/ev:all exposures;po:summit/c,Ld; **SAXIRUDO**/2887/E;po:E-ridge/s; **SEDUATRA**/2884-85/SE,S;po:SE-face/Lr-s; **SESLOVAT**/2880-81/SW,W;po:W-face of NNW-ridge/r; **SILEEXSC**/2889/po:NNW-ridge/c; **TARAALPI**(cf)/2887-88/SE;po:SE-face/Ls; **TRISDIST**/2888-89/SW,W;po:NNW-ridge/r!; **TRISSPIC**/2890/ev:all exposures;po:summit/c;

Dritter Leiterkopf (3105m): 47°3'21"N 12°44'18"E / Glocknergruppe, Hohe Tauern / Date: 20 08 1993 / Interval: 15am;
S: CERAUNIF/3103/po:SE-face (area of the uppermost E-ridge)/s; **DRABFLAD**/3104/E;po:5m S/s; **HUTCALPI**(cf)/3090/SE;po:S-ridge/r!; **POA_ALPI**/3105/E;po:summit-ridge/r-s; **SAXIANDR**/3098/po:flat saddle at the rounded E-ridge/Ls; **SAXIBIFL**/3097-98/N;po:N-face/r; **SAXIMOSC**/3102/E;po:S-ridge + uppermost E-ridge/r-s; **SAXIOPPO**/3105/po:ridge-area, uppermost SE-face/c-d; **SAXIRUDO**/3105/SE;po:uppermost SE-face/c-d;

Appendix 2:

Definition of species codes (nomenclature according to Ehrendorfer 1973): ACHIATRA = *Achillea atrata* L., ACHIMOSC = *Achillea moschata* Wulf., AGROALPI = *Agrostis alpina* Scop., AGRORUPE = *Agrostis rupestris* All., ALCHFLAB = *Alchemilla flabellata* Buser, ANDRALPI = *Androsace alpina* (L.)Lam., ANDRHELV = *Androsace helvetica* (L.)All., ANTECARP = *Antennaria carpatica* (Wahlenb.)Bluff & Fing., ANTHALPI = *Anthoxanthum alpinum* A.& D. Löve, ARABALPI = *Arabis alpina* L., ARABCOER = *Arabis caerulea* All., ARABPUMI = *Arabis pumila* Jacq., ARENCILI = *Arenaria ciliata* L.emend.L., ARNIMONT = *Arnica montana* L., ARTEGENI = *Artemisia genipi* Web., ARTEMUTE = *Artemisia mutellina* Vill., AVENFLEX = *Avenella flexuosa* (L.)Parl., AVENVERS = *Avenochloa versicolor* (Vill.)Holub, CAMPCENI = *Campanula cenisia* L., CAMPCOCH = *Campanula cochleariifolia* Lam., CARDRESE = *Cardamine resedifolia* L., CARECURV = *Carex curvula* All., CAREPARV = *Carex parviflora* Host, CARRRUPE = *Carex rupestris* All., CERALATI = *Cerastium latifolium* L., CERAPEDU = *Cerastium pedunculatum* Gaudin, CERAUNIF = *Cerastium uniflorum* Clairv., DOROCLUS = *Doronicum clusii* (All.)Tausch, DOROGLAC = *Doronicum glaciale* (Wulf.)Nyman, DRABDUBI = *Draba dubia* Suter, DRABFLAD = *Draba fladnizensis* Wulf., DRABHOPP = *Draba hoppeana* Rchb., DRABLADI = *Draba ladina* Br.-Bl., DRABTOME = *Draba tomentosa* Clairv., DRYAOCTO = *Dryas octopetala* L., ELYNMYOS = *Elyna myosuroides* (Vill.)Fritsch, EMPEHERM = *Empetrum hermaphroditum* Hagerup, ERIGUNIF = *Erigeron uniflorus* L., ERITNANU = *Eritrichum nanum* (L.)Schrad.ex Gaudin, EUPHMINI = *Euphrasia minima* Jacq.ex DC., FESTALPI = *Festuca alpina* Suter, FESTHAL. = *Festuca halleri* agg., FESTINTE = *Festuca intercedens* (Hackel)Lüdi ex Becherer, FESTPUMI = *Festuca pumila* Chaix, GENTACAU = *Gentiana acaulis* L.s.str., GENTBAVA = *Gentiana bavarica* L.var.*subacaulis* Cust., GENTBRAC = *Gentiana brachyphylla* Vill., GENTNANA = *Gentianella nana* (Wulf.)Pritch., GENTORBI = *Gentiana orbicularis* Schur, GENTPROS = *Gentiana prostrata* Haenke, GENTPUNC = *Gentiana punctata* L., GEUMMONT = *Geum montanum* L., GEUMREPT = *Geum reptans* L., GNAPSUPI = *Gnaphalium supinum* L., HIERALPI = *Hieracium alpinum* L., HIERGLAN = *Hieracium glanduliferum* Hoppe, HOMOALPI = *Homogyne alpina* (L.)Cass., HUPESELA = *Huperzia selago* (L.)Bernh.ex Schrank&Mart., HUTCALPI = *Hutchinsia alpina* (L.)R.Br., HUTCBREV = *Hutchinsia alpina* ssp. *brevicaulis* (Hoppe)Arc., JUNCJACQ = *Juncus jacquinii* L., JUNCTRIF = *Juncus trifidus* L., JUNICO_A = *Juniperus communis* L.ssp.*alpina*(Neilr.)Celak., LEONALPI = *Leontopodium alpinum* Cass., LEONHELV = *Leontodon helveticus* Mérat emend.Widd., LIGUMUT1 = *Ligusticum mutellinoides* (Cr.)Vill., LINAALPI = *Linaria alpina* (L.)Mill., LLOYSERO = *Lloydia serotina* (L.)Rchb., LUZUALPI = *Luzula alpino-pilosa* (Chaix)Breistr., LUZULUTE = *Luzula lutea* (All.)DC., LUZUSPIC = *Luzula spicata* (L.)DC., MINUBIFL = *Minuartia biflora* (L.)Schinz & Thell., MINUGERA = *Minuartia gerardii* (Willd.)Hayek, MINURECU = *Minuartia recurva* (All.)Schinz & Thell., MINUSEDO = *Minuartia sedoides* (L.)Hiern, MOERCILI = *Moehringia ciliata* (Scop.)DT., NARDSTRI = *Nardus stricta* L., OREODIST = *Oreochloa disticha* (Wulf.)Lk., OXYRDIGY = *Oxyria digyna* (L.)Hill, OXYTCAMP = *Oxytropis campestris* (L.)DC., PEDIASPL = *Pedicularis aspleniifolia* Floerke ex Willd., PEDIKERN = *Pedicularis kerneri* DT., PHYTGL_P = *Phyteuma globulariifolium* ssp.*pedemontanum* (R.Schulz)Becherer, PHYTGLOB = *Phyteuma globulariifolium* s.str. Sternb.&Hoppe, PHYTHEMI = *Phyteuma hemisphaericum* L., PINUCEMB = *Pinus cembra* L., POA_ALPI = *Poa alpina* L., POA_LAXA = *Poa laxa* Haenke, POA_MINO = *Poa minor* Gaudin, POLYVIVI = *Polygonum viviparum* L., POLYVULG = *Polypodium vulgare* L., POTEAURE = *Potentilla aurea* L., POTECRAN = *Potentilla crantzii* (Cr.)Beck ex Fritsch, POTEFRIG = *Potentilla frigida* Vill., PRIMGLUT = *Primula glutinosa* Wulf., PRIMHIRS = *Primula hirsuta* All., PRIMINTE = *Primula integrifolia* L.emend.Gaudin, PRIMLATI = *Primula latifolia* Lapeyr., PRIMMINI = *Primula minima* L., PULSVERN = *Pulsatilla vernalis* (L.)Mill., RANUGLAC = *Ranunculus glacialis* L., RHODFERR = *Rhododendron ferrugineum* L., SALIHERB = *Salix herbacea* L., SALIRETU = *Salix retusa* L., SALISERP = *Salix serpillifolia* Scop., SAXIADSC = *Saxifraga adscendens* L., SAXIANDR = *Saxifraga androsacea* L., SAXIAPHY = *Saxifraga aphylla* Sternb., SAXIBIFL = *Saxifraga biflora* All., SAXIBRYO = *Saxifraga bryoides* L., SAXICAES = *Saxifraga caesia* L., SAXIEXAR = *Saxifraga exarata* Vill., SAXIMOSC = *Saxifraga moschata* Wulf., SAXIOPPO = *Saxifraga oppositifolia* L., SAXIPANI = *Saxifraga paniculata* Mill., SAXIRUDO = *Saxifraga rudolphiana* Hornsch.ex Koch, SAXISEGU = *Saxifraga seguieri* Spreng., SEDUALPE = *Sedum alpestre* Vill., SEDUATRA = *Sedum atratum* L., SEMPMONT = *Sempervivum montanum* L., SENEIN_C = *Senecio incanus* L.ssp.*carniolicus*(Willd.)Br.-Bl., SESLOVAT = *Sesleria ovata* (Hoppe)Kern., SIBBPROC = *Sibbaldia procumbens* L., SILEACAU = *Silene acaulis* (L.)Jacq., SILEEXSC = *Silene exscapa* All., SOLDPUSI = *Soldanella pusilla* Baumg., TANAALPI = *Tanacetum alpinum* (L.)C.H.Schultz, TARAALPI = *Taraxacum alpinum* Hegetschw. s. l., TRISDIST = *Trisetum distichophyllum* (Vill.)PB., TRISSPIC = *Trisetum spicatum* (L.)K.Richter, VACCGAUL = *Vaccinium gaultherioides* Bigelow, VACCVITI = *Vaccinium vitis-idaea* L., VEROALPI = *Veronica alpina* L., VEROBELI = *Veronica bellidioides* L.

MONITORING OF EASTERN AND SOUTHERN SWISS ALPINE TIMBERLINE ECOTONES

CONRADIN A. BURGA & ROGER PERRET

Department of Geography, University of Zurich, Winterthurerstrasse 190, CH-8057 Zurich

Keywords: Alpine vegetation patterns, climatic change, environmental parameters, forest and tree limit, vegetation monitoring

Nomenclature: Binz & Heitz (1990)

Abstract

Firstly, general features of the upper limit of the Swiss Alpine forest and tree limits have been discussed. There are so far numerous investigations and definitions of these different boundaries. The timberline ecotone is here defined as the altitudinal interval between the forest limit and the tree species or lethal limit. The most important question is which are the crucial parameters affecting the upper forest and tree limit. Furthermore, there is the difficulty of distinction of natural and anthropozoogenic factors influencing the present timberline ecotone. Finally, physico-geographical and geological parameters which form the definitive "facies" of the timberline ecotone have to be taken into account.

To investigate the timberline ecotone of the Eastern and Southern Swiss Alps, six sites were studied (mapping of the present vegetation and single trees; investigations of palaeoecological and dendrochronological aspects). Some results of the comparison of the six investigation areas are: 1. The main factors affecting the timberline ecotone are radiation/altitude, average summer temperatures, extreme climatic events, topography/exposure, water supply and anthropozoogenic impact. 2. Not all Alpine plant communities are well suited for tree regrowth and reforestation. 3. The climate-induced natural and/or anthropozoogenic vegetation change would first be conspicuous on extreme exposed sites. A rise of the tree limit above its present potential natural altitude can only be expected under warmer and moister conditions similar to e.g. the hypsithermal.

Monitoring of these test regions will help to check the models and to ascertain further climatically induced changes within the timberline ecotone.

Burga & Kratochwil (eds.), BIOMONITORING, 179-194

1. Introduction - General features of previous investigations at the timberline ecotone in the Alps

The *upper forest limit* is generally the most striking altitude boundary of vegetation in the Alpine mountain region. It is however not always clear what is meant by the term "forest" (geobotanical and forestry definitions of "forest"). In general, tree groups must have a certain density (canopy density of about 40%), a minimum height (about 3 to 5 metres), and a minimum area (about 100 to 2'000 square metres) to be considered as forest (cf. e.g. Bachmann 1964 or Pott & Hüppe 1992). Forests are also characterized by even climatic conditions, by characteristic soil formation processes and by a forest flora adapted to these conditions. The term "tree" and thus the altitude of the *upper tree limit* are not easy to determine, either. The definition propounded by Holtmeier (1967, 1974), Ellenberg (1996) and many other authors, however, appears to be suitable in this sense (cf. also former contributions to the Alpine forest limit and tree limit by Imhof 1900, Brockmann-Jerosch 1919 and Däniker 1923). According to this definition a tree is characterized as a plant penetrating the winter snow cover with its top branches. Since the average height of winter snow cover varies from place to place, it is not possible to indicate a generally valid average for a minimum tree height. Under the protective snow cover, small individual bush-shaped trees climb far above the tree limit. The highest of these small types of trees, usually only a few decimetres high, then form the so-called *tree species limit* or *lethal limit.* The latter is not a clearly defined altitude boundary but a boundary which is closely connected with local climatic and edaphic factors (exposure, wind, duration of snow cover, water supply, boulders, rock content of substrate etc.).

The natural upper forest limit is mainly a *climatically determined* boundary (temperature line). The numerous investigations on this subject by e.g. Michaelis (1934a,b); Larcher (1957, 1963); Aulitzky (1963); Tranquillini (1966, 1967, 1974, 1976, 1979); Baig et al. (1974); Holtmeier (1974, 1985, 1986, 1987, 1989); Baig & Tranquillini (1980) and Körner (1999) have shown this in the Alps and in other mountain areas of the world. All these authors came to the conclusion that both the forest limit and the tree limit are determined primarily by the effects of short growing seasons and the difficult late winter water balance situation associated with desiccation damage on frozen soil. The same is also valid for other mountain areas (cf. e.g. Wardle 1968 or Schwarz 1983). Furthermore, different authors have tried to establish *heat sums* to determine the forest limit: according to Turner & Blaser (1977) or Ellenberg (1996), for a possible tree growth at the Alpine forest limit at least 100 - 110 days with an average air temperature of 5°C are needed. Further minimal heat sums for the forest limit are given by Gross (1989). The 10°C July isotherm is also often used as a rough approach to the altitude of the forest limit (cf. e.g. Brockmann-Jerosch 1919; Holtmeier 1974; Tuhkanen 1980). Since the Alpine forest limit however is formed by different trees (Central Alps: principally Swiss stone pine and larch; Prealps: spruce), e.g. Aulitzky & Turner (1982) propose that such heat sums and temperature limit values must always be related to the tree species concerned. Moreover, the climatic conditions in the various regions of the Alps (continental climate in the Central Alps, oceanic climate in the Prealps and the Southern Alps) and thus the daily and annual temperature courses are different, so that temperature averages will only be a very rough approach to the temperature requirements of the trees forming the forest limit (cf. e.g. Brockmann-Jerosch 1919). Apart from insufficient heat sums and/or low temperatures during the growing season, additional factors, especially traumatic events, may be directly or indirectly responsible for the altitude of the upper forest limit. As to the main factors causing the timberline ecotone, besides regional climatic tree limits and forest limits also local (high snow accumulation, wind exposure, cold air), edaphic (rocky surfaces, talus deposits, changes in dryness or wetness), orographic (avalanche trails, rocky sites or fluvial erosion) as well as anthropogenic (forest clearings, alpine pastures) tree limits and forest limits can be differentiated (cf. e.g. Köstler & Mayer 1970). Today, the course of the upper forest limit (= *actual* or *current* forest limit) in the Alps is, with the exception of too steep or impassable slopes for pasturing, now almost everywhere *anthropozoogenically* determined or influenced, and does not reflect the physiologically-topographically possible *potential natural climatic* forest limit. The mainly anthropozoogenically caused transition zone between the present forest limit and the present tree limit, which is characterized by the appearance of individual trees and groups of trees, is often designated as "Kampfzone" ("fighting zone") (cf. e.g. Öchslin 1950). The altitudinal interval between the forest limit and the tree species or lethal (physiologic) limit is here considered as the *timberline ecotone* (cf. Figure 1, which shows also the general altitudinal zonation of Swiss forest vegetation). The structure of the upper forest limit in the Alps is frequently characterized by such a transition zone, with a possible extension of 300 or more metres of altitude (cf. e.g. Landolt 1984 or Ozenda 1985). The timberline ecotone often consists of dwarf shrub heath or of populations of *Pinus mugo* ssp. *mugo* or *Alnus viridis*. In this context the question is discussed whether the forest limit and the tree limit under natural conditions would coincide. Opinions on this question are contradictory. For example Scharfetter (1938); Schiechtl (1967); Nägeli (1969); Kral (1973) and Ellenberg (1996) suppose that the forest limit and the tree limit under natural conditions would coincide and form a closed boundary. This point of view was substantiated notably by Klötzli (1991, 1993) in certain landscapes of the earth which are still anthropogenically unaffected, e.g. the Argentinean Nahuel Huapi National Park, the west coast of the Southern island of New Zealand (Westland) and some steep fjords of the Norwegian coast where the forests reach up to the tree limit and form a *closed* boundary. Dense forests which have been regarded as very close to a natural state and reach up to the natural forest limit were even found in the Central Alps by Nägeli (1969) in the Sertigtal, by Schiechtl (1967) in the Kaunertal and Matschertal, and by Tranquillini (1979) in the Radurscheltal. According to Holtmeier (1985) there are, however, in many anthropogenically unaffected mountain regions also transition zones between the natural forest limit and the tree limit (e.g. in some Central Asiatic mountain regions and notably in many mountain regions of Western and Northern America). A natural structure of the forest limit as one can expect to find in the Swiss Central Alps can be seen for example in the Southern Canadian Rocky Mountains (Banff) or in the Japanese Alps (Hondschu, Hokkaido, own observations and investigations by Yanagimachi & Ohmori 1991). Depending on local climatic differences, the relief, the recent climate history and the physiological properties and needs of the trees at the forest limit, both a forest limit which coincides with the tree limit, or a transition zone between the forest limit and the tree limit is possible (Klötzli 1991, 1993; Holtmeier 1993).

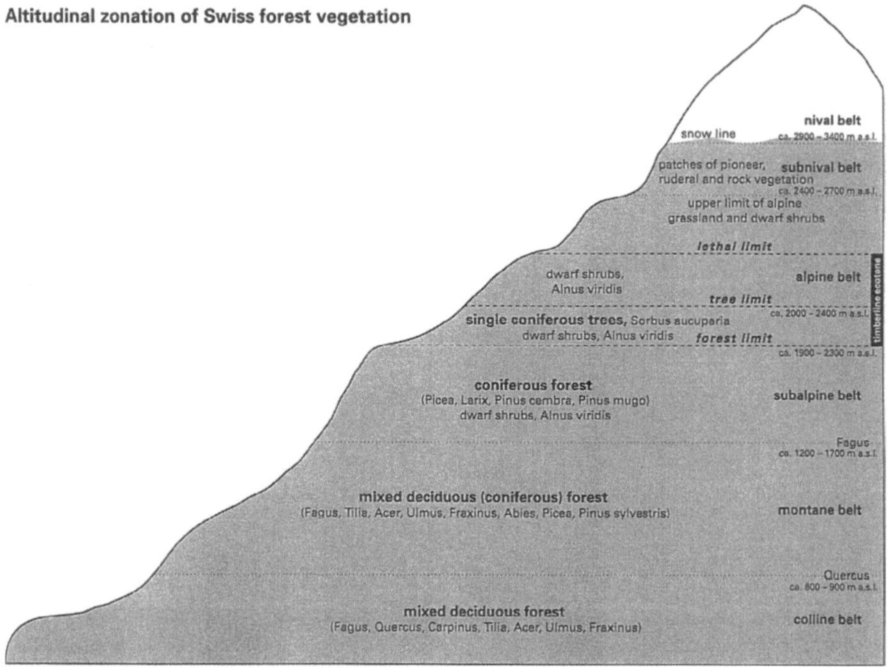

Figure 1. Altitudinal zonation of Swiss forest vegetation (scheme).

There are various possibilities to define the approximate position of the potential natural forest limit in areas with anthropozoogenically caused forest limit depressions. Usually the upper limit of dense *Rhododendron ferrugineum* populations is used as a reference point (cf. e.g. Eblin 1901; Hager 1916; Nägeli 1969; Landolt 1984). Further indicators are mentioned by Holtmeier (1974): relic trees far above the present forest limit; tree stumps and trunks above the present forest limit which bear marks of chopping and cutting; traditions and field names indicating areas which were previously forested (cf. e.g. Rikli 1909; Furrer 1955, 1957; Mayer 1955). A further possibility has been suggested by Hager (1916) using forest flowering plants such as *Luzula sylvatica* or *Pyrola minor* as potential forest indicators. A first map of the forest limits of Switzerland was published by Imhof (1900). A more modern cartographic overview of the altitude of the potential natural forest limit in Switzerland based on observations of the highest located trees and groups of trees has been given by Landolt (1984). This map indicates the potential natural forest limit in the Prealps at around 2'000 metres, in the Central Alps between 2'200 and 2'400 metres and in the Southern Alps between 2'000 and 2'200 metres. The higher average forest limits in the Central Alps compared with the Northern and Southern Alps can be explained by the "Massenerhebungseffekt" ("mass elevation effect") (cf. e.g. Brockmann-Jerosch 1919). The latter is caused by the higher average altitude which leads to higher temperatures due to the higher thermic capacity during the day and the summer. The low precipitation in the Central Alps, reducing the duration of snow cover, and increasing radiation due to the lower average proportion of cloud cover also have a beneficial effect. Furthermore, *exposure* significantly influences the altitude of the potential natural forest limit: since south-exposed slopes have higher radiation sums and a higher heat retention capacity than north-exposed ones, their tree limits and forest limits mostly are clearly higher (according to Landolt 1984 the difference is approx. 100 metres of altitude).

2. Study area (Eastern and Southern Swiss Alps)

The area studied to investigate the timberline ecotone of the Eastern and Southern Swiss Alps is shown in Figure 2. The following six sites have been included in this study:

1. Weisstannen valley (Canton of Saint Gallen)/Prealps (ca. 1'700 to ca. 2'200 metres a.s.l.; Perret 2001)
2. Upper Rhine valley (Canton of Grisons)/Northern Central Alps (ca. 1'600 to ca. 2'300 metres a.s.l.; Schuhmacher 1998)
3. Morteratsch valley (Canton of Grisons, Upper Engadine Valley)/Central Alps (ca. 1'900 to ca. 2'500 metres a.s.l.; Fischer 1999)
4. S-charl valley (Canton of Grisons, Lower Engadine Valley)/Central Alps (ca. 1'800 to ca. 2'400 metres a.s.l.; Gelmi 1996)
5. Poschiavo valley, Val di Campo (Canton of Grisons)/Southern Central Alps (ca. 1'900 to ca. 2'400 metres a.s.l.; Schudel 1999)
6. Ticino Alps, Alpe di Gesero/Val Morobbia (Canton of Ticino)/Southern Alps (ca. 1'700 to ca. 2'200 metres a.s.l.; Caminada 1998)

The sites marked with a triangle in Figure 2 are considered as further modelling areas. Moreover, the ongoing projects "Ecocline" (Theurillat et al. 1996) and "FOREST" (FOREST conference 1997) dealing with the timberline ecotone in the Western Swiss Alps are indicated.
Table 1 gives an overview of the main physico-geographical parameters (geology, precipitation, continentality, altitude of the potential natural tree limit) and vegetational criteria like dominant tree species of the timberline ecotone, dominant plant communities of the upper subalpine and lower alpine belt in N- and S-exposure under anthropozoogenic and natural conditions.

3. Methods

3.1. GEOBOTANICAL (PLANT-SOCIOLOGICAL) MAPPING

The present vegetation of the timberline ecotone of the different investigation sites (cf. Figure 2) has been documented during the years 1994 - 2000 by plant-sociological mapping (scales 1:10'000 and 1:5'000). Furthermore, single trees within the timberline ecotone have been recorded according to e.g. Walder (1983). The mapped altitudinal intervals reach from ca. 300 - 500 metres below the potential natural tree limit up to the upper krummholz limit or up to the upper *Rhododendron ferrugineum, R. hirsutum* or *Alnus viridis* limit. The vegetation records mainly concentrate on a detailed mapping of subalpine forests, and subalpine and alpine dwarf shrub heath and grassland communities on the basis of associations. For each species, the performance was estimated by the usual Braun-Blanquet cover-abundance scale. Mapping of the trees includes all plants considered as trees according to the definition given above, i.e. trees which are higher than the average snow cover during winter.

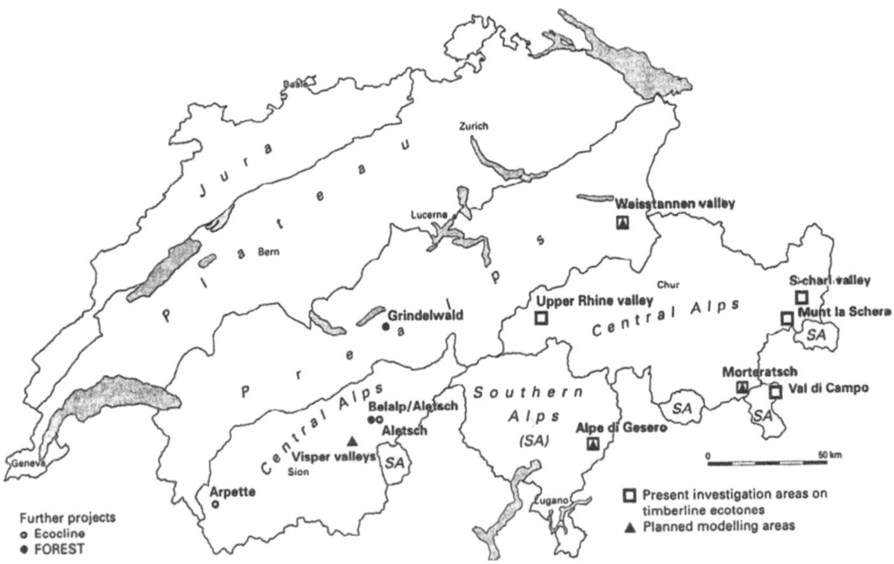

Figure 2. Investigation sites on Swiss Alpine timberline ecotones.

3.2. DENDROCHRONOLOGY

Climatic variations in the past are reflected in the age pattern, growth form and rate of increment of the trees. Dendrochronological investigations at the upper Alpine tree limit are therefore very useful to establish past conditions near this limit. Subfossil wood found above the current tree limit constitutes a record of range-limit dynamics and climatic variability throughout the Holocene.

Table 1. Overview of the investigated sites.

Locality / **Criteria**	**Weisstannen valley**	**Upper Rhine valley**	**Morteratsch valley**	**S-charl valley**	**Poschiavo valley (Val di Campo)**	**Ticino Alps (Alpe di Gesero)**
Geology	mainly mixed rocks (Flysch), particularly rather acid rocks (Verrucano)	mainly plutonic rocks: gneiss, particularly Verrucano	plutonic rocks: mainly granodiorites, granites and diorites	plutonic rocks (granites) and basic rocks (limestones)	plutonic rocks: mainly granites and amphibolithes	mainly plutonic rocks: granites, gneiss
Approx. precipitation in 1'800 m a.s.l.[1]	ca. 2'200 mm	ca. 1'800 mm	ca. 1'000 mm	ca. 800 mm	ca. 1'400 mm	ca. 2'200 mm
Continentality angle after Gams (1931)	ca. 39°	ca. 45°	ca. 61°	ca. 66°	ca. 52°	ca. 39°
Average altitude of the potential natural treeline	ca. 2'000 m	ca. 2'150 m	ca. 2'400 m	ca. 2'350 m	ca. 2'300 m	ca. 2'150 m
Altitudinal difference of the potential natural treeline between S- und N-exposed slopes	ca. 100 m	more or less no differences	more or less no differences	ca. 100 m	more or less no differences (20 m)	ca. 50 m
Dominant tree species of the timberline ecotone	spruce (mainly stunted forms), in N-exposure also *Sorbus aucuparia* potential: sporadic Swiss stone pine	spruce potential: also Swiss stone pine and sporadic larch	Swiss stone pine, sporadic larch	on acid rocks: Swiss stone pine and larch on limestones: Mountain Pine	larch, sporadic Swiss stone pine	mainly spruce, particularly larch, sporadic *Sorbus aucuparia*
Dominant plant communities of the upper subalpine belt: N-exposure (a: anthropozoogenic; b: potential)	a: Nardion, Poion alpini, Rhododendro-Vaccinietum, Empetro-Vaccinietum b: Alnetum viridis, Vaccinio-Piceion, Rhododendro-Vaccinietum	a: Rhododendro-Vaccinietum, Empetro-Vaccinietum, Nardion b: Vaccinio-Piceion, Alnetum viridis, Larici-Pinetum cembrae	a: Rhododendro-Vaccinietum, Nardion b: Larici-Pinetum cembrae	a (acid): Rhododendro-Vaccinietum, Junipero-Arctostaphyletum (upper part) b (acid): Larici-Pinetum cembrae, Festucetum variae; (basic): Erico-Pinion mugo	a: Nardion, particularly Poion alpini b: Larici-Pinetum cembrae, Rhododendro-Vaccinietum, Nardion	a: Rhododendro-Vaccinietum, Alnetum viridis, Nardion b: Junipero-Laricetum, Alnetum viridis

[1] After Atlas der Schweiz, sheet 12 - sums of precipitation 1901-1940, and other sources (cf. References)

Table 1 continued

Locality / Criteria	Weisstannen valley	Upper Rhine valley	Morteratsch valley	S-charl valley	Poschiavo valley (Val di Campo)	Ticino Alps (Alpe di Gesero)
Dominant plant communities of the upper subalpine belt: S-exposure (a: anthropozoogenic; b: potential)	a: Nardion, Poion alpini, Junipero-Arctostaphyletum callunetosum b: Vaccinio-Piceion	a: Poion alpini, Nardion b: Vaccinio-Piceion, Larici-Pinetum cembrae, Festucetum variae	a: Junipero-Arctostaphyletum b: Larici-Pinetum cembrae	a (basic): no anthropozoogenic plant communities; (acid): Junipero-Arctostaphyletum b (basic): Erico-Pinion mugo; (acid): Larici-Pinetum cembrae	a: Nardion, larch-forest pastures b: Larici-Pinetum cembrae, Rhododendro-Vaccinietum, Nardion, particularly Festucetum variae	a: Junipero-Arctostaphyletum, Nardion b: Junipero-Laricetum, Vaccinio-Piceion, Festucetum paniculatae
Dominant plant communities of the lower alpine belt: N-exposure (a: anthropozoogenic; b: potential)	a: Nardion, Caricetum ferrugineae, Poion alpini, Rhododendro-Vaccinietum (lower part) b: Caricetum ferrugineae, Rhododendro-Vaccinietum (lower part), Alnetum viridis (lower part)	a: Empetro-Vaccinietum, Rhododendro-Vaccinietum (lower part) b: no investigated area without anthropogenic impact	a: Nardion, Caricetum sempervirentis b: Nardion	a + b (acid): Caricetum curvulae, Junipero-Arctostaphyletum (lower part) b (basic): not recorded	a: no anthropozoogenic plant communities in the investigated area b: Empetro-Vaccinietum, particularly Rhododendro-Vaccinietum (lower part), Cetrario-Loiseleurietum (upper part), Caricetum curvulae (upper part)	Only the lower part was investigated a: no anthropozoogenic plant communities in the investigated area b: Festucetum variae
Dominant plant communities of the lower alpine belt: S-exposure (a: anthropozoogenic; b: potential)	a: Nardion, Poion alpini, Junipero-Arctostaphyletum callunetosum (lower part) b: Festucetum violaceae, Junipero-Arctostaphyletum callunetosum (lower part)	Only the lower part was investigated a: Nardion b: Empetro-Vaccinietum, Festucetum variae	a: Nardion, Junipero-Arctostaphyletum (lower part) b: Nardion, Junipero-Arctostaphyletum (lower part), Empetro-Vaccinietum	a + b (acid): Caricetum curvulae, Festucetum variae, Junipero-Arctostaphyletum (lower part) b (basic): Elynetum, Caricetum firmae	a: no anthropozoogenic plant communities in the investigated area b: Festucetum variae, Junipero-Arctostaphyletum (lower part), particularly Festucetum violaceae	Only the lower part was investigated a: no anthropozoogenic plant communities in the investigated area b: Festucetum paniculatae, Caricetum curvulae, Festucetum variae

Different dendroecological studies have been carried out in Alpine areas to investigate the relationship between climate and tree ring width and wood density (cf. e.g. Kienast 1985; Lingg 1986; Cherubini 1993, 1996). In the Swiss Alps, long tree ring chronologies spanning over several hundred to several thousands of years were published by Bircher (1982); Renner (1982); Schär & Schweingruber (1987); Zumbühl & Holzhauser (1988); Holzhauser & Zumbühl (1996) and Holzhauser (1997).

3.3. PALAEOECOLOGY

There are different approaches to investigate former tree limits and forest limits using pollen and plant macrofossil analyses. Peat bogs located at the upper limit of the timberline ecotone are valuable archives in most cases. The analysis of surface samples (moss cushions, cf. e.g. Welten 1950; Heim 1970 or Burga 1984) and pollen analyses showed that the *amount of non-arboreal pollen* (NAP-value) of moss cushions close to the investigated peat bog is usually higher in the case of forested than of non-forested areas. This allows the calculation of a *"critical NAP-value"* for the altitude of the forest limit. Changes of the NAP-values then allow the evaluation of past positions of the forest limit. Nevertheless, the NAP-values for the altitude of the forest limit can differ depending on the vegetation above the forest limit (open tree areas, populations of *Alnus viridis* or *Pinus mugo* ssp. *mugo*, dwarf shrub heath or alpine grassland). For the evaluation of the altitude of former forest limits it is therefore necessary first to determine the potential natural vegetation and thus also the potential natural forest limit in the vicinity of each investigated peat bog. In addition, surface samples (pollen content in samples of moss cushions) have to be analyzed to establish characteristic NAP-values and pollen spectra for the different vegetation units (cf. e.g. Jochimsen 1986 or the study at the San Bernardino Pass area by Burga 1984).

Absolute pollen analyses including pollen concentrations (pollen grains per cm^3) or influx values (pollen grains per cm^2 and year) can also give evidence of the altitude of former forest limits. It is generally accepted that higher pollen concentrations and influx values of the trees forming the forest limit indicate higher forest limits, due to the higher pollen production of the trees at the forest limit (mainly mountain pine and spruce, partly Swiss stone pine but not larch) in comparison to most of the plants above the forest limit. Like NAP-values, absolute pollen analyses can only be used for the determination of former forest limits if anthropogenic influences on the vegetation close to the investigated peat bogs can be excluded. Just like climatically caused depressions of the forest limits, forest clearings usually result in higher NAP-values and lower pollen concentrations and pollen influx values of the trees forming the forest limit.

Macrofossil finds of trees (e.g. wood, needles, seeds, conifer cones, stomata etc.) forming the forest limit and the tree limit provide a further possibility to determine the position of former forest limits (cf. e.g. Ammann & Wick 1993; Tinner & Ammann 1996; Bauerochse & Katenhusen 1997; Burga & Perret 1998). In particular, pollen samples may contain *stomata*. As soon as the latter are found in pollen samples, it can be concluded that trees were present close to the investigated peat bog area. In most cases it can be decided, based on the stomata frequency, whether the peat bog was surrounded by individual trees, groups of trees or by a forest. Stomata and macrofossils are mainly found in limnic sediments (thereby rapidly reaching the anaerobic zone), whereas they are usually completely decomposed in slowly growing peat bogs in the forest limit range (cf. e.g. Lang 1994). The absence of stomata in peat bog sediments does *not in all cases* reflect a lower tree limit and forest limit in relation to the altitude of the investigated peat bog, since complete decomposition of the conifer needles may have occurred.

4. First results and future perspectives

The timberline ecotone investigations so far include six selected areas of the Eastern and Southern Swiss Alps with different ecological boundary conditions (cf. Figure 2). In a first step, status quo studies carried out by students (see above and the References) as theses or doctoral theses should provide results as to the "facies" and the variability of the timberline ecotones (cf. Introduction and Table 1). In all studies the present vegetation and plant communities, the occurrence of the uppermost single trees and/or stunted forms have been mapped. Moreover, life forms, plant biodiversity and geomorphological processes have been included.
The comparison of the six investigated areas in the Eastern and Southern Swiss Alps has provided the following main results (cf. Table 1):

- The main factors influencing the timberline are radiation/altitude/summer main temperatures, topography/exposure, water supply of the soil and anthropozoogenic impact.
- The continentality angle seems to be directly connected with the altitude of the potential natural tree limit. But compared with the Northern Alps, in the Southern Alps the tree limit is located higher at the same continentality angle. This is because the continentality angle depends exclusively on the precipitation sums and the altitude (cotangent of the annual precipitation sums divided by the altitude) and therefore only indirectly on the temperature. Thus, the higher Southern Alpine tree limit depends on the higher average air temperature during the growing season at the same altitude compared with the Northern Alps.
- The altitudinal difference of the potential natural tree limit between S- and N-exposed slopes is most distinct in clear oceanic or continental localities (Weisstannen and S-charl valley, Alpe di Gesero/Ticino Alps, cf. Table 1). The difference varies between 50 and 100 metres. The altitudinal difference of the potential natural tree limit between S- and N-exposed slopes in the oceanic areas could be due to the dominating position of *Picea abies* in the timberline ecotone. In contrast to more continental areas where *Pinus cembra* and *Larix decidua* dominate in the forest limit range, *Picea abies* seems to react to the different radiation sums in the various exposures. So, for *Picea abies* the solar radiation during the growing season represents probably an important factor for tree growth. Besides, also geomorphological parameters like slope inclination and stability, rock fall activity, snow avalanches etc. locally strongly influence the altitude and the pattern of the timberline ecotone.
- Dominant trees of the investigated timberline ecotones are *Picea abies* (especially Prealps and Southern Alps), *Larix decidua, Pinus cembra* and *Pinus mugo* s.l.

(Central Alps). Less frequent are *Pinus sylvestris, Abies alba, Sorbus aucuparia, S. aria, Salix appendiculata, Betula pubescens, B. pendula, Acer pseudoplatanus* and *Populus tremula.* Among the deciduous trees, *Sorbus aucuparia*, which occurs from the colline belt up to the tree limit, has to be especially mentioned. *Sorbus aucuparia* is most abundant in oceanic areas of the Prealps (e.g. Weisstannen valley) and the Southern Alps (e.g. Alpe di Gesero), but can also be found in the Central Alps, mainly in moist places where the Alnetum viridis occurs. *Pinus mugo* s.l. grows as pioneer mainly on protalus ramparts and debris cones. On "better sites" it is replaced by *Picea abies, Larix decidua* or *Pinus cembra.*

- The sharp change of lithology from carbonate to siliceous rocks can lead to a distinct change of the timberline ecotone type from e.g. the Erico-Pinetum mugo to the Larici-Pinetum cembrae (cf. e.g. S-charl valley).
- Among the dominant plant communities a shift can be recorded between the more oceanic Northern and Southern Alps and the continental Central Alps (cf. Eggenberg 1994) due to the different precipitation regime and snow cover: While in the Northern and Southern Alps the dwarf shrub communities Empetro- and Rhododendro-Vaccinietum do not only occur on N-exposed slopes, but partly also in E- and W-exposures, they are restricted to N-exposures in the Central Alps. The same is to mention for the Alnetum viridis. On the other side, the Junipero-Arctostaphyletum typicum is only found in the Central Alps on S-exposed slopes and is absent in the Northern and Southern Alps (however here we have the subassociation callunetosum that occurs on less extreme locations).
- The dominant plant communities and the snow cover of the upper subalpine belt influence clearly the possibility of the establishment of trees or forest. Thus e.g. the Junipero-Arctostaphyletum (callunetosum) and Empetrum-Vaccinietum dwarf shrub communities as well as the Festucetum violaceae and the Hypochoeris uniflora-Nardion strictae grass communities provide suitable seed beds for tree growth, in contrast to a dense grass cover of e.g. *Calamagrostis villosa* (Vaccinio-Piceetum calamagrostietosum), *Nardus stricta* (Aveno-Nardetum) or *Festuca varia* (Festucetum variae) or a dense tall herb cover (Adenostylo-Cicerbitetum, Alnetum viridis).
- Where *Picea abies* dominates at the forest limit (Northern and Southern Alps, oceanic part of the Central Alps), the upper limits of the *Rhododendron ferrugineum-* and *Alnus viridis*-populations in N- and NE-exposures are often significantly higher than the climatic forest limit. Thus the former theory that the upper limits of the *Rhododendron ferrugineum-* and *Alnus viridis*-populations correspond to the potential natural forest limit is not valid for these regions.
- The potential natural timberline ecotone is expected to be very narrow (only 0 - 50 metres of altitude), at least on more or less regularly inclined slopes without creeping snow, snow slides, avalanches, debris flows etc. The trees near the potential natural tree limit are clearly smaller and often show clusters of stunted trees by layering *(Picea abies)* or by seed deposits of *Nucifraga caryocatactes.*

- Present observations show, compared with ancient photographs and vegetation maps (e.g. Upper Rhine valley, Hager 1916), for most regions of the Alps increased forest surfaces. This is mainly due to artificial reforestation and quick reforestation of abandoned subalpine pastures (cf. e.g. Alpe di Gesero) and only in exceptional cases to the rise of the temperatures since the end of the Little Ice Age around 1850.

Which reactions of the tree limit to global warming can be expected?
Two main problems can be distinguished in this sense:

1. Distinction between anthropozoogenic impacts and natural effects on the timberline ecotone. Among the natural effects, positive and negative feedbacks can be expected. One negative feedback could be due to ice melting and permafrost recession which might entail slope destabilisation and subsequently increasing erosion of potential forest sites. Besides, tree growth (reforestation) in more structured areas with warmer climate initially only can occur on favourable places, such as ridges, undulations and other convex surfaces. The formation of stunted tree zones thereby often can be observed in areas with spruce forming the forest limit (regarding the formation and structure of stunted tree zones cf. Kuoch & Amiet 1970). The stunted trees are able to improve locally the bad microclimatic conditions, and tree growth can gradually advance. As a result of snow cover over a long period, creeping snow, fungus attack, soils or vegetation communities unfavourable for tree regeneration, low seed production and propagation of the trees forming the tree and the forest limit and other influences, however, many areas remain unforested for a long time. Only if the climatic conditions stay more or less stable over an extended period, the establishment of a "closed" forest limit can be expected. But it should always be taken into account that the topographic variability may also lead to an open forest limit with an upper transition zone during longer periods of climatic stability (e.g. concave surfaces are rather moist or wet and therefore unsuitable sites for reforestation). Moreover, the increased atmospheric CO_2 concentration, the increased nitrogen emissions, the effects of the tropospheric O_3 and heavy metal dust must be considered. According to newer results, the first two effects would lead to an additional increase of tree growth (cf. e.g. Innes 1991; Häsler 1992; Beniston & Innes 1998), in contrast to the last two factors. It can be expected that at the transition zone between the oceanic Prealps and the continental Central Alps (e.g. Upper Rhine valley) the first signs of vegetation change due to global warming are recognized, i.e. under moister conditions, oceanic plant species would shift to the Central Alps. A climate-induced vegetation change would first be conspicuous on extreme exposed, i.e. sensitive sites. The local conditions of small sites are crucial for the successful establishment of trees.

2. Knowledge of the time-lag of vegetation reaction or the rates of vegetation changes related to climatic change. Are they within a time-scale of several decades or centuries? Holocene vegetation and climate history/palaeoecology provide information on the magnitude of such processes (cf. Burga & Perret 1998). During the Holocene, the Swiss Alpine tree limit fluctuated about ± 100 metres related to the present potential position. But as to the reaction time of vegetation and thus also of the tree limit after a climatic change there is more or less no information. A rise of the tree limit above its present potential natural altitude can only be expected

under warmer and moister climatic conditions, similar to e.g. the Mid-Holocene temperature optimum (cf. Burga 1988; Burga & Perret 1998).

For further research, it is planned to use geographical information systems (GIS) to put together data of the mentioned vegetation and tree mappings, dendrochronological investigations and additional studies (e.g. solar radiation or seed dispersal within and above the timberline ecotone) of the six investigated sites in the Swiss Alps. For example, these data will be used for models simulating shifts of the tree limit and the forest limit under different climate scenarios in the Swiss Alpine test regions. Monitoring of these test regions will help to check the modelling results and to ascertain further climatically induced changes (mainly extreme climatic events) of the tree and forest limit.

Acknowledgements

We are grateful to the Swiss Science Foundation for the support of the timberline project in the Eastern Swiss Prealps (PhD thesis of R. Perret, project number 3100-039296.93).

References

Ammann, B. & Wick, L. 1993. Analysis of fossil stomata of conifers as indicators of the alpine tree line fluctuations during the Holocene. pp. 175-185. In: Frenzel, B. (ed), Oscillations of the Alpine and Polar Tree Limits in the Holocene. Paläoklimaforschung 9. Fischer, Stuttgart.

Aulitzky, H. 1963. Bioklima und Hochlagenaufforstung in der subalpinen Stufe der Inneralpen. Schweizerische Zeitschrift für Forstwesen 114: 1-25.

Aulitzky, H. & Turner, H. 1982. Bioklimatische Grundlagen einer standortsgemässen Bewirtschaftung des subalpinen Lärchen-Arvenwaldes. Mitteilungen der Eidgenössischen Anstalt für das forstliche Versuchswesen 58: 327-580.

Bachmann, P. 1964. Die heutige Waldgrenze im Goms. Institut für Orts-, Regional- und Landesplanung der ETH Zürich.

Baig, M.N., Tranquillini, W. & Havranek, M. 1974. Cuticuläre Transpiration von *Picea abies*- und *Pinus cembra*-Zweigen aus verschiedener Seehöhe und ihre Bedeutung für die winterliche Austrocknung der Bäume an der alpinen Waldgrenze. Centralblatt für das gesamte Forstwesen 91: 195-211.

Baig, M.N. & Tranquillini, W. 1980. The effects of wind and temperature on cuticular transpiration of *Picea abies* and *Pinus cembra* and their significance in desiccation damage at the alpine timberline. Oecologia 47: 252-256.

Bauerochse, A. & Katenhusen, O. 1997. Holozäne Landschaftsentwicklung und aktuelle Vegetation im Fimbertal (Val Fenga, Tirol/Graubünden). Phytocoenologia 27: 353-453.

Beniston, M. & Innes, J.L. (eds) 1998. The Impacts of Climate Variability on Forests. Lecture Notes in Earth Sciences 74.

Binz, A. & Heitz, C. 1990. Schul- und Exkursionsflora für die Schweiz. 19th ed. Schwabe & Co., Basel.

Bircher, W. 1982. Zur Gletscher- und Klimageschichte des Saastales. Physische Geographie 9: 1-233.

Brockmann-Jerosch, H. 1919. Baumgrenze und Klimacharakter. Beiträge zur geobotanischen Landesaufnahme 6.

Burga, C.A. 1984. Aktuelle Vegetation und Pollengehalt von Oberflächenproben der obermontanen bis subalpinen Stufe am Bernhardin-Pass (Graubünden/Schweiz). Jahresberichte der Naturforschenden Gesellschaft Graubünden 101: 53-99.

Burga, C.A. 1988. Swiss vegetation history during the last 18'000 years. New Phytologist 110(4): 581-602.

Burga, C.A. & Perret, R. 1998. Vegetation und Klima der Schweiz seit dem jüngeren Eiszeitalter. Ott, Thun.

Caminada, M. 1998. Vegetationsgeographische Untersuchungen zum aktuellen Zustand des Waldgrenz-Ökotons und seiner Dynamik in den insubrischen Alpen (Val Morobbia). (Unpublished thesis University of Zurich.)

Cherubini, P. 1993. Studio dendrocronologica su *Pinus pinea* L. in due differenti stazioni sulla costa mediterranea in Toscana (Italia). Dendrochronologia 11: 87-99.

Cherubini, P. 1996. Spatiotemporal growth dynamics and disturbances in a subalpine spruce forest in the Alps: a dendrochronological reconstruction. Canad. J. of Forest Res. 26: 991-1001.

Däniker, A. 1923. Biologische Studien über Baum- und Waldgrenze, insbesondere über die klimatischen Ursachen und deren Zusammenhänge. Vierteljahrsschr. Naturf. Ges. Zürich 68: 1-102.

Eblin, B. 1901. Die Vegetationsgrenzen der Alpenrose als unmittelbare Anhalte zur Festsetzung früherer bzw. möglicher Waldgrenzen in den Alpen. Schweizerische Zeitschrift für Forstwesen 52: 133-157.

Eggenberg, S. 1994. Dynamik und Vegetation an der Waldgrenze und die Abhängigkeit ihrer Zusammensetzung vom Klima. Doctoral thesis University of Bern.

Ellenberg, H. 1996. Vegetation Mitteleuropas mit den Alpen. 5th ed. Ulmer, Stuttgart.

Fischer, T. 1999. Waldgrenzökoton und Wiederbewaldungsdynamik im Gebiet des Morteratschgletschers. (Unpublished thesis University of Zurich.)

FOREST-Infotagung 1997. Proceedings of the conference held at Bern (Switzerland) on 7 November 1997, incl. brochure.

Furrer, E. 1955. Probleme um den Rückgang der Arve in den Schweizer Alpen. Mitteilungen der Schweizerischen Anstalt für das forstliche Versuchswesen 31: 669-705.

Furrer, E. 1957. Das Schweizerische Arvenareal in pflanzengeographischer und forstgeschichtlicher Sicht. Ber. Geobot. Inst. Rübel Zürich 29: 16-23.

Gelmi, D. 1996. Vegetationsgeographische Untersuchungen zur Waldgrenze im Unterengadin (Val S-charl). (Unpublished thesis University of Zurich.)

Gross, M. 1989. Untersuchungen an Fichten der alpinen Waldgrenze. Dissertationes Botanicae 139.

Hager, K. 1916. Verbreitung der wildwachsenden Holzarten im Vorderrheintal (Kt. Graubünden). Erhebungen über die Verbreitung der wildwachsenden Holzarten in der Schweiz. 3. Lieferung. Stämpfli, Bern.

Häsler, R. 1992. Beobachtungen in den Aufforstungen "Stillberg" bei Davos. Eine Übersicht über mögliche Auswirkungen des Treibhauseffekts auf Gebirgswälder. Bündnerwald 5/92: 40-46.

Heim, J. 1970. Les relations entre les spectres polliniques récents et la végétation actuelle en Europe occidentale. Univ. Louvain, Lab. Palyn. et Phytosoc.

Holtmeier, F.-K. 1967. Zur natürlichen Wiederbewaldung aufgelassener Alpen im Oberengadin. Wetter und Leben 19: 195-202.

Holtmeier, F.-K. 1974. Geoökologische Beobachtungen und Studien an der subarktischen und alpinen Waldgrenze in vergleichender Sicht. Erdwissenschaftliche Forschung 8.

Holtmeier, F.-K. 1985. Die klimatische Waldgrenze - Linie oder Übergangssaum (Ökoton). Ein Diskussionsbeitrag unter besonderer Berücksichtigung der Waldgrenzen in den mittleren und hohen Breiten der Nordhalbkugel. Erdkunde 39: 271-285.

Holtmeier, F.-K. 1986. Die obere Waldgrenze unter dem Einfluss von Klima und Mensch. Abhandlungen des Westfälischen Museums für Naturkunde 48: 395-412.

Holtmeier, F.-K. 1987. Der Baumwuchs als klimaökologischer Faktor an der oberen Waldgrenze. Münstersche Geographische Arbeiten 27: 145-151.

Holtmeier, F.-K. 1989. Ökologie und Geographie der oberen Waldgrenze. Berichte der Reinhold-Tüxen-Gesellschaft 1: 15-45.

Holtmeier, F.-K. 1993. Timberlines as indicators of climatic changes: problems and research needs. pp. 211-222. In: Frenzel, B. (ed), Oscillations of the Alpine and Polar Tree Limits in the Holocene. Paläoklimaforschung 9.

Holzhauser, H. 1997. Fluctuations of the Grosser Aletsch Glacier and the Gorner Glacier during the last 3200 years: new results. pp. 35-54. In: Frenzel, F., Boulton, G.S., Gläser, B. & Huckriede, U. (eds), Glacier Fluctuations during the Holocene. Paläoklimaforschung 24.

Holzhauser, H. & Zumbühl, H.J. 1996. To the history of the Lower Grindelwald Glacier during the last 2800 years - palaeosols, fossil wood and historical pictorial records - new results. Z. Geomorph. N.F., Suppl.-Bd. 104: 95-127.

Imhof, E. 1900. Die Waldgrenze in der Schweiz. Gerland's Beiträge zur Geophysik 4(3): 241-330.

Innes, J.L. 1991. High-altitude and high-latitude tree growth in relation to past, present and future global climate change. The Holocene 1(2): 168-173.

Jochimsen, M. 1986. Zum Problem des Pollenfluges in den Hochalpen. Dissertationes Botanicae 90.

Kienast, F. 1985. Tree ring analysis, forest damage and air pollution in the Swiss Rhone valley. Land Use Policy 2: 74-77.

Klötzli, F. 1991. Alpine Vegetation: stabil und natürlich? pp. 70-83. In: Müller, J.P. & Gilgen, B. (eds), Die Alpen - ein sicherer Lebensraum? Ergebnisse der 171. Jahresversammlung der SANW 1991 in Chur. Desertina, Disentis.

Klötzli, F. 1993. Dornpolster und Kissenpolster - zwei divergierende Adaptionen. pp. 155-162. In: Brombacher, C., Jacomet, S. & Haas, J.N. (eds), Festschrift Zoller. Dissertationes Botanicae 196.

Körner, C. 1999. Alpine Plant Life. Functional Plant Ecology of High Mountain Ecosystems. Springer, Berlin.

Köstler, J.N. & Mayer, H. 1970. Waldgrenzen im Berchtesgadener Land. Jahrbuch des Vereins zum Schutze der Alpenpflanzen und -tiere 35: 1-35.

Kral, F. 1973. Zur Waldgrenzdynamik im Dachsteingebiet. Jahrbuch des Vereins zum Schutze der Alpenpflanzen und -tiere 38: 71-79.

Kuoch, R. & Amiet, R. 1970. Die Verjüngung im Bereich der oberen Waldgrenze der Alpen mit Berücksichtigung von Vegetation und Ablegerbildung. Schweizerische Anstalt für das forstliche Versuchswesen 46: 161-342.

Landolt, E. 1984. Unsere Alpenflora. 5th ed. SAC, Zürich.

Lang, G. 1994. Quartäre Vegetationsgeschichte Europas. Fischer, Jena.

Larcher, W. 1957. Frosttrocknis an der Waldgrenze und in der alpinen Zwergstrauchheide auf dem Patscherkofel bei Innsbruck. Veröffentlichungen Ferdinandeum Innsbruck 37: 49-81.

Larcher, W. 1963. Zur spätwinterlichen Erschwerung der Wasserbilanz von Holzpflanzen an der Waldgrenze. Berichte des naturwissenschaftlichen medizinischen Vereins Innsbruck 53: 125-137.

Lingg, W. 1986. Dendroökologische Studien an Nadelbäumen im alpinen Trockental Wallis (Schweiz). Eidgenössische Anstalt für das forstliche Versuchswesen 287: 1-81.

Mayer, K.A. 1955. Frühere Verbreitung der Holzarten und einstige Waldgrenze im Kanton Wallis. IV. Oberwallis. Mitteilungen der Schweizerischen Anstalt für das forstliche Versuchswesen 31: 563-668.

Michaelis, P. 1934a. Ökologische Studien an der alpinen Baumgrenze. IV. Zur Kenntnis des winterlichen Wasserhaushaltes. Jahrb. Wiss. Botan. 80: 169-247.

Michaelis, P. 1934b. Ökologische Studien an der alpinen Baumgrenze. V. Osmotischer Wert und Wassergehalt während des Winters in den verschiedenen Höhenlagen. Jahrb. Wiss. Botan. 80: 337-362.

Nägeli, W. 1969. Waldgrenze und Kampfzone in den Alpen. HESPA 19: 1-44.

Öchslin, M. 1950. Der Kampfzonenwald. Mitteilungen der Naturforschenden Gesellschaft Schaffhausen 23: 273-279.

Ozenda, P. 1985. Die Vegetation der Alpen im europäischen Gebirgsraum. Fischer, Stuttgart, New York.

Perret, R. 2001. Geobotanische Untersuchungen zur heutigen und potentiell-natürlichen Struktur der Waldgrenzökotone im Weisstannental (östliche Schweizer Voralpen, Kt. St. Gallen). Doctoral thesis University of Zurich. (In prep.)

Pott, R. & Hüppe, J. 1992. Vegetationskundliche und pollenanalytische Studien zur oberen Waldgrenze im Fimbertal (Silvretta). Mitteilungsblatt der Hannoverschen Hochschulgemeinschaft 19: 49-76.

Renner, F. 1982. Beiträge zur Gletschergeschichte des Gotthardgebietes und dendroklimatologische Analysen an fossilen Hölzern. Physische Geographie 8.

Rikli, M. 1909. Die Arve in der Schweiz. Neue Denkschriften der Schweizerischen Naturforschenden Gesellschaft 44.

Schär, E. & Schweingruber, F. 1987. Nacheiszeitliche Stammfunde aus Grächen im Wallis. Schweizerische Zeitschrift für das Forstwesen 138: 497-515.

Scharfetter, R. 1938. Pflanzengesellschaften der Ostalpen. Wien.

Schiechtl, H.M. 1967. Die Physiognomie der potentiellen natürlichen Waldgrenze und Folgerungen für die Praxis der Aufforstung in der subalpinen Stufe. Mitteilungen der forstlichen Bundes-Versuchsanstalt Mariabrunn 75: 5-55.

Schudel, K. 1999. Waldgrenzökoton im Val da Camp. (Unpublished thesis University of Zurich.)

Schuhmacher, K. 1998. Das Waldgrenzökoton im Wandel. Vegetationsgeographische Untersuchungen im Vorderrheintal. (Unpublished thesis University of Zurich.)

Schwarz, R. 1983. Simulationsstudien zur Theorie der oberen Baumgrenze. Erdkunde 37: 1-11.

Theurillat, J.-P., Felber, F., Geissler, P., Guisan, A. & Gobat, J.-M. 1996. Le projet "Ecocline" et le programme prioritaire "Environnement". Bull. Murithienne 114: 151-162.

Tinner, W. & Ammann, B. 1996. Treeline Fluctuations Recorded for 12,500 Years by Soil, Pollen, and Plant Macrofossils in the Central Swiss Alps. Arctic and Alpine Research 28: 131-147.

Tranquillini, W. 1966. Über das Leben der Bäume unter den Grenzbedingungen der Kampfzone. Allgemeine Forstzeitung 77: 127-132.

Tranquillini, W. 1967. Über die physiologischen Ursachen der Wald- und Baumgrenze. Mitteilungen der forstlichen Bundes-Versuchsanstalt Mariabrunn 73: 457-487.

Tranquillini, W. 1974. Der Einfluss von Seehöhe und Länge der Vegetationszeit auf das cuticuläre Transpirationsvermögen von Fichtensämlingen im Winter. Berichte der Deutschen Botanischen Gesellschaft 87: 175-184.

Tranquillini, W. 1976. Water relations and alpine timberline. Ecological Studies 19: 473-491.

Tranquillini, W. 1979. Physiological ecology of the alpine timberline. Tree existence at high altitudes with special reference to the European Alps. Ecological Studies 31: 1-137.

Tuhkanen, S. 1980. Climatic parameters and indices in plant geography. Acta Phytogeographica Suecica 67.

Turner, H. & Blaser, P. 1977. Mikroklima, Boden und Pflanzen an der oberen Waldgrenze. Eidgenössische Anstalt für das forstliche Versuchswesen 173.

Walder, U. 1983. Ausaperung und Vegetationsverteilung im Dischmatal. Mitteilungen der Eidgenössischen Anstalt für das forstliche Versuchswesen 59: 81-212.

Wardle, P. 1968. Engelmann Spruce *(Picea engelmannii* ENGEL.) at its upper limits on the Front Range, Colorado. Ecology 49: 483-495.

Welten, M. 1950. Beobachtungen über den rezenten Pollenniederschlag in alpiner Vegetation. Ber. Geobot. Inst. Rübel Zürich 22: 48-56.

Yanagimachi, O. & Ohmori, H. 1991. Ecological status of *Pinus pumila* scrub and the lower boundary of the Japanese alpine zone. Arctic and Alpine Research 23: 424-435.

Zumbühl, H.J. & Holzhauser, H. 1988. Alpengletscher in der Kleinen Eiszeit. Sonderheft zum 125jährigen Jubiläum des SAC. Die Alpen 67: 129-322.

OBSERVED CHANGES IN VEGETATION IN RELATION TO CLIMATE WARMING

GABRIELE CARRARO[1], PIPPO GIANONI[1], ROBERTO MOSSI[1], FRANK KLÖTZLI[2] & GIAN-RETO WALTHER[2]

[1]Dionea SA, Office for forestry engineering, environmental consulting and landscape planning, CH-6600 Locarno, Switzerland; [2]Geobotanical Institute ETH, Zürichbergstrasse 28, CH-8044 Zürich, Switzerland

Keywords: Chorological tension, forest ecosystems, Insubria, laurophyllisation, neophytes, vegetation ecology

Abstract

The authors analyse the changes in forest vegetation in the north and south of the Alps by comparing 300 vegetation relevés of the forties and sixties with relevés of the nineties. The resulting changes in species composition and vegetation structure can be explained by various factors, one of which is the milder winter climate of the last three decades.

At these examples, the connection between similar bicolimates and the concept of the "chorological tension", the authors make some prognoses of laurophyllisation in western Eurasia.

Introduction

In 1987, forest stands with an amazing richness in laurophyllous species were found in the Canton of Ticino. This vegetation represents the most thermophilous forest type on common soils in the southern part of the Alps (Carraro & Gianoni 1987; Gianoni et al. 1988).

The investigation of a probable reference value of thermophilous forest structures, especially focusing on potential climatic changes, is subject of this project.

The research project is mainly based on three different topics:

- the comparison of old relevés from 1940/60 with the present situation on the northern and the southern side of the Alps
- the investigation of the phenomena on a regional and local scale (Locarnese)
- the comparison with similar ecosystems and the analysis and evaluation of literature.

The research project focuses mainly on the vegetation of the colline and submontane belt in the southern part of the Swiss Alps. In this area the elevated thermic gradient provoked a new type of vegetation with particular aspects of the laurophyllous type. In this sense the northern part of the Alps can be considered as a comparable example for a colder climate.

Burga & Kratochwil (eds.), BIOMONITORING, 195-205

The following questions were addressed:

- Do any natural forest formations exist, in which effects related to the warming of the atmosphere can be detected?
- To what degree can these effects be distinguished from other common phenomena of the development of vegetation under human influence?
- Which criteria should be used to set up a list of species or vegetation structures as bioindicators?
- What are the conclusions for land management, forest management and the reciprocal concessions considering socio-cultural aspects?

1. Methods

The principal work of this research programme consisted in the resurvey of existing phytosociological relevés in the northern and in the southern part of the Swiss Alps. Determining the influence of the climate in the natural context is an extremely complex task and can only be done approximately, taking into account the knowledge already gathered by other comparisons of old and new relevés in Switzerland (Kissling et al. 1988; Kissling 1989; Kuhn 1990; Werdenberg & Hainard 1989).
Nevertheless, a second level of analysis was applied by observing the distribution of thermophilous vegetation on a regional mesoclimatic scale in order to facilitate the description of the effect of one single factor. The temperature data of 10 meteorological stations, situated in typical sites on the slopes forming the Locarnese basin, were collected during the winter of 1994/95.
The comparison of typical climatic data of specific similar, just slightly warmer bioclimates with the Central European data - as presented by Klötzli (1988) - clearly shows the importance of the mean temperature and the winter temperature as limiting factors for some vegetation types.

2. Results

There are different scientific approaches to the topics vegetation change and climate warming, such as the construction of models based on palynological analysis, the application of geographic information systems or laboratory experiments with a physiological approach concentrated on particular species.
These theoretical approaches are justified because the diverse factors that interact in nature hardly provide useful data without the respective treatment.
On the other hand it is evident that these models can only be calibrated and ameliorated using direct observations on a 1:1-scale in the same ecosystem; these finally prove the predictions of a theoretical model.
The comparison of phytosociological relevés only allows a qualitative assessment for some examples of forest vegetation: little disturbed stands and stands close to the corresponding climax.

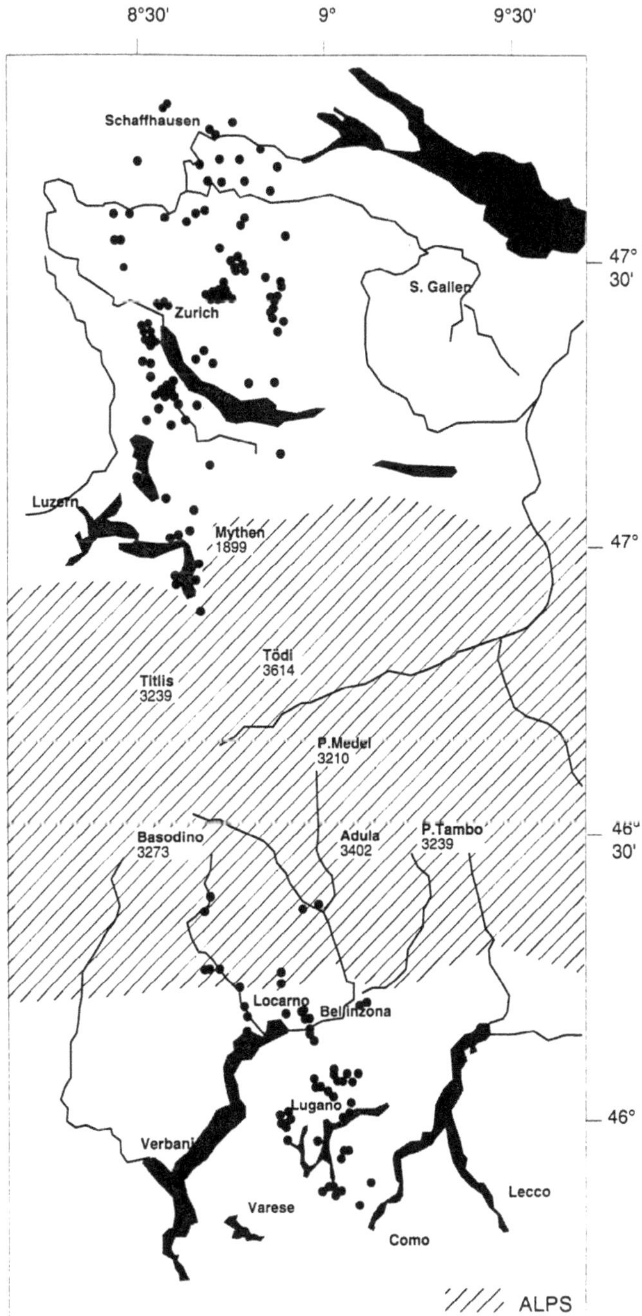

Figure 1. Distribution of the relevés on the northern and southern side of the Swiss Alps, resurveyed in 1994.

If the plants are subdivided into groups of species with different strategies, the composition tends to become less diverse, more shade-tolerant and eutrophic. This situation builds up new functional structures in both regions, the southern and the northern side of the Alps.
This can mainly be explained by the shading of the tree layer, by more fertile forest soils, which have been less intensively managed in this century (compared with past centuries), and by the parallel atmospheric nitrogen input.
The consistent reduction in the number of species and the various substitutions (internal dynamics) are characteristic; these phenomena seem to be more accentuated in forests further from their climax (succession).

- Establishing of climax (*Fagus, Aruncus, Carex sylvatica, Luzula, Calamagrostis, Prenanthes*) and/or nitrophilous species (*Acer pseudoplatanus, Geum, Circaea, Carex pendula, Lamium*)
- Reduced number of some mesophilous species, typical of interior forest gaps (*Fragaria, Festuca heterophylla, Salvia glutinosa, Aegopodium*) or resulting directly from management practices (*Quercus, Carpinus, Picea*)
- Strong reduction in acido- and heliophilous species (*Vaccinium, Carex pilulifera, Hieracium murorum, Pteridium, Solidago virga-aurea*)
- Increase in neophytes (*Robinia pseudacacia, Impatiens parviflora, Prunus serotina*)
- Slight decrease in montane species (*Abies, Lonicera alpigena, Sambucus racemosa, Astrantia, Galium rotundifolium, Saxifraga cuneifolia, Phyteuma scheuchzeri, Valeriana tripteris*)
- Increase in thermophilous climax species (*Tamus communis, Evonymus latifolia*)
- Increase in all evergreens (laurophyllous species) in higher degree in direct vicinity of the settlements up to a distance of 200-300 metres (*Cotoneaster div. spec., Lonicera div. spec., Prunus laurocerasus, Trachycarpus fortunei, Laurus nobilis*). Analogically, all the autochthonous evergreen species have established themselves in a higher colline belt (*Hedera, Taxus, Ilex, Vinca*).

According to literature the synusia of laurophyllous species must be considered as a structural element typical of the mature states of the studied formations and definitely stands for a degree of major maturity of the investigated forest phytocoenosis.
Synusia rich in diverse laurophyllous species, from shrubs to trees, have been found on the coast of the Biscayan Gulf (at a distance of ca. 1000 kilometres!) in a form almost identical to the one described from various parts of Insubria.
The phenomenon has an own functional peculiarity (winter photosynthesis) and a certain constancy in the various areas exposed to the same climate.
Even without the climate warming of the last decades a similar phenomenon could have occurred either with autochthonous species - as testified by stadia rich in *Taxus* and *Ilex* - or with exotic species as already testified in the beginning of the last century but limited to some parts of the warmest areas of Insubria.

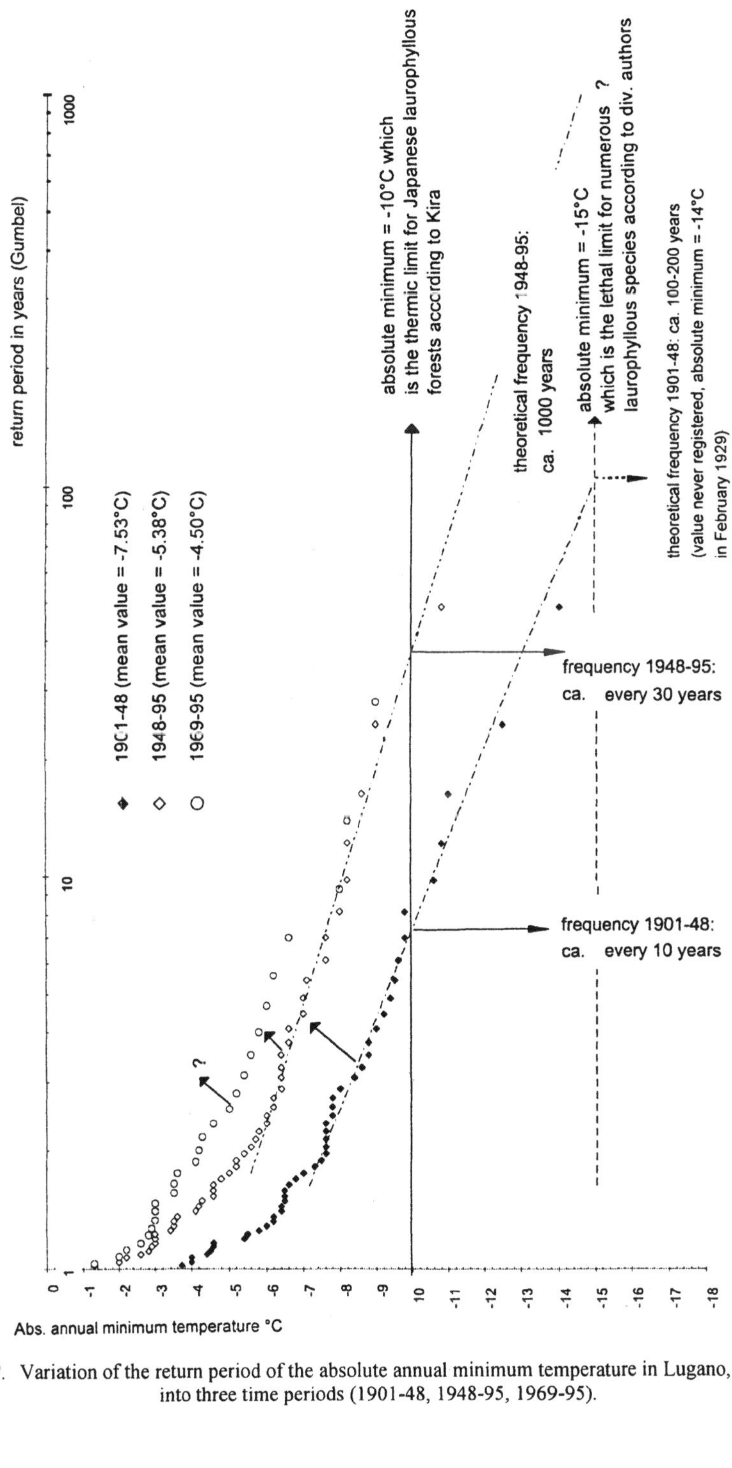

Figure 2. Variation of the return period of the absolute annual minimum temperature in Lugano, subdivided into three time periods (1901-48, 1948-95, 1969-95).

As an interpretative instrument of the structural change and the penetration by neophytes of the Central European forests, analogies and divergencies between the forest structure in situ or in other climatically analogous regions elsewhere in the world (Japan, China, south-western Caucasus, western Pyrenees) have been investigated.
This allows a first statement: the strong expansion of evergreen species observed in the understorey coincides with present structures in warmer to subtropical climates, which seem to be better qualified as transition zone towards the climax of Laurisilvae.
Taking into account the ecotonal areas which differentiate the above-mentioned formations from the deciduous biome, a principally thermic order ("mild winters") can be observed. Therefore, the phenomenon could also represent a connection with the on-going climatic modifications.
In fact, it had been shown in the Insubrian zone that the slight warming of the winter climate during the last decades may play a crucial role; for confirmation the local mesoclimate was analysed. The differences in days with frost, temperature minima and absolute minima of various forest stations coincide with the density and vitality of thermophilous evergreen species of an equal situation relative to chorology, fertility and distance to settlements.
The order of the deviations of the investigated climatic parameters is - in absolute and in relative terms - comparable with the change in mean January temperature (+ 1-2°C), number of days with frost (- 40%) and absolute minima (+ 2-3°C) as recorded in the second half of this century.
With a certain reliability it can be stated that in this context the massive spreading of synusia of laurophyllous species is a climate-dependent phenomenon.

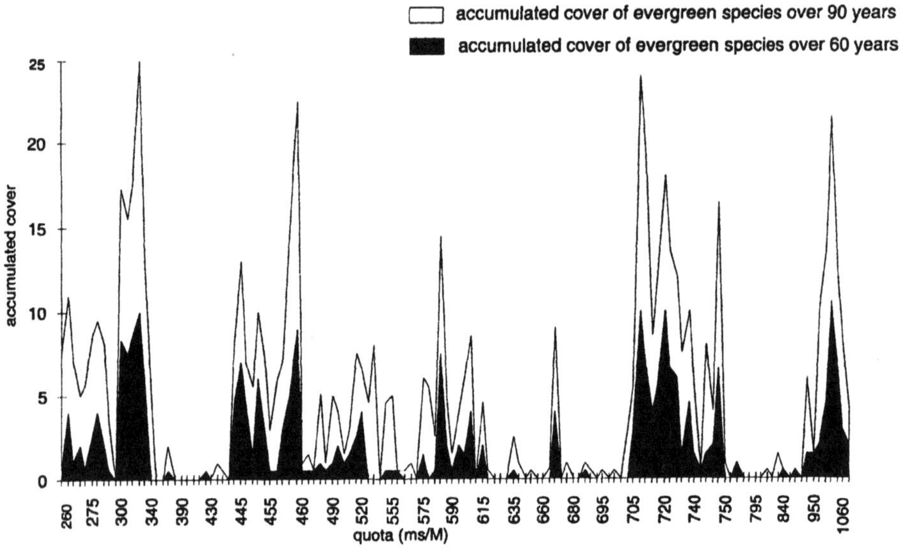

Figure 3. Variation of cover values of evergreen species on the southern side of the Swiss Alps (Hedera, Ilex, Taxus, Daphne laur., Ruscus, Vinca, Asplenium ad-ni.; Laurus, Ligustrum luc., Cotoneaster, Trachycarpus, Lonicera jap., Prunus laur., Niburnum) - 132 relevés.

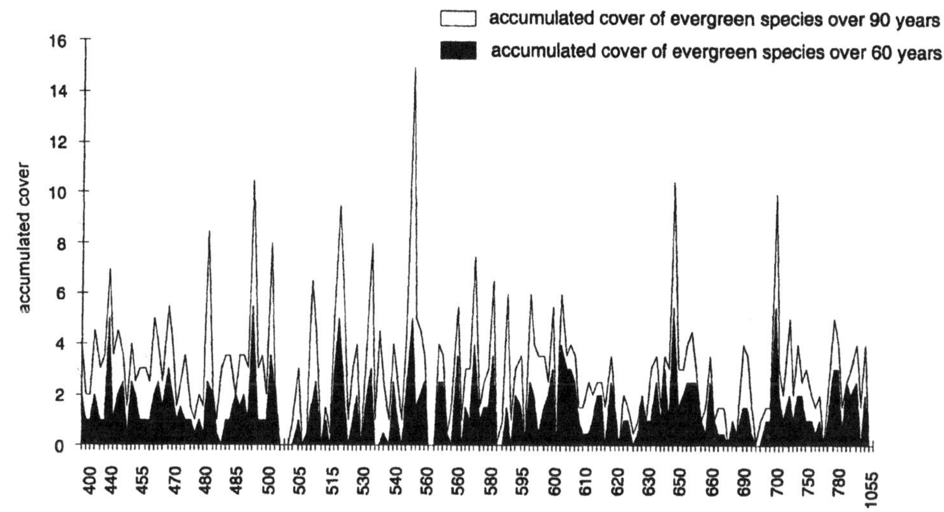

Figure 4. Variation of cover values of evergreen species on the northern side of the Swiss Alps (Hedera, Ilex, Taxus, Daphne laur., Ruscus, Vinca, Asplenium ad-ni.; Laurus, Ligustrum luc., Cotoneaster, Trachycarpus, Lonicera jap., Prunus laur., Niburnum) - 173 relevés.

A key for interpreting the changes in forest vegetation

The colline forest framing the south-Alpine lakes and in a broader sense the forest zone of western Europe exposed to an oceanic climate demonstrate consistent structural differences, caused by the reduced number of evergreen elements which are able to occupy a well-defined ecological niche: namely, to exploit the possibility to photosynthesize during the cold season, like the big conifers in western America or - as the best example - the components of the Macaronesian, Chinese or Japanese Laurisilvae do. The major part of the exotic elements in propagation, originating from the last-mentioned area and from the ecotonal zone in contact with deciduous forests, are rare, inefficiently established or completely absent in the Central European vegetation.

How would the vegetation change when exposed to a climatic change? Into which situation would the autochthonous vegetation be modified? The knowledge gained so far (ecophysiology of the species and of the vegetation formations) indicates the necessity to consider a further integrative dimension - as an assessment factor for a particular forest vegetation - which, in turn, indicates the tendency to converge towards other formations. This tendency does not only concern mature stands (tendency to evolve towards the proper climax), but also more complex structures (convergence towards functional structures typical of the bioclimate).

Therefore, a first component of this dimension must be expected in the distance from the forest climax, recognizable on the local and/or regional scale. The second component can be approached theoretically, on a larger scale up to the level of hemispheres, by comparing different formations: some - from a historical, or better from a chorological background - establish themselves more or less efficiently on the species or ecophysiological level, which means that they also appear more or less completely on the structural level.

To validate the potential of change caused by the variation of any parameter it is important to know the "chorological tension" already present or likely to develop in the investigated vegetation, which means the structural difference to a climax found in a given climatic situation:

- based on examples of climax forests present on the regional level
- considering the differences between the last-mentioned and other formations frequently found in the same hemisphere, in situations characterized by a more completed chorological heritage.

The importance of the concept of the "chorological tension" of a given type of vegetation includes obviously the unacceptable effect of very aggressive neophytes in insular or relatively "more insular" systems compared to the provenance zone of the hosts. In the present vegetational context - also on a global scale - it becomes crucial to consider this dimension.

In Insubria the absence of synusia with laurophyllous species is more obvious than anywhere else. This deficiency results to a large extent from the last glaciation which destroyed the thermo-hygrophilous aspect of subtropical forests that were present in Europe in the Tertiary, 600'000 years BP. Additionally, the Mediterranean elements are less competitive in humid-temperate climate migrating from the South (Etesian climate, sclerophyllous biome). However, man exploited these elements rather intensively from the Neolithic Period to the Middle Ages (Burga & Perret 1998).

Considering the thermo-hygrophilous ecotypes of *Abies*, *Ilex*, *Rhododendron* and *Taxus* or - in situations of the sub-Mediterranean ecotones - thermo-hygrophilous ecotypes of *Laurus*, *Smilax* and *Arbutus*, functionally competitive ecotypes exist in the examined ecological niches. Taking into account that their fragmented distribution is partly - if not totally - due to human impact since the Neolithic Period (fire, pasturing, clear cutting) their absence now creates space for various competitive species with a potential for a higher variance than anywhere else. In this sense the Insubrian forest is a highly indicative example of a system with high "chorological tension": it is modified already at its present stage and tends to modify mainly with a climate warming of only 1-2°C, in a way that climatic change stresses or produces "ex-novo" the existence of other structural and ecological configurations, the "chorological tension" increases. Presently this crucial factor for strong structural changes may be determined by conclusions gained from analogies. This is the case for the Insubrian zone compared with bioclimates and structures of warm hyperoceanic zones like the Laurisilvae (Japan) and for the coast of the Bay of Biscay.

With the same approach in Central Europe a "chorological tension" can be shown as well, although to a lower degree: the in situ vegetation is closer to the expressed vegetation of the same hemisphere exposed to the same climate, the structural changes and the penetration by neophytes are less evident.

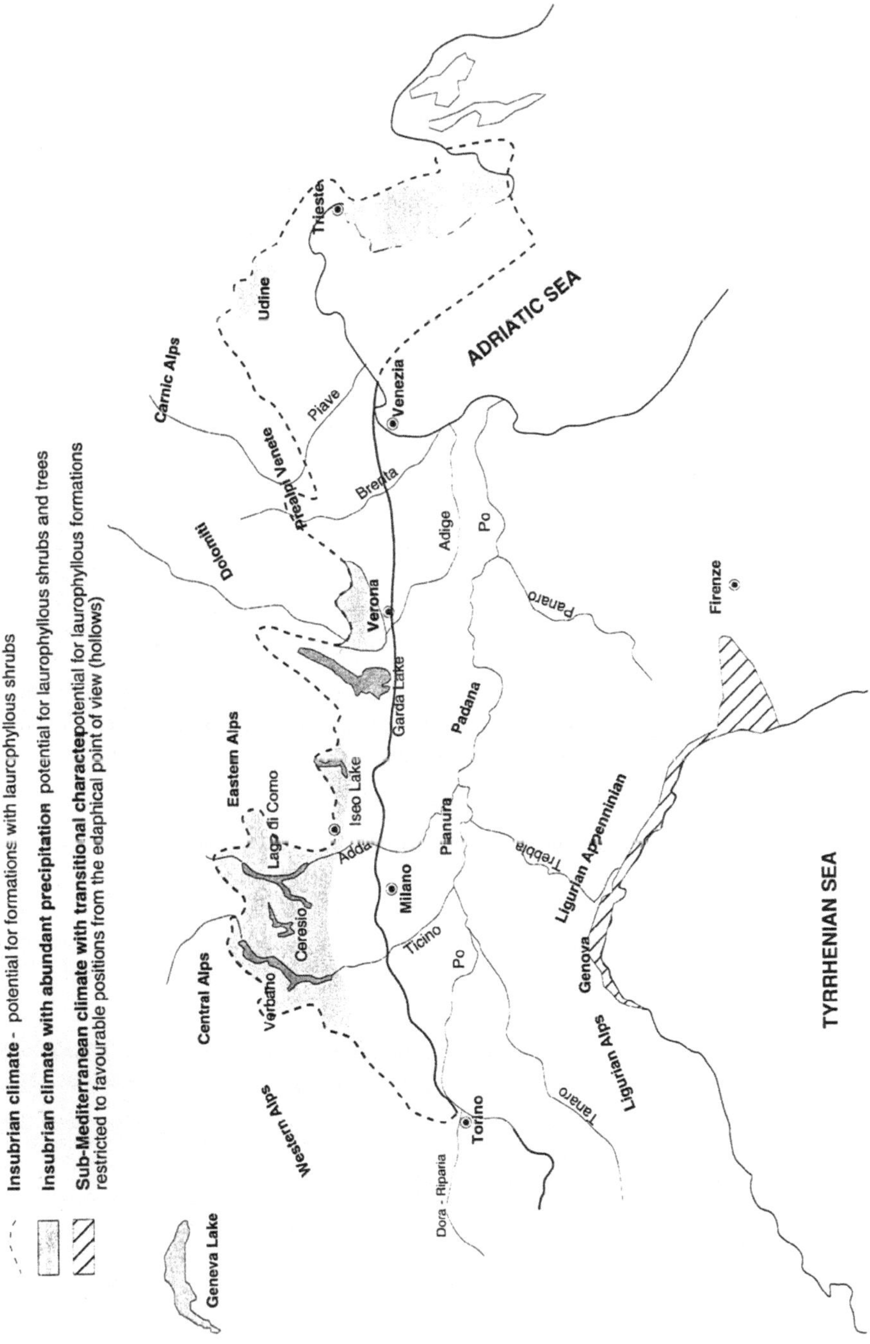

Figure 5. Areas with potential laurophyllisation in the sub-Mediterranean region.

3. Conclusions

The investigations have proved the importance of a factor which is able to provoke a considerable structural change: it is indispensable not only for validating the presence of vacant or suboccupied ecological niches with particular regard to climate and vegetation, but also for deriving the best strategies for conservation or recomposition and the most appropriate management practices.

Concerning forest ecosystems the following suggestions on important issues can be made:

- to conserve and protect mature stages and to actively favour autochthonous laurophyllous species with thermophilous ecotypes
- to conserve the genetic variability of populations, to ameliorate the biological network if damaged; to conserve unbroken forest populations with reduced proportion of external forest edges
- to prevent the spread of aggressive neophytes and to regulate the transcontinental introduction of plant species and living material
- to conserve the present population of exotic laurophyllous species
- to promote European laurophyllous species in silviculture and landscape
- to promote the conservation of laurophyllous biomes already mentioned in this report as strategic areas for the biodiversity of western Eurasia (Caucasus, Balkans, Pyrenees, extra-zonal relicts in the Mediterranean basin).

The following scientific issues should be emphasized:

- to ameliorate and refine the working tools to promote vegetation science in parallel fields
- to include quantitative efforts referring to population dynamics (biological values, state, critical mass, dynamics of diaspores, etc.) in the prognosis and to assess the quality and presence of biological bridges for particular formations or species (of strategic and invasive values)
- to study in parallel other European regions and areas in other continents, introducing related concepts of structural and functional analogies and of "chorological tension"
- to examine the opportunity to introduce antagonists of invasive species from the original area (biological control)
- to accumulate knowledge of existing equilibria in the principal ecotonal zones of the continental Laurisilvae - deciduous forest and laurophyllous-sclerophyllous forest
- to accumulate knowledge of the presence and the population genetics of some Mediterranean species with peri- and oro-Mediterranean distribution (thermo-hygrophilous ecotypes)
- to encourage the application of geographical information systems

- to establish and connect an observation and information network, particularly of species which show an "aggressive" and "invasive" behaviour under analogical climatic conditions on an international level
- to establish a long-term vegetation monitoring network.

References

Burga C.A. & Perret R. 1998. Vegetation und Klima der Schweiz seit dem jüngeren Eiszeitalter. Ott Verlag, Thun.

Carraro G. & Gianoni G. 1987. Studio sulla presenza delle diverse specie di quercia nostrana in Ticino e della loro importanza per la selvicoltura a dipendenza della stazione e tenendo conto del fenomeno d'ibridazione. Lavoro di diploma ETHZ: 1-198. (Polycopy.)

Gianoni G., Carraro G. & Klötzli F. 1988. Thermophile, an laurophyllen Pflanzenarten reiche Waldgesellschaften im hyperinsubrischen Seenbereich des Tessins. Ber. Geobot. Inst. ETH Stiftung Rübel, Zürich 54: 164-180.

Klötzli F. 1988. On the global position of evergreen broad-leaved non-ombrophilous forest in the subtropical and temperate zones. Veröff. Geobot. Inst. Stiftung Rübel ETH Zürich 98: 169-196.

Kissling P. 1989. Changement floristique depuis 1950 dans les forêts des Alpes suisses. Bot. Helv. 99(1): 27-43.

Kissling P., Kuhn N. & Wildi O. 1988. Le relevé mérocénotique et son application à l'étude du changement floristique en forêt. Bot. Helv. 98(1): 39-75.

Werdenberg K. & Hainard P. 1989. Modifications de la composition floristique dans la forêt genevoise et pollution atmosphérique par l'azote. Le lien est-il réel? Une réponse à l'Institut fédéral de recherches forestières. Sausurrea 19: 57-66.

LAUROPHYLLISATION - A SIGN OF A CHANGING CLIMATE?

GIAN-RETO WALTHER

Geobotanical Institute ETH Zürich, Zürichbergstrasse 38, CH-8044 Zürich, Switzerland

Keywords: Evergreen broad-leaved forest, global change, Insubria, laurineous species, long-term monitoring, vegetation shift

Abstract

Evergreen broad-leaved plants have been cultivated in Switzerland for more than 200 years. Only in the last decades about a dozen of these species succeeded in escaping from the gardens and in spreading into the surrounding forests. Meanwhile, they have grown up to the tree layer and must be considered presently as naturalized. Former deciduous forest transformed into evergreen deciduous broad-leaved mixed forest. As temperature is the major climatic determinant for species distribution, particularly limiting the distribution of evergreen towards deciduous vegetation, the possible link between these vegetation shifts and climatic change is discussed.

1. Introduction

In 1987 two forest engineers investigated forest stands rich in *Quercus* species in southern Switzerland as to their phytosociological status and the occurrence of introgression within the indigenous *Quercus* species (*Quercus petraea*, *Q. robur*, *Q. pubescens*, *Q. cerris*) (Carraro et al. 1987). They discovered species groups that did not fit into the sociological pattern described by Antonietti (1968). These deviating formations have been called "Lauro-Quercetum castanosum (prov.)" because they are dominated by evergreen broad-leaved species (Gianoni et al. 1988).
More recently, exotic species such as *Cotoneaster* spp., *Lonicera* spp. and *Prunus laurocerasus* have been found much more frequently in northern Switzerland and locally, they are even naturalized (Landolt 1993).
Within the National Research Program 31 (Climatic Changes and Natural Disasters) a project focused on vegetation changes with possible relations to a changing climate (Carraro et al. 1999). In addition a supplementary project investigated the occurrence of forest stands rich in evergreen broad-leaved - so-called laurineous or laurophyllous - species in southern Switzerland (Walther 1995). All these findings resulted in a first overview of vegetation changes in Swiss lowland forests, given by Klötzli et al. (1996), and discussed at the international conference on "Recent shifts in vegetation boundaries of deciduous forests, especially due to general global warming", held in Switzerland in March 1997 (Klötzli & Walther 1999a).

Burga & Kratochwil (eds.), BIOMONITORING, 207-223

2. Impact of climatic change on vegetation

There has been a general upward trend in global mean annual temperature, from about 14.5°C in 1866, when reliable records began, to around 15.4°C in 1995 - the warmest year on record (Bright 1997). This trend has continued and even become more pronounced in the following years. Between May 1997 and September 1998, there has been a string of 16 consecutive months where each monthly mean global temperature broke all previous records (Karl et al. 2000). The climatic changes predicted for most of the temperate forest regions over the period 1990 - 2050 are increases in both summer and winter temperatures of 1 - 2°C, which are likely to lead to shifts in vegetation belts and significant alterations in the composition of forests (IPCC 1995). In Switzerland, an increase in the annual mean temperature of 0.8°C (average value) has been measured at the permanent meteorological stations in the period from 1961 to 1990. The Swiss meteorological stations are scattered all over the country and located at different altitudes, but there was not a single station with a deviating trend (Aschwanden et al. 1996). In the first ten years of the period (1961 - 1970) the increase amounted only to +0.1°C, in the last decade (1981 - 1990), however, the temperature rose by +1.5°C.

Most of the research into the impact of climatic changes on vegetation in Switzerland - as everywhere else - has either focused on changes in physiological processes due to increased aerial CO_2-concentrations or on modelling. CO_2-experiments have shown that an increased CO_2-uptake may lead to changes in photosynthetic activity, leaf-, wood- and root-biomass as well as its proportions, and the activity of soil bacteria and mycorrhiza. The behaviour of species varies individually and thus their relative abundance is influenced (e.g. Körner 1997 and Fuhrer & Bader 1996). Models suggest that the zonal vegetation of Switzerland is very sensitive to climate warming (Brzeziecki et al. 1995) and therefore changes in the species composition are expected (Fischlin 1995), as well as an upward movement of forest vegetation belts in the Alps (Kienast et al. 1996). But as yet, only few field studies investigating fluctuations in natural or semi-natural habitats have been performed. These indicate that in the alpine and nival belt an upward movement of species is an overall trend (Grabherr et al. 1994). Also the displacement of the arctic tree line (Lescop-Sinclair & Payette 1995) and, more generally spoken, changes of both forest growth rates and the position of the latitudinal and altitudinal tree lines are occurring (Innes 1994). As the latter stated, it seems likely that climatic change is involved, since the growth at and the movement of the natural timber line are primarily controlled by temperature. Apart from a forest range shift, the vegetation composition (predominance of species) may change significantly (Andrasko 1990). In experimentally heated plots a shifting dominance within montane meadow vegetation could be observed (Harte & Shaw 1995). Finally, vegetation shifts, including the spread of exotic evergreen broad-leaved species, have been demonstrated in studies in the colline and submontane belt with a change in the species composition of up to 60 % (Klötzli et al. 1996). These changes do not only suggest differences in average temperatures, but also in nutrient and average light conditions, as well as a drop in the pH values. However, a link between the process called "laurophyllisation" and the climatic change seems highly probable.

3. The process called "laurophyllisation"

Laurophyllisation defines the process of the spreading of evergreen broad-leaved (laurophyllous) species into deciduous forest and thus represents a biome shift from deciduous to evergreen broad-leaved forest. There are different definitions of the term "evergreen broad-leaved forest" (discussed e.g. in Song 1999). Species of the laurophyllous type have evergreen broad-leaved foliage but - in contrast to sclerophyllous species - they are not adapted to drought.

The tertiary forests in Central Europe were rich in evergreen broad-leaved species, however, most of the representatives of the laurophyllous type were eliminated during the glaciations (Mai 1995). At present there are only few relict evergreen broad-leaved woody species left in Central Europe (e.g. *Ilex aquifolium*, *Hedera helix*). There is more diversity in laurophyllous relict species in climatically more favoured areas like the Black Sea coast in Georgia, where more of the evergreen broad-leaved woody species have survived the extreme climatic events of the past (see e.g. Zazanashvili 1999).

Southern Switzerland (Ticino) - especially the region surrounding the lakes (e.g. Lago Maggiore and Lago di Lugano) called Insubria - has for Switzerland unique climatic conditions with an annual mean temperature of ca. 12°C, annual mean precipitation of 1500 to 2000 mm, scattered throughout the year but concentrated on few days with heavy rainfall, and relatively mild winter temperatures. It is appreciated as recreation area and as a popular holiday destination. The climate has allowed to cultivate a lot of exotic thermophilous species from all over the world in the house gardens. The favourites have been evergreen broad-leaved species from south-eastern USA and from eastern Asia, including the palm *Trachycarpus fortunei.* The botanical garden of Brissago Island maintains a collection of more than 2000 species, amongst them many subtropical plants from, e.g., East Asia, Australia, South Africa and South America. A closer look at the climatic position of Insubria shows its vicinity to areas with native evergreen broad-leaved forests (Figure 1) (cf. also Klötzli 1988).

The white dots represent meteorological stations. A code is assigned to each station (cf. Table to Figure 1) and the absolute minimum temperature is indicated. The stations belonging to the same geographical region are encircled by ellipses.

The Ticino - particularly the warmest part represented by the stations of Locarno and Lugano - is situated in an ecotonal position, right on the present boundary of laurophyllous vegetation.

A comparison of the climatic situation in areas with deciduous broad-leaved vegetation, represented by Locarno (Switzerland), and evergreen broad-leaved vegetation, represented by Yokohama (Japan), with climate diagrams (Walter & Lieth, 1960-67) is shown in Figure 2.

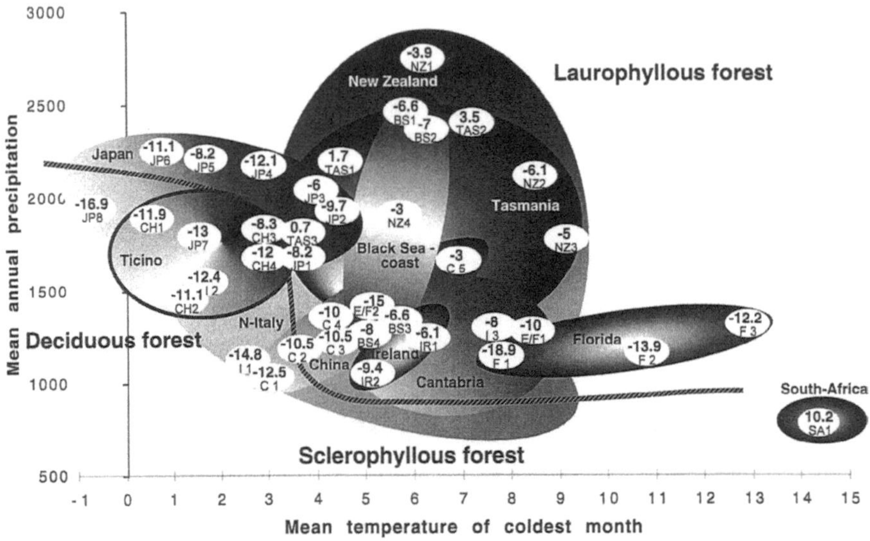

Region of the world	Code	Meteorological station	mean temp. of coldest month*	mean annual precipitation*	absolute min.temp.*	mean daily min.temp.*	Biome type*
Ticino	CH1	Rivera-Bironico	0.7	1981	-11.9		V2
	CH2	Bellinzona	1.4	1589	-11.1		V2
	CH3	Lugano	2.8	1725	-12.0	-1.8	V2
	CH4	Locarno	2.8	1890	-8.3		V2
Northern Italy	I 1	Treviso	2.8	1056	-14.8	-0.2	V1
	I 2	Domodossola	1.4	1594	-12.4	-2.6	V2
	I 3	Genova	7.7	1295	-8.0	5.9	V(VI)
Cantabria	E/F1	Bagnères de Bigorre	5.0	1383	-15.0	-0.2	V(VI)
	E/F2	Biarritz	8.6	1255	-10.0	5.1	V1
Black Sea - coast	BS1	Rize	5.7	2510	-6.6	3.6	V(IV)
	BS2	Batum	6.4	2404	-7.0	3	V
	BS3	Giresun	5.7	1330	-6.6	3.9	V(IV)
	BS4	Zonguldak	5.0	1261	-8.0	1.9	V
SW-Ireland	IR1	Valentia	6.4	1303	-6.1	5	V(VI)1
	IR2	Cork	5.0	1047	-9.4	2.7	V(VI)2
China	C 1	Chinkiang	2.8	1026	-12.5		V2
	C 2	Wuhu	3.5	1183	-10.5		V1
	C 3	Hankow	4.3	1252	-10.5		V1
	C 4	Kiukiang	4.3	1410	-10.0		V1
	C 5	Wenchow	7.1	1673	-3.0		II(V)

Table continued

Japan	JP1	Yokohama	3.6	1747	-8.2	-0.4	VI(V)2
	JP2	Sakai	4.3	1960	-9.7	0.4	V3b
	JP3	Hamamatsu	4.3	2018	-6.0	1	V2
	JP4	Fushiki	2.9	2232	-12.1	-1.3	VI(V)1
	JP5	Kamo	1.4	2240	-8.2		VI(V)1
	JP6	Toyama	0.7	2284	-11.1	-2.8	VI(V)1
	JP7	Niigata	1.4	1780	-13.0	-1.5	VI(V)1
	JP8	Sakata	-0.7	1964	-16.9		VIIa
South Africa	SA1	East London	14.3	808	10.2	2.8	V
Florida	F 1	Columbia	7.7	1172	-18.9	2.1	V(2)
(SE-USA)	F 2	Charleston	10.7	1168	-13.9	6.6	V(II)
	F 3	Jacksonville	13.1	1323	-12.2	7.4	V(II)
Tasmania	TAS1	Waratah	4.3	2190	1.7		V(4)
	TAS2	Zeehan	7.1	2470	3.5		V(4)
	TAS3	Moina	3.6	1795	0.7		V(4)
New Zealand	NZ1	Hokitika	6.1	2760	-3.9	1.1	V3
	NZ2	Waihi	8.6	2145	-6.1	4.3	V1
	NZ3	Whangamata	9.2	1796	-5.0	5	V1
	NZ4	Onepoto	5.7	1922	-3.0	3.3	V(VI)1

* according to Walter & Lieth (1960-67)

Figure 1. The position of the Ticino in a global context, shown in a diagram with mean annual precipitation and mean temperature of the coldest month.

The two graphs are quite similar, the main difference being in the winter temperatures. It is commonly supposed that the temperatures in this season - especially the low extremes - are a crucial factor determining the boundary with deciduous forest (Box 1994, Woodward 1987, Song 1983, Jäger 1968). For species with wintergreen leaves, the critical temperatures for leaf survival may be the lowest ones in winter (Box 1996, Woodward 1987). Woodward & Williams (1987) consider a minimum temperature of -15°C as critical in controlling the poleward spread of vegetation that is dominated by evergreen broad-leaved species. Also in Eastern Asia the periodical or episodical occurrence of minimum temperatures of -10°C is considered to be the limiting factor for evergreen broad-leaved tree species (Miehe pers. comm.).

In view of its climate, it is not surprising that in the Ticino the cultivation of adult plants, even of the frost-hardier species from warm-temperate regions, was possible. But until recent times regeneration could not be observed, although the exotic species have been introduced and cultivated in the gardens for more than two hundred years (Table 1).

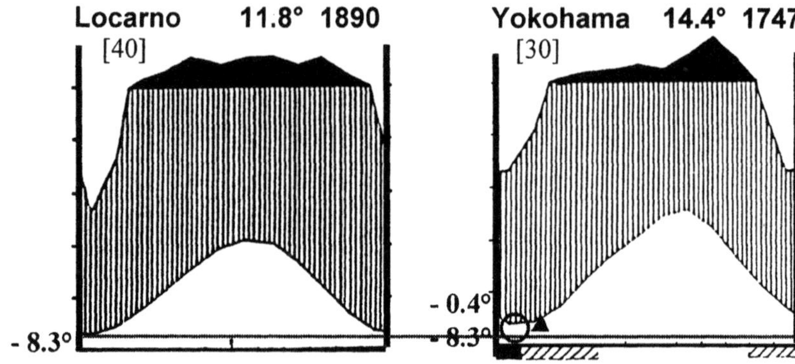

Figure 2. Comparison of the climate diagrams of areas with deciduous broad-leaved vegetation and with evergreen broad-leaved vegetation.

First observations of escaped garden cultivars were reported from azonal sites like gorges and sheltered places, with some individuals able to endure the natural conditions in the wild. Then more and more species succeeded in germinating from seeds and in growing up to young plants, but they were still restricted to gardens. In Ellenberg & Rehder (1962), Leibundgut (1962) and Ellenberg & Klötzli (1972) there are no records on the colonization of zonal forest stands by exotic laurophyllous species. But in the last few decades the picture has changed. A regular and sustained colonization by a whole set of laurophyllous species has taken place (Figure 3). At particular sites some individuals have already grown up to the upper tree layer (cf. also Walther 1995, Klötzli et al. 1996, Klötzli & Walther 1999b).

Figure 4 shows the current appearance of the forest in particular areas of southern Switzerland; a dozen or so species must be considered as naturalized.

In core areas there are laurophyllous species which have reached the upper tree layer (*Cinnamomum glanduliferum*), followed by others in the lower tree layer (*Ligustrum lucidum*, *Laurus nobilis, Ilex aquifolium*). The tree layer shows a mosaic of evergreen and deciduous species, whereas the shrub layer is dominated by rejuvenating laurophyllous tree species (*Cinnamomum glanduliferum*, *Trachycarpus fortunei*, *Laurus nobilis*, *Prunus laurocerasus*) and various evergreen and semi-evergreen shrub species (*Elaeagnus pungens*, *Ruscus aculeatus*, *Lonicera japonica*). Indigenous deciduous species are very much under pressure. There is a second stage of laurophyllisation with evergreen broad-leaved species having not (yet) reached the tree layer but growing vigorously in the shrub layer. These two categories are restricted to the areas surrounding the lakes (cf. Figure 5), suggesting that the influence of lakes may be significant (see also Oberdorfer 1964, p. 184). Outside the area influenced by the lake, there are still some exotic laurophyllous plants scattered within the native forest complex, but not in a dominating position and restricted to the shrub layer (Figure 5).

Table 1. The occurrence of laurophyllous species in a historical view.

Time-scale	Species name	Report	Location		Source
1576	*Prunus laurocerasus*	introduction to Italy		Introduction	Jäggli 1924
1671	*Trachycarpus fortunei*	first botanical garden with exotic species	Isola Madre (established by Prince Borromeo)		Pfister 1977 (Schröter 1936)
1688	*Liliodendron tulipifera*	introduction from North America			Schröter 1936
1730	*Camellia japonica*	introduction from Eastern Asia			Schmid 1956
2nd half of 18th century	*Eriobotrya japonica*	introduction from Japan			Schmid 1956
1833	*Olea europaea*	escaped	Gandria	Spread on azonal sites	Christ 1882
1858	*Jubaea spectabilis*	sowing of seed (grown up to 17 m at the time of publication)	Isola Madre (Lake Major)		Schröter 1936
1882	*Agave americana*	rejuvenating group	Minusio and Madonna dal Sasso		Christ 1882
	Ruscus hypoglossum	gives the impression of an escaping species	Isola Madre		Kny 1882
1891	*Eucalyptus amygdalina*	introduction (grown up to about 30 m at the time of publication)	Botanical garden of Brissago Island		Schröter 1936
1896	*Lonicera japonica*	subspontaneous occurrence	Trevano near Lugano		Becherer 1960
1904	*Laurus nobilis*	subspontaneous occurrence	Morcote and Gandria		Bettelini 1904
	Buxus sempervirens	subspontaneous occurrence	Casoro and Aldesago		
1907	*Laurus nobilis* *Agave americana*	naturalized escaped	between Gandria and Oria		Rickli 1907
1910	*Dicksonia antarctica*	introduction of 12 stems	Villa Carlotta (Lake Como)		Schröter 1936
1920	*Phyllostachys cf. bambusoides*	escaped	moist gorges near Lugano and on slopes near Gandria		Voigt 1920
	Trachycarpus fortunei	subspontaneous occurrence	near Gandria		
1926	*Eucalyptus globulus*	escaped, killed by frost, but resprouting	Castagnola - Gandria		Schröter 1926
	Pittosporum tobira	cultivated / escaped (?)	Lugano - Melide		
	Acacia dealbata	everywhere killed by frost, except on Brissago Island			

Table 1 continued

1936	*Trachycarpus fortunei*	seedlings near adult trees	Brissago Island	Spread on azonal sites	Schröter 1936
	Cinnamomum camphora	seedlings near adult trees	gardens and parks		
	Elaeagnus pungens	rarely self-sowing	Insubria		
	Ruscus hypophyllum	escaping	Insubria		
1950	*Aucuba japonica* *Prunus laurocerasus* *Trachycarpus fortunei*	subspontaneous occurrence in moist and shady gorges	Ronco - Brissago		Guide to the botanical gardens of the canton Ticino
1956	*Eriobotrya japonica*	seedlings	house gardens		Schmid 1956
	Ruscus hypophyllum	escaping "here and there"			
	Cinnamomum glanduliferum	seedlings	house gardens		
	Camellia japonica	rejuvenating	house gardens		
	Prunus laurocerasus	often escaping			
	Aucuba japonica	escaping "here and there"			
	Elaeagnus pungens	seedlings	house gardens		
	Laurus nobilis	spontaneous-like occurrence	warmer sites		
	Ligustrum lucidum	seedlings	house gardens		
	Trachycarpus fortunei	rejuvenating vigorously	Brissago Island and in moist gorges on the southern slopes above Ronco		
	Osmanthus fragrans	seedlings	house gardens		
	Arundinaria japonica	rejuvenating	Brissago Island and in moist gorges on the southern slopes above Ronco		
1960	*Lonicera japonica*	often escaped and naturalized		Spread on zonal sites	Becherer 1960
1961	*Trachycarpus fortunei* *Arundinaria japonica*	easily escaping			Schmid 1961
1964	*Lonicera japonica*	escaping	near settlements		Oberdorfer 1964
	Laurus nobilis *Buxus sempervirens*	subspontaneous occurrence in Salvio-fraxinetum taxotosum cover value: +			
1968	*Laurus nobilis*	subspontaneous occurrence in Erisithalo-Ulmetum (prov.) and Helleboro Ornetum (prov.) cover value: **r** resp. +			Antonietti 1968
	Lonicera japonica	subspontaneous occurrence in Erisithalo-Ulmetum (prov.) cover value: +			

Table 1 continued

1975		subspontaneous occurrence in **herb** layer of Querco-Fraxinetum;			Zuber 1979
		cover values:			
	(Chamaerops humilis) =*Trachycarpus fortunei*	+	(herb layer)		
	Elaeagnus pungens	+	(herb layer)		
	Laurus nobilis	+ resp. **1**	(herb layer)		
	Lonicera japonica	+ resp. **1**	(herb layer)		
	Prunus laurocerasus	+	(herb layer)		
1988		subspontaneous occurrence in **shrub** and **lower tree** layer of Querco-Fagetea;			Gianoni et al. 1988
		cover values:			
	Laurus nobilis	up to **4**	(lower tree layer)		
	Elaeagnus pungens	up to **4**	(shrub layer)		
	Prunus laurocerasus	+ resp. **1**	(shrub layer)		
	Trachycarpus fortunei	up to **2**	(shrub layer)		
	Lonicera japonica	up to **4**	(shrub layer)		
	(Cinnamomum camphora) = *C. glanduliferum*	+	(shrub layer)		
	Arundinaria japonica	+	(shrub layer)		
1995		subspontaneous occurrence up to the **upper tree** layer of Cruciato glabrae-Quercetum castanosum		Spread on zonal sites	Walther 1995
		cover values:			
	(Cinnamomum camphora) = *C. glanduliforum*	up to **3**	(upper tree layer)		
	Laurus nobilis	up to **2**	(lower tree layer)		
	Ligustrum lucidum	**1**	(lower tree layer)		
	Elaeagnus pungens	up to **3**	(shrub layer)		
	Trachycarpus fortunei	up to **3**	(shrub layer)		
	Prunus laurocerasus	up to **2**	(shrub layer)		
	Lonicera japonica	up to **1**	(shrub layer)		
	Buxus sempervirens	**1**	(shrub layer)		
	Acacia dealbata	**1**	(shrub layer)		
	Aucuba japonica	+	(shrub layer)		
	Cotoneaster sp.	+	(shrub layer)		
	Pittosporum tobira	+	(herb layer)		
1996	*Cinnamomum glanduliferum*	naturalized	Insubria		Klötzli et al. 1996
	Laurus nobilis	naturalized	Insubria		
	Ligustrum lucidum	subspontaneous	Solduno near Locarno		
	Elaeagnus pungens	naturalized	Insubria		
	Trachycarpus fortunei	naturalized	Insubria		
	Prunus laurocerasus	naturalized	southern CH		

Table 1 continued

	Lonicera japonica	subspontaneous naturalized	northern CH Insubria	Spread on zonal sites	
	Pittosporum tobira	subspontaneous	particular areas in Insubria		
	Acacia dealbata	subspontaneous	particular areas in Insubria		
	Eriobotrya japonica	subspontaneous	particular areas in Insubria		
	Aucuba japonica	subspontaneous	particular areas in Insubria		
	Buxus sempervirens	subspontaneous	particular areas in Insubria		
	Lonicera pileata	naturalized subspontaneous	southern CH northern CH		
	Lonicera henryi	subspontaneous	northern CH		
	Viburnum tinus	subspontaneous	particular areas in Insubria		
	Cotoneaster sp.	subspontaneous	southern & northern CH		
1998	*Cinnamomum glanduliferum*	naturalized	Insubria		Carraro et al. 1999
	Elaeagnus pungens	naturalized	Insubria		
	Trachycarpus fortunei	naturalized	Insubria		
	Laurus nobilis	naturalized	Insubria		
	Prunus laurocerasus	naturalized	southern CH		
	Lonicera japonica	naturalized	Insubria		
	Lonicera pileata	naturalized subspontaneous	southern CH northern CH		
	Ligustrum lucidum	subspontaneous	Solduno near Locarno		
	Pittosporum tobira	subspontaneous	particular areas in Insubria		
	Acacia dealbata	subspontaneous	particular areas in Insubria		
	Eriobotrya japonica	subspontaneous	particular areas in Insubria		
	Aucuba japonica	subspontaneous	particular areas in Insubria		
	Buxus sempervirens	subspontaneous	particular areas in Insubria		
	Lonicera henryi	subspontaneous	northern CH		
	Viburnum tinus	subspontaneous	particular areas in Insubria		
	Cotoneaster sp.	subspontaneous	southern & northern CH		
	Camellia japonica	subspontaneous	house gardens & parks		

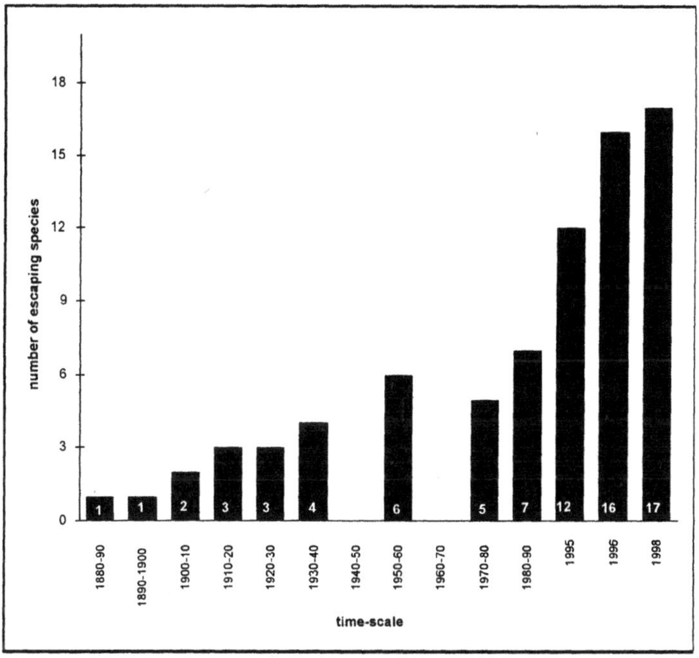

Figure 3. Time-scale for the development of laurophyllisation.

The larger the distance to the core area, the smaller the diversity of exotic laurophyllous species. At the margin areas of the occurrence of laurophyllisation, the winter-hardiest species are left, such as *Prunus laurocerasus*, *Lonicera* spp. and *Cotoneaster* spp., together with the indigenous evergreen species, e.g. *Ilex aquifolium* and *Hedera helix*. The same species occur on the northern side of the Alps in the Swiss Midlands. Also in this area exotic evergreen garden plants succeeded in escaping from the gardens into the surrounding forests. However, the picture is not (yet) as advanced as it is in southern Switzerland. It may be similar to the state reported in early publications on the occurrence of exotic species in southern Switzerland in azonal sites (Table 1), although in the gardens more and more exotic species have been cultivated in recent years, such as *Trachycarpus* and *Araucaria*. They survived the recent winters without protection and undamaged.

Figure 4. Winter picture of a forest rich in laurophyllous species.

The occurrence of exotic laurophyllous species coincides nicely with the sensitivity of the individual species to low temperatures (Table 2).
The differences in the table may be explained as a consequence of the different sites, the samples of the various species were collected from.

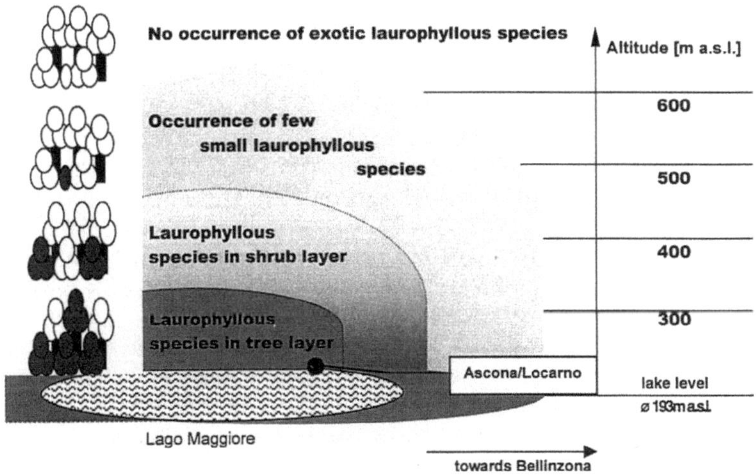

Figure 5. Scheme of the occurrence of laurophyllous species.

Table 2. Sensitivity of laurophyllous species to low temperatures (in °C). (s) seedling, (a) adult plant

Species name	**Sakai & Larcher (origin area of species)**	**Larcher 1964 (Lake Garda)**	**Walder 1985 (Lake Major)**
Eucalyptus globulus	-5	-6	
Acacia dealbata	-5	-10	-9 – -11
Phoenix canariensis	-7.5$_{(s)}$ – -10.5$_{(a)}$	-9	-9
Nerium oleander	-6 – -8	-11	-9 – -11
Eucalyptus viminalis	-8		
Cinnamomum camphora	-10		-8
Chamaerops humilis	- 11.5$_{(a)}$	-12	-11
Cinnamomum glanduliferum		-13	-10
Laurus nobilis	-10 – -14	-12	-11 – -14
Olea europaea	-10 – -14	-13	< -10
Viburnum tinus	-10 – -14	-13	< -10
Osmanthus fragrans	-13		< -10
Trachycarpus fortunei	-12$_{(s)}$ – -14$_{(a)}$	-14	< -11 – -14
Eriobotrya japonica		-14	< -10
Pittosporum tobira			< -10
Ligustrum japonicum	-15		
Aucuba japonica	-15		< -10
Camellia japonica	-15 – -20		-11
Magnolia grandiflora	-20	-16	< -10
Prunus laurocerasus		-11 – -17	
Ilex aquifolium	< -15		
Hedera helix	< -15		

4. Plants as bioindicator of the temperature regime

It is generally accepted that climate exerts the predominant control on the world distribution of zonal vegetation (Walter & Breckle 1983-91) and that temperature is the major climatic determinant of species distribution (Woodward 1987). "The ranges and growth of temperate tree species in wetter, maritime and high-latitude regions often can be related to the length of the growing season, measured in degree-days, and to absolute minimum temperatures in more continental areas" (IPCC 1995). In Japan, the region dominated by evergreen trees rapidly changes to a region dominated by broad-leaved deciduous trees, once the minimum temperature drops below a certain value (Woodward 1987). The knowledge of the close relationships between forest vegetation and climate allows to interpret the existing forest type in terms of climatic conditions (Jahn 1975). The knowledge of the way in which vegetation responded to past climatic changes may also give insights into the likely response of vegetation to future climatic changes induced by the "greenhouse effect" (Huntley 1990).
Plant functional types may be seen as predictors of transient responses of vegetation to global change (Chapin et al. 1996). Evergreen broad-leaved species in particular may act as bioindicators of subtle differences in the low-temperature regime of a region. In Central Europe the -0.5°C isotherm of the coldest month is interpreted as representing the northern limit for *Ilex aquifolium* (Iversen 1944). Dierschke (1993) suggests to consider climbing ivy (*Hedera helix*) as a possible indicator of milder winter conditions. The increasing number of evergreen broad-leaved species sensitive to low temperatures leads to the hypothesis that the conditions for thermophilous species have ameliorated. Therefore, the first criterion for possible bioindicators of climatic change according to de Groot et al. (1995), "climate sensitivity", has already been fulfilled. In addition the following criteria are listed: habitat constraints, position within distribution range, dispersal capacity, functional position in the ecosystem and suitability for monitoring. The laurophyllous species fulfil a number of these criteria. In consequence they can well be included in a list of possible bioindicators of climatic change.

5. Discussion and conclusions

As the altitudinal belts are linked to particular environmental conditions and especially to different temperature regimes, Ozenda & Borel (1995) expect these belts to shift upwards with increasing temperature. Scherzinger (1996) indicates a latitudinal shift of 100 to 200 km northwards and an altitudinal one of 150 to 200 m upwards. At a regional scale, variations in climate cause variations in the relative boundaries between forest ecosystems (WMO 1994). We might be confronted with the - at least partial - replacement of vegetation belts. Especially in situations as presented in Switzerland, there is an opening of a new ecological niche at the lower end of the vegetation zonation. The climatic conditions at low elevations with the characteristically very humid Insubrian climate are not suitable for typical Mediterranean vegetation which occurs in the south of Switzerland. Furthermore, models suggest that there will be no significant changes in the precipitation regime in these areas (Brzeziecki et al. 1995). Carraro et al. (1999) stressed the concept of the "chorological tension" with the impossibility of unoccupied niches in natural habitats. Laurophyllous species seem to be

best adapted to these novel conditions, and they have started to establish a new type of forest vegetation. According to Bright (1997) it is likely that non-native plants will profit from new climatic conditions.
Such a drastic change in former site parameters has the consequence that formerly well adapted, autochthonous and natural forest communities may become stands inappropriate to the site (Scherzinger 1996). Vice-versa, stands which were less suited to the site may become the best adapted to the new conditions. The expected movement of species requires a reassessment of the concept of exotic species and invasion (Kutner & Morse 1996). It is likely that with climatic change the former definition will become increasingly outdated and controversial as species move to new areas or are introduced to sites other than those where they presently occur.
Vegetation shifts do not only occur in the human-influenced border situation of evergreen to deciduous broad-leaved vegetation as it is presented in southern Switzerland; similar shifts have also been reported from evergreen broad-leaved forests in east China (Song 1999). In Japan (Fujiwara pers. comm.) and Georgia (Nakhutsrishvili pers. comm.) new research projects have been launched to investigate fluctuations at the margin of evergreen to deciduous vegetation.
Ecotonal areas are expected to be most responsive to climatic change (Brühlheide in: Klötzli & Walther 1999). Therefore, these areas are crucial for the early detection of long-term ecosystem changes in flora and fauna. The detection of shifts as a consequence of climatic change may be expected first in ecotonal areas and, when vegetation shifts do occur in such areas, we find ourselves confronted with the question whether this is already a sign of climatic change.

References

Andrasko, K. 1990. Global warming and forests: an overview of current knowledge. Unasylva 163: 3-11.

Antonietti, A. 1968. Le associazioni forestali dell'orizzonte submontano del Canton Ticino su substrati pedogenetici ricchi di carbonati. Mitt. EAFV 4: 88-221.

Aschwanden, A., Beck, M., Häberli, Ch., Haller, G., Keine, M., Roesch, A., Sie, R. & Stutz, M. 1996. Klimatologie der Schweiz (Jahrgang 1996). Klimatologie 1961-1990, Heft 2, Band 1 von 4. Bereinigte Zeitreihen. Die Ergebnisse des Projektes KLIMA 90 Band 1: Auswertungen. SMA Zürich.

Becherer, A. 1960. Die Flora des Tessins und des Comerseegebietes im Lichte der neueren Erforschung. Bauhinia 1(3): 261-281.

Bettelini, A. 1904. La flora legnosa del Sottoceneri. Diss. Università Zurigo.

Box, E.O. 1994. Eastern North America: Natural Environment and Sampling Strategy. pp. 21-60. In: Miyawaki A., Iwatsuki K. & Grandtner M.M. (eds), Vegetation in eastern North America. University of Tokyo Press.

Box, E.O. 1996. Plant functional types and climate at the global scale. J. Veg. Sci. 7: 309-320.

Bright, Ch. 1997. Tracking the ecology of climate change. pp. 78-94. In: Brown L.R., Flavin C. & French H. (eds), State of the world 1997. Worldwatch, New York.

Brzeziecki, B., Kienast, F. & Wildi, O. 1995. Modelling potential impacts of climate change on the spatial distribution of zonal forest communities in Switzerland. J. Veg. Sci. 6(2): 257-268.

Carraro, G., Gianoni, G. & Gianola, G. 1987. Studio sulla presenza delle diverse specie di quercia nostrana in Ticino e della loro importanza per la selvicoltura a dipendenza della stazione e tenendo conto del fenomeno d'ibridazione. M. Sc.-Thesis, ETH Zürich: 1-198. (Polycopy.)

Carraro, G., Gianoni P., Klötzli, F., Mossi, R. & Walther, G.-R. 1999. Observed changes in vegetation in relation to climate warming. Final report NFP 31. vdf, Zürich.

Chapin, F.S. III, Bret-Harte, M.S., Hobbie, S.E. & Zhong, H. 1996. Plant functional types as predictors of transient responses of arctic vegetation to global change. J. Veg. Sci. 7: 347-358.

Christ, H. 1882. Das Pflanzenleben der Schweiz. 2nd. ed. Schulthess, Zürich.
de Groot, R.S., Ketner, P. & Ovaa, A.H. 1995. Selection and use of bio-indicators to assess the possible effects of climate change in Europe. J. Biogeograph. 22: 935-943.
Dierschke, H. 1994. Pflanzensoziologie. UTB, Ulmer, Stuttgart.
Ellenberg, H. & Rehder, H. 1962. Natürliche Waldgesellschaften der aufzuforstenden Kastanienflächen im Tessin. Schweiz. Zeitschr. Forstw. 113: 128-142.
Ellenberg, H. & Klötzli, F. 1972. Waldgesellschaften und Waldstandorte der Schweiz. Schweiz. Anst. Forst. Versuchsw. 48(4).
Fischlin, A. 1995. Assessing sensitivities of forests to climate change: experiences from modelling case studies. pp. 145-147. In: Guisan A., Holten J.I., Spichiger R. & Tessier L. (eds), Potential ecological impacts of climate change in the Alps and Fennoscandian mountains. Ed. Conserv. Jard. Bot. Genève.
Fuhrer, J. & Bader, S. 1996. Klimaänderung und Grünland. Nationales Forschungsprogramm 31. Klimaänderung und Naturkatastrophen. info 10: 11-12.
Gianoni, G., Carraro, G. & Klötzli, F. 1988. Thermophile, an laurophyllen Pflanzenarten reiche Waldgesellschaften im hyperinsubrischen Seenbereich des Tessins. Ber. Geobot. Inst. ETH, Stiftung Rübel, Zürich 54: 164-180.
Grabherr, G., Gottfried, M. & Pauli, H. 1994. Climate effects on mountain plants. Nature 369: 1-448.
Harte, J. & Shaw, R. 1995. Shifting dominance within a montane vegetation community: results of a climate-warming experiment. Science 267: 876-880.
Huntley, B. 1990. European post-glacial forests: compositional changes in response to climatic change. J. Veg. Sci. 1: 507-518.
Innes, J.L. 1994. Climatic sensitivity of temperate forests. Env. Poll. 83: 237-243.
IPCC 1995. Climate change 1995. Impacts, Adaptations and Mitigation of Climatic Change: Scientific-Technical Analyses. Contribution of Working Group II to the Second Assessment Report of the Intergovernmental Panel on Climate Change.
Iversen, J. 1944. *Viscum*, *Hedera* and *Ilex* as climate indicators. Geologiska Foreningens i Stockholm Forhanlingar 66: 463-483.
Jäger, E. 1968. Die klimatischen Bedingungen des Areals der Dunklen Taiga und der sommergrünen Breitlaubwälder. Ber. Dtsch. Bot. Ges. 81(8): 397-408
Jäggli, M. 1924. Cenni sulla Flora Ticinese. Grassi & Co., Bellinzona.
Jahn, G. 1975. Die Vegetation als Klima-Indikator. pp. 355-362. In: Dierschke H. (ed) 1977, Vegetation und Klima. Cramer, Vaduz.
Karl, T.K., Knight, R.W. & Baker, B. 2000. The record breaking global temperatures of 1997 and 1998: Evidence for an increase in the rate of global warming? Geophysical Research Letters 27(5): 719-722.
Kienast, F., Wildi, O., Brzeziecki, B., Zimmermann, N. & Lemm, R. 1996. Klimaänderung und mögliche langfristige Auswirkungen auf die Vegetation der Schweiz. Nationales Forschungsprogramm 31. Klimaänderung und Naturkatastrophen. info 11: 11-12.
Klötzli, F. 1988. On the global position of the evergreen broad-leaved (non-ombrophilous) forest in the subtropical and temperate zones. Veröff. Geobot. Inst. ETH, Stiftung Rübel, Zürich 98: 169-196.
Klötzli, F., Walther, G.-R., Carraro, G. & Grundmann, A. 1996. Anlaufender Biomwandel in Insubrien. Verh. Ges. f. Ökol. 26: 537-550.
Klötzli, F. & Walther, G.-R. (eds) 1999a. Recent shifts in vegetation boundaries of deciduous forests, especially due to general global warming. Proceedings to the conference, Mte. Verità Schriftenreihe. Birkhäuser, Basel.
Klötzli, F. & Walther, G.-R. 1999b. Recent vegetation shifts in Switzerland. pp. 15-29. In: Klötzli F. & Walther G.-R. (eds), Recent shifts in vegetation boundaries of deciduous forests, especially due to general global warming. Proceedings to the conference, Mte. Verità Schriftenreihe. Birkhäuser, Basel.
Kny, L. 1882. Die Gärten des Lago Maggiore. Garten-Zeitung. Parey, Berlin.
Körner, Ch. 1997. Auswirkungen des erhöhten Kohlendioxidgehaltes der Atmosphäre auf die alpinen Ökosysteme. Nationales Forschungsprogramm 31. Klimaänderung und Naturkatastrophen. info 12: 6-7.
Kutner, L.S. & Morse, L.E. 1996. Reintroduction in a changing climate. pp. 23-48. In: Falk D.A., Millar C.I. & Olwell M. (eds), Restoring Diversity. Island Press, Washington D.C.
Landolt, E. 1993. Über Pflanzenarten, die sich in den letzten 150 Jahren in der Stadt Zürich stark ausgebreitet haben. Phytocoenologia 23: 651-663.

Larcher, W. 1964. Winterfrostschäden in den Parks und Gärten von Arco und Riva am Gardasee. Veröff. Museum Ferdinandeum Innsbruck 43: 153-199.
Leibundgut, H. 1962. Waldbauprobleme in der Kastanienstufe Insubriens. Schweiz. Zeitschr. Forstw. 113: 164-188.
Lescop-Sinclair, K. & Payette, S. 1995. Recent advance of the arctic treeline along the eastern coast of Hudson Bay. J. Ecol. 83: 929-936.
Mai, D.H. 1995. Tertiäre Vegetationsgeschichte Europas. Fischer, Stuttgart.
Oberdorfer, E. 1964. Der insubrische Vegetationskomplex, seine Struktur und Abgrenzung gegen die submediterrane Vegetation in Oberitalien und in der Südschweiz. Beitr. naturk. Forsch. SW-Deutschl. 23(2): 41-187.
Parco botanico del cantone Ticino. undated. Isole di Brissago. Guida ufficale.
Pfister, M. 1977. Sonnenstube Tessin. Ringier, Zürich.
Rickli, M. 1907. Zur Kenntnis der Pflanzenwelt des Kts. Tessin. Ber. Schweiz. Bot. Ges. 17: 27-63.
Sakai, A. & Larcher, W. 1987. Frost survival of plants. Ecol. Stud. 62. Springer, Berlin, Heidelberg.
Scherzinger, W. 1996. Naturschutz im Wald. Qualitätsziele einer dynamischen Waldentwicklung. Ulmer, Stuttgart.
Schröter, C. 1926. Exkursion der Volkshochschule ins Tessin. 1-9. (Polycopy).
Schröter, C. 1936. Flora des Südens. Rascher, Zürich.
Schmid, E. 1956. Flora des Südens. 2nd. ed. Rascher, Zürich.
Song, Y. 1983. Die räumliche Ordnung der Vegetation Chinas. Tuexenia 3: 131-157.
Song, Y. 1999. The historical shift of the evergreen broad-leaved forest in East-China. pp. 253-272. In: Klötzli F. & Walther G.-R. (eds), Recent shifts in vegetation boundaries of deciduous forests, especially due to general global warming. Proceedings to the conference, Mte. Verità Schriftenreihe. Birkhäuser, Basel.
Voigt, A. 1920. Beiträge zur Floristik des Tessins. Mitt. bot. Mus. Univ. Zürich 85: 332-357.
Walder, M. 1985. Die Auswirkungen des Kälteeinbruchs im Januar 1985 auf wärmeliebende Pflanzen in einem Tessiner Garten. Schweiz. Beitr. zur Dendrologie 35: 189-199.
Walter, H. & Breckle, S.-W. 1983-91. Ökologie der Erde. Band 1-4. Fischer, Stuttgart.
Walter, H. & Lieth, H. 1960-67. Klimadiagramm-Weltatlas. Fischer, Jena.
Walther, G.-R. 1995. Ausbreitung und Grenzen laurophyller Arten im Südtessin. M. Sc.-Thesis, ETH Zürich: 1-37 + Annex (Polycopy).
WMO 1994. Climate variability, agriculture and forestry. Report of the CAgM-IX Working Group on the study of climate effects on agriculture including forests, and of the effects of agriculture and forests on climate. Technical note No. 196, WMO-No. 802. Secretariat of the World Meteorological Organization, Geneva.
Woodward, F.I. 1987. Climate and plant distribution. Cambridge University Press, London.
Woodward, F.I. & Williams, B.G. 1987. Climate and plant distribution at global and local scales. Vegetatio 69: 189-197.
Zazanashvili, N. 1999. On the Colkhic vegetation. pp. 181-197. In: Klötzli F. & Walther G.-R. (eds), Recent shifts in vegetation boundaries of deciduous forests, especially due to general global warming. Proceedings to the conference, Mte. Verità Schriftenreihe. Birkhäuser, Basel.
Zuber, R.K. 1979. Untersuchungen über die Vegetation und die Wiederbewaldung einer Brandfläche bei Locarno (Kanton Tessin). Beiheft zu Zeitschr. Schweiz. Forstverein 65.

CHANGES OF PLANT COMMUNITY PATTERNS, PHYTOMASS AND CARBON BALANCE IN A HIGH ARCTIC TUNDRA ECOSYSTEM UNDER A CLIMATE OF INCREASING CLOUDINESS

INGO MÖLLER[1], CHRISTOPH WÜTHRICH[2] & DIETBERT THANNHEISER[1]

[1]*Institute of Geography, University of Hamburg, Bundesstrasse 55, D-20146 Hamburg, Germany;* [2]*Department of Geography, University of Basel, Spalenring 145, CH-4055 Basel, Switzerland*

Keywords: Arctic, carbon balance, climatic change, modeling, plant communities, scenario technique

Abstract

Climate models predict a pronounced warming in the Arctic which will be accompanied by an increasing cloudiness during the vegetation period in summer. This study determines which consequences can be expected regarding the spreading of plant communities, their phytomass and the spatial carbon balance in a high arctic catchment under a climate with increased cloudiness after 40-50 years. Using the present and the most likely future distribution patterns of the prevailing plant communities, changes in phytomass and CO_2 fluxes were calculated. The calculations showed that an alteration in the distribution of the plant communities is much less effective in changing the carbon balance of the high arctic tundra than a change of important prevailing ecological conditions, such as light or length of the snow-free period.

1. Introduction

In the last years numerous studies dealt with the effects of an increasing CO_2-content of the atmosphere or with effects of higher temperatures on the arctic vegetation (Oberbauer et al. 1986; Wookey et al. 1995). Climate models predicted a pronounced warming in the Arctic (IPCC 1996) which is supported for the High Arctic by meteorological data (Steffensen 1982; Førland et al. 1997). The warming of the arctic islands since the beginning of this century has caused an immigration of plant species, as Skye (1989) could show for two species on Hopen island in the south of Svalbard. In the last few years, the possibility of a further spreading of thermophilous species (e.g. *Cassiope tetragona, Empetrum nigrum ssp. hermaphroditum* and *Betula nana*) in relation with increased warming on Spitsbergen was pointed out several times (e.g. Wüthrich 1991; Thannheiser 1994; Elvebakk & Spjelkavik 1995).

Burga & Kratochwil (eds.), BIOMONITORING, 225-242

On the other hand, it is assumed today that arctic warming is strongest during winter and that in summer an increase in cloudiness has to be expected additionally (Maxwell 1992). This would have other ecological consequences than the well-studied increase of temperature and atmospheric CO_2 concentration because increased cloudiness leads to reduced light intensities (photon flux densities), temperature and evaporation on the earth's surface. Since many plants in the Arctic do not grow under light-saturated conditions (Tieszen 1973), less light directly reduces plant primary production, resulting in a shift in the tundra carbon balance towards carbon loss (Wüthrich et al. 1997a and b). The west coast of Spitsbergen is strongly exposed to humid westerly winds from the North Atlantic and therefore often shows cloudy and foggy weather during the vegetation period with late snow melting and a poor plant species inventory (Möller et al. 1998). Many of the plants in this area (e.g. *Racomitrium sp.*) are adapted to foggy weather or to a long-lasting snow cover (e.g. *Phippsia algida*).
In the extraordinarily humid summer of 1996 we observed permanent cloudiness and low temperatures in the coastal plain of Eidembukta (Thannheiser et al. 1998). The plants displayed a very low vitality. The carbon balance of this high arctic tundra showed remarkable carbon losses from soil to atmosphere, since most of the communities did not exceed their light compensation points (Wüthrich et al. 1998). In this paper we present our spatial modeling of vegetation changes under an "increasing cloudiness" scenario at the harsh west coast of Svalbard. The aim of the paper is to determine which changes can be expected regarding the spreading and the composition of plant communities, their phytomass and the spatial carbon balance within a clearly defined high arctic catchment, if with climate change such foggy and cloudy summers as in 1996 become the "average summer" on Svalbard.

2. Methods

2.1. SELECTION OF THE STUDY AREA

The study catchment with an area of 58.45 ha is located north of Eidembukta, in the southern part of Svartfjellstranda next to the sea (south of St. Jonsfjorden, approximately 78°22'N, 12°43'E). It is situated on a shallow coastal plain at Forlandsundet which is typical of Western Spitsbergen (Figure 1). In the course of the last glacial period the coastal plain was formed under glacial impact. During the Holocene it came under marine influence which determined its shape. As a result of the glacio-isostatic rebound during the deglaciation, the plain was uplifted above the sea level. Nowadays single beach terraces separate extended plains from slight depressions. The highest beach ridges of a pronounced beach ridge series are found at 60 m above sea level. The study area is divided by brooks and brooklets which are permanently fed from snow patches during the summer months (Figure 2).
The coastal plain is exposed to the predominant west winds, which cause precipitation in the form of rain, drizzle and coastal fog nearly every day during the short vegetation period (Førland et al. 1997). A permanent cloud cover as well as the damp fogs reduce the duration of bright sunshine and therefore the solar radiation input. Consequently, the temperature sum is lower than in comparable areas of the inner fiord zone of Spitsbergen,

which has a more continental climate and is mostly free of fog. The study area Eidembukta has been chosen for an examination of a present tundra ecosystem because it has been exposed for many years to such conditions as little light, much drizzle and a short vegetation period, which can be expected for other areas on the Svalbard Archipelago with a regionally increasing cloudiness as well.

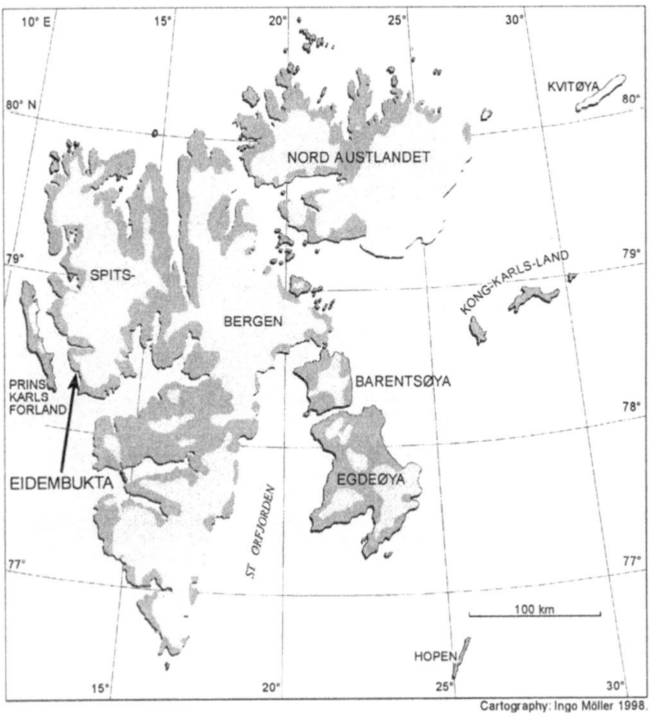

Figure 1. Location of the catchment Eidembukta on Spitsbergen, Svalbard.

2.2. STUDIES OF THE VEGETATION

In the catchment only 58 different vascular plant species could be registered. Particularly among the cryptogams, species dominate that are adapted to fog as e.g. *Stereocaulon* and *Racomitrium* species (Möller et al. 1998). We assume that the unfavourable light conditions entail a poor species diversity of the vegetation and a low primary production. To document the vegetation inventory and its organization, 115 relevés were drawn up according to the floristic-sociological method of Braun-Blanquet (1964), in stands which were uniform in floristic composition and structure over an area larger than the minimal area. The nomenclature of the species followed the Norsk Flora (Elven 1994) for vascular plants, Frey et al. (1995) for mosses [except *Tomenthypnum nitens* (Hedw.) Loeske] and Krog et al. (1980) and Wirth (1995) for lichens. For setting up of the vegetation units the noted syntaxa were used, but the phytocenoses with the neutral term 'community' do not have any association rank due to a missing synsystematical synopsis. Each community is represented by a partial constancy table, the taxa are listed with their constancy class in each community. Some non-significant taxa with low constancy were omitted. Constancy class I means present in less than 20 % of the records, II means present in 20-40 %, III present in 40-60 %, IV 60-80 % and V 80-100 %.

Figure 2. Mosaic of moss tundra vegetation (DST and DSI), snow bed vegetation (PA and CP) and vegetation of wet sites (DC) in the upper third of the catchment Eidembukta.

To obtain accurate information about the spatial distribution of the vegetation units, a vegetation map was prepared, which is based both on conventional vegetation mapping and on computer-aided image processing and interpretation of a scanned colour-infrared aerial photograph, as described by Möller et al. (1998). After the scanning of the aerial photograph the pixel size (and the spatial resolution of the study) was 2.5 x 2.5 m^2. At the end of the image classification, the number of wrongly assigned pixels was about 6 %. For the present vegetation map (Figure 3) small vegetation units - like stands of the **Deschampsietum alpinae**, the *Dryas octopetala* community and *Skua hummocks*, which extend over few square decimetres or metres only - were not classified and reproduced. The vegetation map contains an equal-area representation of the vegetation units which is true to scale. This map allowed to determine the area proportions for each plant community and to upscale from point-measurements to the catchment scale by GIS-techniques. It provides the basis on which the "increasing cloudiness scenario" could be calculated quantitatively and spatially.

2.3. DETERMINATION OF THE PHYTOMASS AND CARBON FLUXES

Carbon content of the vegetation

Mid-season above-ground and below-ground phytomass of the plant communities in the study area was determined according to the "harvesting method" (cf. Steubing & Fangmeier 1992). For each dominant and characteristic community three ¼ square metre plots were cut out, sorted in the field and air-dried. The following work in the laboratory included the cleaning of the samples, the sorting out of the individual plant species and at last their drying for 10 hours at 80°C. Carbon pools in the phytomass were calculated using dry matter data and supposing a carbon content of 50 % of the dry plant material.

CO_2-Fluxes

A mobile CO_2 measurement system was used for measuring carbon fluxes of the different plant communities in the catchment. The measurement system consisted of a pump, an infrared gas analyzer (IRGA, LICOR LI6252, Lincoln, USA), a recorder, a transparent measurement chamber, a reference chamber, a dew point sensor, temperature-sensing elements inside and outside the chamber and a datalogger. Artificial light allowed the measurement of photosynthetic activity under controlled light conditions (100 µmol photons m^{-2} s^{-1}). Photosynthetically active radiation (PAR) was measured using a LICOR LI-189 light meter. The whole system was transportable on a big-wheeled hand-cart and powered by a 12 V lead battery which was periodically recharged using a solar panel. The whole instrumental set-up is described in detail in Wüthrich et al. (1998).

2.4. SCENARIO

The "increasing cloudiness scenario" is based on the present spatial distribution of the plant communities and a maximum-likelihood distribution after 40-50 years of climate change with weather conditions as we described for 1996. The present-day and the expected overall extension of the individual plant communities were calculated by summarizing the assigned pixels in the classified aerial photograph. Attempts to calculate a vegetation map containing the future distribution of the vegetation via logical mathematical terms were abandoned because the results were not satisfying. Only rarely is a certain plant community replaced by another. In principle the plant communities are not vicariant. Furthermore, the micro-relief existing in a certain habitat has such a strong effect on the spreading of the vegetation that it has to be included in the map of the future distribution of the plant communities. For these reasons, we have drawn the scenario map of the plant communities manually on the basis of the present-day distribution. In this way, the empirical knowledge on the micro-relief and on the ecology of the plant communities could be best taken into account.

In our scenario we expect (as in the summer of 1996) an extended snow cover period, a shorter and cooler vegetation period, lower evaporation, increasing wetland- and stream-dominated areas, higher soil humidity and greater overland- and subsurface-flows. Depending on these prevailing conditions the snow bed- and moss communities would slightly extend their distribution, while areas covered by polar desert communities or dwarf-shrub communities would decrease or disappear. The highest increases in spatial coverage are expected for the wetland- and stream-dominated vegetation. Using present and future distribution patterns of the prevailing plant communities, we calculated changes in phytomass and CO_2-fluxes resulting from this scenario. In the Discussion (Chapter 4) some results of the analysis of the soil organic matter are included which were reported before (Wüthrich et al. 1998).

3. Results

3.1. PRESENT AND FUTURE VEGETATION UNITS

The plant cover on the coastal plain of Svartfjellstranda is strongly characterized by the maritime influence of Forlandsundet and gives today a monotonous impression. The micro-relief, the edaphic and the microclimatic conditions differentiate the various plant combinations. Different snow thickness - and thereby different frost protection and varying length of the snow-free period - has the greatest impact. Eighteen vegetation units (associations, communities and their subunits; Table 1) were differentiated. The vegetation units are grouped into five vegetation types of nearly homogenous structure and nearly uniform synecology. The partial constancy tables (Table 2 and 3) reveal the floristic composition of the phytocenoses. As shown in Figure 3 and Table 1, the phytocenoses of the snow beds and the moss tundra have the widest spreading in the catchment area and form a nearly coherent vegetation cover. In contrast, the other communities occur only in small patches.

The present vegetation types and plant communities

Snow is the characterizing element of the vegetation in the study area. The coastal plain is strongly influenced by snow drifts in late winter. Although abundant rain in early summer promotes the melting of the snow, snow remainders are found even in late summer. Due to the limited vegetation period of four to six weeks and the late warming, the snow bed communities dominate the coastal plain. They are located on the terrace surfaces, particularly in slight depressions. Within the snow bed vegetation six communities were differentiated (Table 2), which are adjusted to the length of the snow-free period. The *Phippsia algida* community (PA) is covered with snow for the longest time. The *Cetraria delisei* community (CD, with two forms) and the *Cetraria delisei-Salix polaris* community (CS, with two forms as well) are characteristic for the study area and still need moderate snow protection. The characteristic occurrence of *Stereocaulon*-lichens is promoted by the influence of fog. But the grazing by reindeer contributes to their occurrence as well. In contrast to many other lichen species, *Stereocaulon* is indirectly promoted by grazing, because the small thallus can regenerate quickly.

The vegetation of wet sites and brooklets is represented by two phytocenoses only (Table 2). The Deschampsietum alpinae (DA), which is richer in species, occurs in habitats trickled with water and does not extend to more than some square metres. In contrast to this, the *Bryum cryophilum* community (BC) is bound to running water of small and shallow brooklets.

The moss tundra vegetation is spread widely in the study area and represents a nearly coherent vegetation cover. The vegetation type is characterized by the dominance of mosses (particularly *Drepanocladus uncinatus*, *Tomenthypnum nitens* and *Racomitrium canescens*). Table 3 reveals the composition of the *Drepanocladus uncinatus-Salix polaris* community, which is relatively rich in species (DS, with two forms) and the **Tomenthypnetum involuti** (TI), which is poor in species. The *Racomitrium canescens* community (RC) occurs in small patches on terrace edges and on exposed rocks.

Table 1. Vegetation units in the catchment Eidembukta.

Vegetation Unit	Code
Complex of Snow Bed Vegetation and Vegetation of Wet Sites	
Snow Bed Vegetation:	
1. Cetraria delisei-Salix polaris-com., Stereocaulon-Form	**CSS**
2. Cetraria delisei-Salix polaris-com., Typical Form	**CST**
3. Cetraria delisei-community, Stereocaulon-Form	**CDS**
4. Cetraria delisei-community, Typical Form	**CDT**
5. Cerastio regelii-Poetum alpinae	**CP**
6. Phippsia algida-community	**PA**
Water Vegetation and Vegetation of the Wet Sites	
7. Bryum cryophilum-community	**BC**
8. Deschampsietum alpinae	**DA**
Complex of the Moss-, Heath- and Patch Tundra Vegetation:	
Moss Tundra Vegetation:	
9. Drepanocladus uncinatus-Salix polaris-com., Initial Form	**DSI**
10. Drepanocladus uncinatus-Salix polaris-com., Typical Form	**DST**
11. Racomitrium canescens-community	**RC**
12. Tomenthypnetum involuti	**TI**
13. Skua Hummocks	**SH**
Dwarf-Shrub Heath:	
14. Dryas octopetala-community, Stereocaulon-Form	**DOS**
15. Dryas octopetala-community, Typical Form	**DOT**
Patch Tundra Vegetation:	
16. Saxifraga oppositifolia-Salix polaris-com., Stereoc.-Form	**SSS**
17. Saxifraga oppositifolia-Salix polaris-com., Typical Form	**SST**
18. Crustose Lichens-Saxifraga oppositifolia-community	**KS**
Others:	
19. Water	
20. Bare Surfaces	

Table 2. Partial constancy table of the snow bed vegetation and the vegetation of wet sites.

Vegetation type	**Snow Bed Vegetation**						**Wet Site Veg.**	
Plant community	**PA**	**CP**	**CDT**	**CDS**	**CST**	**CSS**	**DA**	**BC**
Number of relevés	5	7	5	5	12	6	5	5
Mean cover (%)	45	38	83	80	85	88	53	73
Mean number of species	16	16	18	20	25	24	19	8
Phippsia algida	V	I	III	II	.	.	V	.
Poa alpina var. alpina	IV	V	V	IV	.	.	IV	IV
Cerastium regelii	V	V	III	I	I	.	V	IV
Salix polaris	.	II	IV	IV	V	V	.	.
Deschampsia alpina	.	I	I	.	.	.	V	.
Saxifraga cernua	V	V	V	V	V	IV	V	IV
Saxifraga oppositifolia	IV	V	IV	IV	V	V	V	II
Saxifraga nivalis	V	III	V	V	V	IV	III	II
Saxifraga cespitosa	IV	IV	IV	V	V	IV	II	I
Oxyria digyna	V	IV	III	V	IV	IV	III	IV
Cochlearia groenlandica	V	V	V	II	III	.	V	IV
Cerastium alpinum	V	II	V	V	V	V	III	.
Luzula arcuata ssp. confusa	.	V	V	II	V	V	.	.
Draba lactea	.	V	.	.	I	III	IV	.
Juncus biglumis	.	I	.	.	I	II	III	.
Saxifraga foliosa	.	.	I	I	III	.	IV	.
Equisetum scirpoides	.	.	.	I	III	II	.	.
Bistorta vivipara	.	.	.	.	IV	II	.	II
Pedicularis hirsuta	.	.	.	.	V	V	.	.
Draba corymbosa	.	.	.	.	IV	III	.	.
Cardamine pratensis ssp. polemonioides	.	.	.	.	.	.	V	.
Ranunculus hyperboreus ssp. arnellii	.	.	.	.	.	.	IV	.
Cetraria delisei	II	I	V	V	V	V	.	.
Stereocaulon alpinum & botryosum	II	III	IV	V	V	V	.	.
Bryum cryophilum	III	.	.	.	.	.	.	V
Cetraria islandica	III	I	IV	III	V	V	.	.
Crustose lichens	III	III	I	III	IV	III	.	.
Distichium capillaceum	V	V	.	II	II	II	V	.
Campylium stellatum	V	V	.	I	II	II	.	.
Bryum spec.	.	V	II	IV	III	II	.	.
Scorpidium scorpioides	.	II	.	.	I	.	V	V
Drepanocladus uncinatus	.	.	V	V	V	V	.	.
Polytrichum juniperinum	.	.	III	V	III	V	.	.
Ditrichum flexicaule	.	.	II	.	III	III	III	.
Polytrichum alpinum	V	III	.	.	III	.	.	.
Ptilidium ciliare	.	.	.	I	V	IV	.	.
Dicranum angustum & spadiceum	.	.	.	I	IV	V	.	.
Hypnum revolutum	.	.	.	.	II	V	II	.
Tomenthypnum nitens	.	.	.	.	I	III	II	.

Table 2 continued

Pogonatum urnigerum	.	.	II	.	.	.	III	.
Tortella fragilis	.	.	.	.	V	I	.	.
Drepanocladus revolvens	.	.	.	.	.	.	III	II
Racomitrium canescens	.	.	.	.	.	V	.	.
Oncophorus wahlenbergii	.	.	.	.	.	IV	.	.
Dicranum muehlenbeckii	.	.	.	.	.	III	.	.
Drepanocladus exannulatus	.	.	.	.	.	.	III	.
Campylium stellatum	.	.	.	.	.	.	III	.
Orthothecium chryseon	.	.	.	.	.	.	III	.
Bryum pseudotriquetrum	.	.	.	.	.	.	III	.
Bryum pallescens & wrightii	.	.	.	.	.	.	.	IV

The present dwarf-shrub heath is characterized by the cushion plant *Dryas octopetala* (DO). It settles only in small patches mostly on terrace edges in the higher third of the study area. The community occurs in a typical and a *Stereocaulon*-form (Table 3). From the sociological point of view, it can be seen as a fragment of the association **Polari-Dryadetum**.

The Polar Desert vegetation is characterized by a sparse and scattered vegetation cover. Only few vegetation islands are found on frequently disturbed sites near the beach and on the top of the highest elevations. The dominant plant in all three communities is *Saxifraga oppositifolia* (Table 3). It is conspicuous that the *Saxifraga oppositifolia-Salix polaris* community (SS, two forms) contains many cryptogams and therefore is rich in species. The *crustose lichens-Saxifraga oppositifolia* community (KS) is found on the tops of the beach ridges. The coverage of this community is less than 50 %.

The vegetation after climate change with increased cloudiness

The future vegetation will be characterized by the following changes: snow bed communities will expand in total. But certain communities will expand to a different extent. The surface areas of the *Phippsia algida* community (PA) and of the typical form of the *Cetraria delisei-Salix polaris* community (CST) will be slightly reduced (Table 4). We assume that the subunits with *Stereocaulon* in the *Cetraria delisei-Salix polaris* community (CS) and in the *Cetraria delisei* community (CD) do not only invade areas of other communities, but also habitats of the typical subunit of the same community (Figure 4). This is a dislocation and a shift of the centre within the communities. Since the brooklet- and wet sites increase in area due to changes in moisture conditions, the corresponding plant communities expand accordingly. However, as they are restricted to certain habitats - the *Bryum cryophilum* community (BC) to brooklets and the **Deschampsietum alpinae** (DA) to wetland areas - the increases in the surface area are moderate. In contrast to the snow bed-, the brooklet- and the wetland vegetation, which profit from an increase in surface areas, the surface area of the moss tundra vegetation will much decrease. As in the *Cetraria delisei* community, a shift of the centre of the community is to be expected in the *Drepanocladus uncinatus-Salix polaris* community with its two subunits. The typical subunit (DST) decreases very strongly, but the initial form (DSI) increases in surface area, because it mainly takes over areas formerly covered by DST. Since the dwarf-shrub heath nowadays grows at its threshold of existence, we expect an extinction of the *Dryas* community under even worse weather conditions. Consequently, it will be superseded. The Polar Desert vegetation will hardly change. However, the decrease of the *crustose lichens-Saxifraga oppositifolia* community (KS) in relation to the increase of the *Saxifraga oppositifolia-Salix polaris* community (SS) is very important but also surprising, because in this case a less productive community (KS) is replaced by a clearly more productive one (SS).

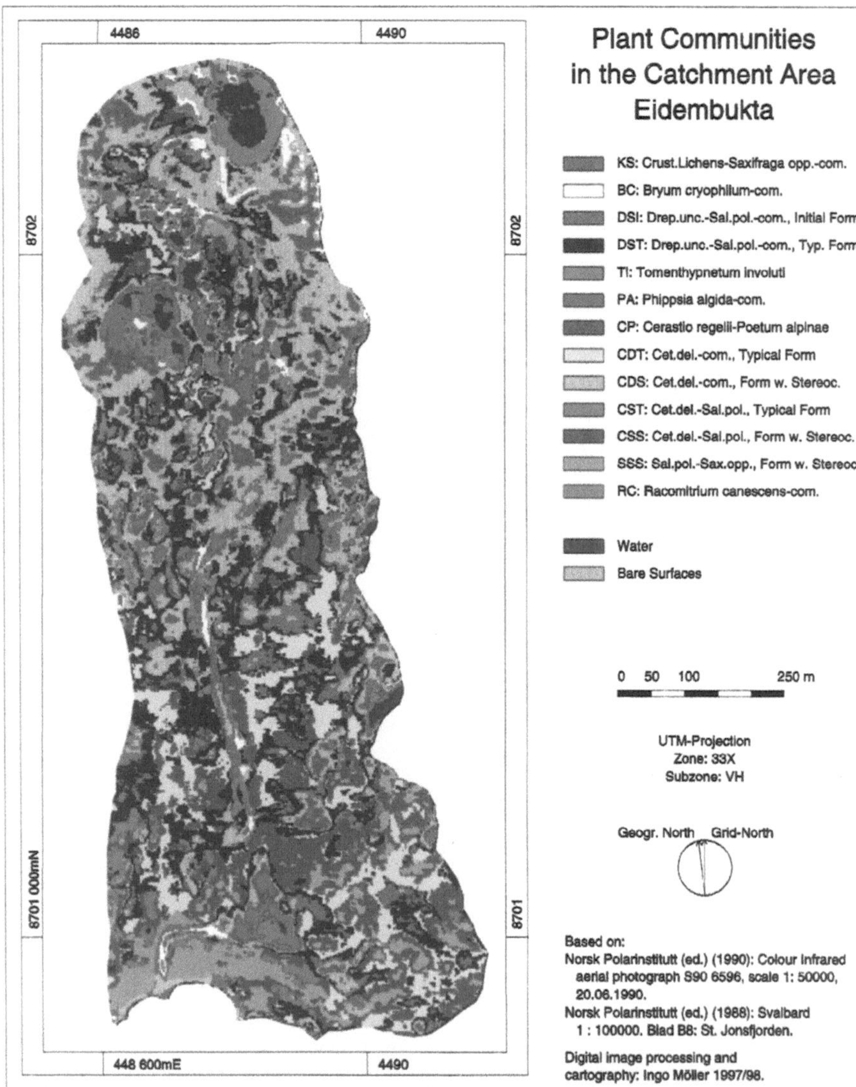

Figure 3. Vegetation map of the catchment Eidembukta, present-day situation.

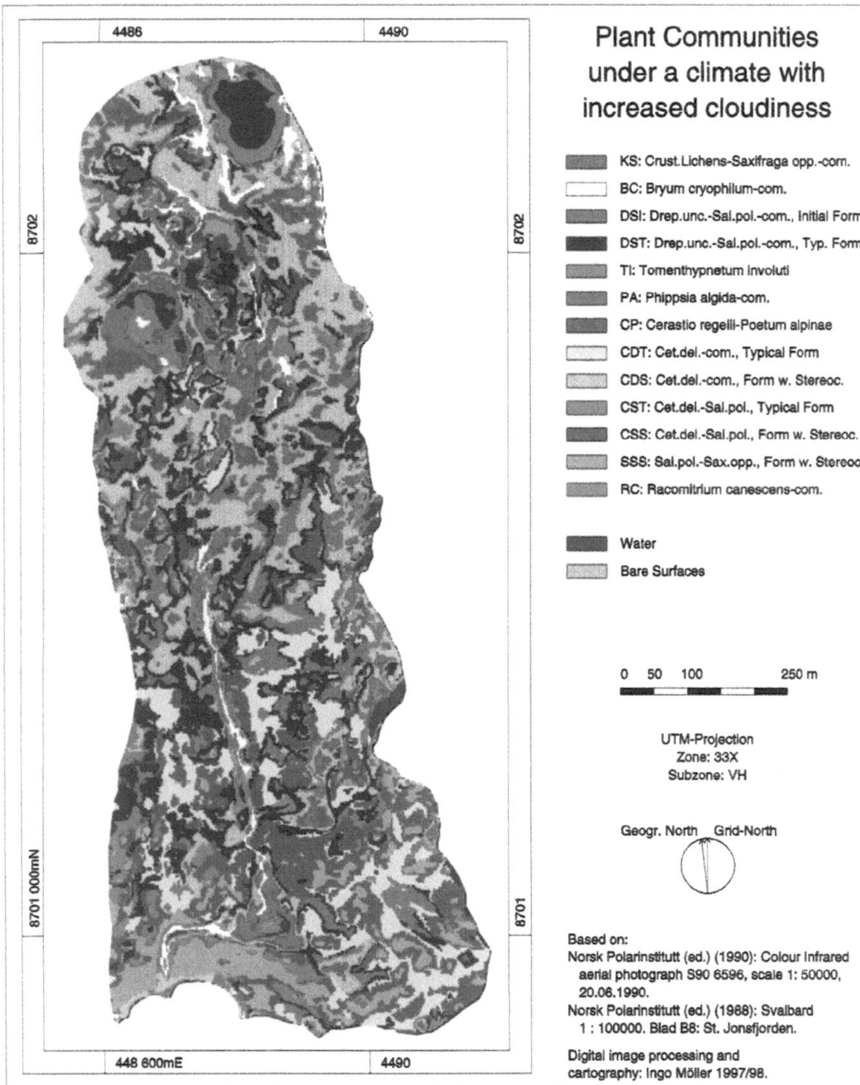

Figure 4. Vegetation map of the catchment Eidembukta, situation "increasing cloudiness scenario", ≈ year 2036.

Table 3. Partial constancy table of the moss tundra- and the polar desert vegetation, and the dwarf-shrub heath.

Vegetation type	**Moss Tundra and Skua Hummocks**					**Polar Desert Veg.**			**D. Shr. Heath**	
Plant community	**TI**	**DST**	**DSI**	**RC**	**SH**	**KS**	**SST**	**SSS**	**DOT**	**DOS**
Number of relevés	2	6	5	5	9	5	5	9	5	5
Mean cover (%)	100	90	93	98	90	65	88	90	83	83
Mean number of species	19	17	18	23	17	17	23	23	21	19
Salix polaris	5	V	V	V	V	V	V	V	IV	IV
Saxifraga oppositifolia	5	V	IV	V	II	V	V	V	IV	IV
Dryas octopetala	.	.	.	.	.	.	.	.	V	V
Cerastium alpinum	5	V	III	IV	V	V	IV	V	II	II
Luzula arcuata ssp. confusa	5	V	IV	V	III	V	V	V	III	III
Bistorta vivipara	.	II	V	II	II	V	V	V	IV	IV
Oxyria digyna	5	V	V	V	IV	II	V	IV	.	.
Stellaria longipes	5	II	I	III	V	III	II	II	.	.
Saxifraga cernua	.	V	IV	IV	IV	V	IV	IV	.	.
Saxifraga cespitosa	.	IV	III	I	III	III	V	V	.	.
Pedicularis hirsuta	.	.	I	I	.	II	IV	IV	II	II
Saxifraga nivalis	.	.	II	I	II	III	IV	IV	.	.
Cochlearia groenlandica	3	I	.	I	III	.	.	.	.	.
Draba lactea	.	I	.	IV	.	.	IV	IV	.	.
Equisetum scirpoides	.	I	.	III	.	.	I	II	.	.
Papaver dahlianum	.	.	.	II	.	IV	II	II	.	.
Draba subcapitata	.	.	.	III	.	.	II	II	.	.
Saxifraga hirculus	.	I	III	.	.	.	.	.	.	.
Poa alpina var. vivipara	.	.	I	.	IV	.	.	.	.	.
Carex rupestris	.	.	.	.	.	.	III	III	.	.
Saxifraga hyperborea	.	.	.	.	III	.	.	.	.	.
Draba pauciflora	.	.	.	.	.	V	.	.	.	.
Tomenthypnum nitens	5	I	II	IV	IV	.	II	II	II	II
Drepanocladus uncinatus	.	V	V	V	IV	.	V	V	I	I
Racomitrium canescens	5	II	.	V	.	I	III	III	V	V
Crustose lichens	.	I	I	.	II	V	V	III	III	III
Stereocaulon alpinum & botryosum	5	.	IV	V	.	II	V	V	V	V
Cetraria islandica	5	V	III	V	II	IV	V	V	V	V
Pleurozium schreberi	5	II	I	III	IV	.	I	I	III	III
Bryum spec.	5	I	IV	.	IV	V	III	III	I	II
Ptilidium ciliare	5	III	I	IV	.	.	III	III	I	I
Thamnolia vermicularis	5	I	.	III	II	.	II	I	V	V
Distichium capillaceum	.	V	III	.	I	II	IV	IV	V	V
Polytrichum alpinum	.	V	V	III	IV	IV	.	.	IV	IV
Cetraria nivalis	.	III	.	IV	II	.	I	I	V	IV
Cetraria delisei	.	.	II	I	.	V	V	V	V	V
Pogonatum urnigerum	5	.	.	.	.	I	I	I	I	I

Table 3 continued

Ditrichum flexicaule	.	II	.	.	II	.	III	III	I	II
Oncophorus wahlenbergii	.	.	I	.	II	.	V	V	V	V
Hylocomium splendens	.	IV	.	IV	III	.	.	.	I	I
Dicranum scoparium & spec.	.	.	.	IV	.	.	III	III	III	III
Polytrichum juniperinum	.	.	.	I	II	.	IV	IV	.	.
Dicranum elongatum	5	.	.	.	.	.	.	.	II	I
Peltigera rufescens	5	I	.	.	.	.	.	.	.	.
Tortula ruralis	5	.	.	.	IV	.	.	.	.	.
Dicranum angustum & spadiceum	.	V	III	.	.	.	.	.	.	.
Drepanocladus revolvens	.	.	I	.	III	.	.	.	.	.
Hypnum bambergeri	.	.	.	.	.	.	.	.	V	V
Alectoria nigricans	.	.	.	.	.	.	.	.	IV	IV

3.2. THE PRESENT AND FUTURE PHYTOMASS-C IN THE CATCHMENT AREA EIDEMBUKTA

The present standing crop and carbon allocation in the catchment

The highest phytomass (3.18 kg m^{-2} was measured in the *Skua hummocks* (SH). This is the consequence of the strong manuring in these habitats by bird excrements. The second highest substance production (1.5 kg m^{-2} was measured in the *Drepanocladus uncinatus-Salix polaris*-moss tundra community (DS), where the phytomass pool is less than half of the phytomass of the *Skua hummocks*. Since the *Racomitrium canescens* community (RC) has an average phytomass of 1.29 kg m^{-2} the moss tundra vegetation includes the vegetation units with the highest phytomass in the study area. A great variability of the phytomass was observable in the communities of the Polar Desert vegetation. While the *crustose lichens-Saxifraga oppositifolia* community (KS) shows a standing crop as low as 0.36 kg m^{-2} the *Saxifraga oppositifolia-Salix polaris* community (SS) has a mean phytomass of 1.46 kg m^{-2} At first glance it is surprising that a Polar Desert habitat showed higher phytomass than the *Dryas octopetala*-dwarf-shrub heath (DO, 1.38 kg m^{-2}. But this can be explained by the fact that the *Dryas octopetala* community grows at its threshold of existence and is fragmentarily developed only. The lowest phytomass was measured within the snow bed communities. In accordance with the shorter snow-free period, the phytomass diminishes from 0.82 kg m^{-2}in the *Cetraria delisei-Salix polaris* community (CS) to 0.15 kg m^{-2}in the **Cerastio regelii-Poetum alpinae** (CP) and 0.13 kg m^{-2} in the *Phippsia algida* community (PA) (Table 4).

Assuming a carbon content of 50 % of the plant material, the carbon reserves in the phytomass of the present coastal tundra are between 0.065-1.59 kg C m^{-2} The above-ground phytomass is clearly less than the below-ground phytomass in most plant communities ($\approx 1:1.9$-2.5). Only in the extremely moss-rich habitats and the pioneer habitats the ratio shifts towards above-ground phytomass ($\approx 1:0.8$-1.2). Extrapolating the point data of the phytomass to the whole catchment by means of the community distribution, a total dry phytomass of $\approx$573.8 t results in the area of 58.45 ha. This corresponds to an average phytomass of about 1 kg m^{-2}terrestrial surface, if water surfaces and bare surfaces without any vegetation were subtracted. Thus, the present mean ratio between above-ground and below-ground phytomass in the whole catchment is 1 : 2.11.

Table 4. Changes of plant community area, total phytomass and seasonal CO_2-loss in a scenario with increased cloudiness.

	Community Area (m^2)		Total Phytomass (t)		CO_2-Loss (t CO_2 /season)	
Plant Community	Today (1996)	Scenario (≈ 2036)	Today (1996)	Scenario (≈ 2036)	Today (1996)	Scenario (≈ 2036)
KS	8'759.4	6'825.9	0.319	0.248	0.086	0.067
TI	15'079.7	13'277.3	2.037	1.793	---	---
DS	163'231.6	138'643.4	24.496	20.806	28.832	24.489
BC	9'194.8	12'846.5	1.2419	1.735	---	---
PA	14'943.9	13'829.7	0.193	0.179	---	---
CP	36'208.2	36'676.3	0.550	0.557	2.183	2.211
CD	180'788.0	189'055.8	14.745	15.419	8.241	8.618
CS	109'293.9	120'352.1	8.914	9.816	15.117	16.661
SS	30'454.4	32'762.4	4.446	4.783	3.938	4.239
RC	3'370.8	3'637.7	0.435	0.469	0.383	0.413
Total	*584'499.0*	*584'499.0*	*57.376*	*55.806*	*58.780*	*56.695*
Difference		*0.0 m²*		*-2.737 %*		*-3.5469 %*

The future standing crop and carbon allocation

Assuming a shift of vegetation distribution as described in Chapter „*The vegetation after climate change with increases cloudiness*", the organic carbon stored in the vegetation will be reduced due to the decrease of the total phytomass. It is plausible that several years of increased cloudiness reduce the phytomass of the present-day plant communities (changing within-community phytomass), because energy reserves in roots, shoots and stems might be depleted after several years of bad weather conditions. However, to extract the effects of within-community changes, we calculated the future phytomass assuming an unchanged phytomass of the individual plant community. The phytomass in the whole catchment area decreases with this scenario from 573.8 t to 558.1 t; a decline of about 2.8 % only. The real future phytomass within the catchment should probably be corrected to a stronger decline, because an average decrease of standing crop has to be assumed under even worse living conditions additionally. The future ratio of above-ground to below-ground phytomass remains with 1 : 2.08 on the same level as the present-day ratio; with a slight shift to the above-ground phytomass. This tendency correlates with the observation that in pioneer and extremely moss-rich habitats the ratio of above-ground to below-ground phytomass shifts towards the above-ground phytomass (see Chapter "*The present standing crop and carbon allocation in the catchment*").

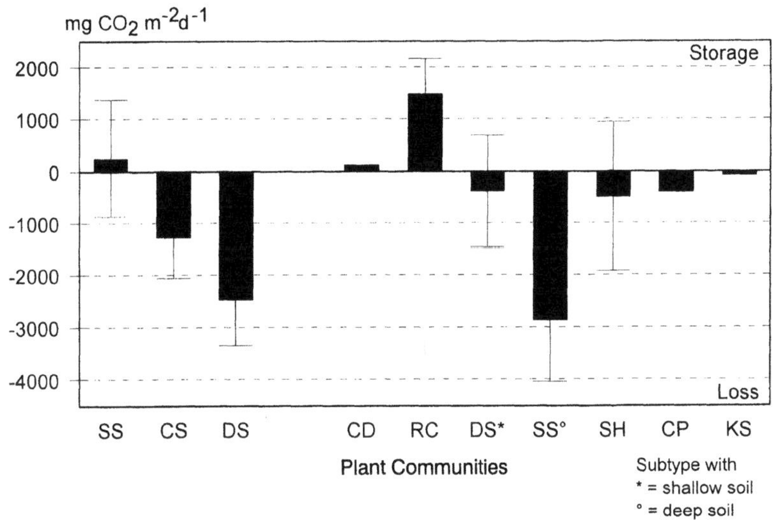

Figure 5. Today's CO_2-fluxes under natural light.

3.3. THE PRESENT AND FUTURE CARBON DIOXIDE FLUXES AND THE CARBON BALANCE

In the summer of 1996, the photosynthetic active radiation (PAR) between 10.00 a.m. and 16.00 p.m. was between 125-520 $\mu mol\ m^{-2}\ s^{-1}$ (mean 340 $\mu mol\ m^{-2}\ s^{-1}$) at the foggy west coast of Spitsbergen (Figure 5). Under these poor light conditions the majority of the measurements showed CO_2-emission to the atmosphere (Wüthrich et al. 1998). Apart from the *Saxifraga oppositifolia-Salix polaris* community (SS) only the *Racomitrium canescens* community (RC) and the *Cetraria delisei* community (CD) were able to store carbon. In these habitats the ecosystem light compensation point was exceeded at a PAR of 260-340 $\mu mol\ m^{-2}\ s^{-1}$. For all other habitats even higher light intensities were necessary to store carbon. Under low light conditions (PAR of 100 $\mu mol\ m^{-2}\ s^{-1}$) which are typical of low angles of incidence of the night-time polar sun and cloudy sky, all the habitats showed marked CO_2-fluxes from soil to atmosphere (carbon loss, see Table 4).

Using these data, the total daily carbon loss of each plant community was calculated assuming 8 hours of daylight ($\approx 340\ \mu mol\ m^{-2}\ s^{-1}$) and 16 hours of low light (100 $\mu mol\ m^{-2}\ s^{-1}$). Then, a rough estimation of seasonal CO_2-loss of each habitat was computed assuming a 60-day period of biological activity (Table 4). The seasonal loss of CO_2 for the classified area of the catchment amounted to 58.779 t CO_2 (Table 4). The average seasonal carbon loss rate of the presented communities using the gas flux pathway was estimated as 301 kg C ha^{-1}. The calculations of the seasonal carbon loss after changes in plant community distribution (but unchanged climate conditions) showed that the seasonal CO_2-emission will be slightly reduced by 3.5 % to 56.694 t CO_2. Accordingly the average seasonal carbon loss rate using the gas flux pathway will be reduced to 292 kg C ha^{-1}.

4. Discussion and Conclusions

In the study catchment Eidembukta the plant cover composed of 58 vascular plant species is strongly characterized by the snow bed- and the moss tundra vegetation. Particularly in slight depressions the vegetation period is limited to four or six weeks. The moss tundra is spread widely in the study area and represents a nearly closed vegetation cover dominated by *Drepanocladus uncinatus*, *Tomenthypnum nitens* and *Racomitrium canescens.* Other communities than the vegetation of wet sites and brooklets (dominated by *Deschampsia alpina* and *Bryum cryophilum*), the dwarf-shrub heath (*Dryas octopetala*) and the Polar Desert vegetation (dominated by *Saxifraga oppositifolia* and *Salix polaris*) occur in small patches only.
We expected in our "increasing cloudiness scenario" an extended snow cover period, a shorter and cooler vegetation period, lower evaporation, increasing wetland and brooklet areas, higher soil humidity and greater overland- and subsurface-flows. Depending on these prevailing conditions the snow bed- and the moss communities would slightly extend their distribution, while areas covered by Polar Desert communities or dwarf-shrub communities would decrease or disappear. The highest increases in spatial coverage are expected for the wetland and brooklet vegetation, which however represent only a small proportion of the catchment area. But a realistic map of the future vegetation distribution does not differ significantly from the present one (compare Figures 3 and 4). Using present and future distribution patterns of the prevailing plant communities, we calculated changes in phytomass and CO_2-fluxes resulting from this scenario.
The "future phytomass" calculation for the catchment of Eidembukta has shown that a changing plant community distribution will entail only slight changes in total phytomass in the catchment. In this study present-day phytomass data are presented, which were measured in an exceptionally cloudy and rainy summer. However, it is plausible that several years of an increased cloudiness might further reduce the phytomass of the present-day plant communities (changing within-community phytomass), because energy reserves in roots, shoots and stems might be depleted after several years with bad weather conditions. It is possible that the C/N ratio and hence the litter quality might change with this scenario as well. These kinds of changes are not included in our scenario modeling.
In the summer of 1996, rainy weather with a permanently cloudy sky caused a strong decrease of carbon assimilation in the plant communities. Despite the midnight sun, all habitats lost considerable amounts of carbon by respiration to the atmosphere. Even under fairly good light conditions only few plant communities were able to store more CO_2 by photosynthesis than was lost by respiration simultaneously. The daily carbon loss expressed as mean C-loss of all measured habitats in the whole catchment is estimated at a rate of 0.581g C m^{-2} in such a chilly and damp summer, of which only 1.31 % leave the system as TOC by rivers (Wüthrich et al. 1998).

This result does not correspond to the relatively high contents of carbon in high arctic soils (5.98 kg C m^{-2}) reported by Wüthrich et al. (1998). So the question arises: Why is so much carbon stored in gelic soils if there is a clear carbon loss even under daylight conditions in the middle of the high arctic summer? Two explanations seem to be plausible:

A. The year of 1996 was "completely out of order". Usually the light climate is much better at Eidembukta, resulting in higher primary productivity which can compensate carbon losses in extremely cloudy and rainy summers.

B. Possibly the present soil carbon pool is a relic of the "Little Ice Age". Although the temperature climate was somewhat harsher in the middle of the last century, there is evidence (e.g. Kutzbach 1987) that the climate was more continental ("cold and arid") at that time. Possibly light was less limiting for the primary producers in the more continental period than it is today. Even the summer temperatures might have been higher due to longer periods with direct sunshine.

At this point we cannot decide which of the two explanations is more realistic. Since local climate effects create large gradients between places as Eidembukta and the next relevant meteorological station (Isfjord Radio, approx. 45 km south), it is impossible to substantiate our observations by long-term micro-meteorological data. Therefore only the low number of plant species and phytoecological extrapolations (e.g. lack of large coherent dwarf-shrub stands) provide some evidence that explanation A is less likely than B.
However, our data show that changing the plant community distribution in the catchment in our "increasing cloudiness scenario" reduced the average seasonal carbon loss rate by 3.5 % only. If we reduce for instance the length of the microbial activity period from 60 to 50 days, the carbon loss will be reduced by 19.6 %! These calculations show that changes of the distribution of the plant communities as well as alterations in the TOC loads of rivers (only 1.31 % of the present C-flux) are much less effective in changing the carbon balance of the high arctic tundra than changes of important prevailing ecological conditions (e.g. reducing the length of the vegetation period or changing PAR between the light compensation point and the light saturation level) for the present plant- and microbial communities. Hence, international tundra research should focus more precisely on the direct and combined effects of light-, temperature- and precipitation changes (particularly influencing the length of the snow-free period and the humidity) on the carbon balance of the prevailing plant (-soil-) communities. In parallel, it is important to investigate the size, turn-over and age of carbon pools in the present plant (-soil-) communities to make realistic predictions on future tundra processes.

References

Braun-Blanquet, J. 1964. Pflanzensoziologie. Grundzüge der Vegetationskunde. 3th ed., Springer., Wien a.o.
Elvebakk, A. & Spjelkavik, S. 1995. The ecology and distribution of Empetrum nigrum ssp. hermaphroditum on Svalbard. Nord. J. Bot. 15(5): 541-552.
Elven, R. 1994. Norsk Flora. Founded by J. & D.T. Lid. 6th ed. Det Norske Samlaget, Oslo.

Førland, E.J., Hanssen-Bauer, I. & Nordli, P.Ø. 1997. Climate statistics and longterm series of temperature and precipitation at Svalbard and Jan Mayen. DNMI Rep., Klima 21: 1-72.

Frey, W., Frahm, J.-P., Fischer, E. & Lobin, W. 1995. Die Moos- und Farnpflanzen Europas. 6. ed., Fischer, Stuttgart a.o. (Kleine Kryptogamenflora, Bd IV).

Gjessing, Y.T. & Øvstedal, D.O. 1976. Energy budget and ecology of two vegetation types in Svalbard. Astarte 8(2): 83-92.

IPCC (ed) 1996. Climate Change 1995. The Science of Climate Change. Univ. Press, Cambridge.

Krog, H., Östhagen, H. & Tönsberg, T. 1980. Lavflora - Norske busk- og bladlav. Universitetsforl., Oslo.

Kutzbach, J.E. 1987. Model simulations of the climatic patterns during the deglaciation of North America. pp. 425-446. In: Ruddiman, W.F. & Wright, H.E. (eds), North America and adjacent oceans during the deglaciation of North America. Geology of North America. Boulder, Colorado.

Maxwell, B. 1992. Arctic climate: potential for change under global warming. pp. 11-34. In: Chapin III, F.S. et al. (eds), Arctic ecosystems in a changing climate: an ecophysiological perspective. Acad. Press, San Diego.

Möller, I., Thannheiser, D. & Wüthrich, C. 1998. Eine pflanzensoziologische und vegetationsökologische Fallstudie in Westspitzbergen. Geoökodynamik XIX(1/2): 1-18.

Oberbauer, S.F., Oechel, W.C. & Riechers, G.H. 1986. Effects of CO_2 enrichment and nutrition on growth, photosynthesis and nutrient concentration of Alaskan tundra plant species. Can. J. Bot. 64: 2993-2998.

Rønning, O.I. 1979. Svalbards flora. Norsk Polarinst. Polarbok, Oslo.

Skye, E. 1989. Changes to the climate and flora of Hopen island during the last 100 years. Arctic 42: 323-332.

Steffensen, E.L. 1982. The climate at Norwegian Arctic Stations. DNMI Rep., Klima, Oslo: 1-44.

Steubing, L. & Fangmeier, A. 1992. Pflanzenökologisches Praktikum. Ulmer, Stuttgart.

Thannheiser, D. 1994. Vegetationsgeographisch-synsoziologische Untersuchungen am Liefdefjord (NW-Spitzbergen). Z. Geomorph. N.F., Suppl. 97: 205-214.

Thannheiser, D. 1996. Spitzbergen. Geogr. Rundschau 48(5): 268-274.

Thannheiser, D., Möller, I. & Wüthrich, C. 1998. Eine Fallstudie über die Vegetationsverhältnisse, den Kohlenstoffhaushalt und mögliche Auswirkungen klimatischer Veränderungen in Westspitzbergen. Verh. Ges. Ökologie 28:475-484.

Tieszen, L.L. 1973. Photosynthesis and respiration in Arctic tundra grasses; field light intensity and temperature responses. Arc. Alp. Res. 5(3): 239-251.

Wirth, V. 1995. Flechtenflora. 2nd. ed. Ulmer, Stuttgart.

Wookey, P.A., Robinson, C.H., Parsons, A.N., Welker, J.M., Press, M.C., Callaghan, T.V. & Lee, J.A. 1995. Environmental constraints on the growth, photosynthesis and reproductive development of Dryas octopetala at a high Arctic polar semi-desert, Svalbard. Oecologia 102: 478-489.

Wüthrich, C. 1991. Landschaftsökologische Umweltforschung: Beiträge zu den Wechselwirkungen zwischen biotischen und abiotischen Faktoren im hocharktischen Ökosystem Spitzbergen. Die Erde 122: 335-352.

Wüthrich, C., Döbeli, C., Schaub, D. & Leser, H. 1994. The pattern of carbon-mineralisation in the high-Arctic tundra (Western and Northern Spitsbergen) as an expression of landscape ecologic environment heterogeneity. Z. Geomorph. N.F., Suppl. 97: 251-264.

Wüthrich, C. & Schaub, D. 1997a. Reaktionspotential unterschiedlicher Moortypen für Änderungen von Licht und Bodenwasserhaushalt. Bulletin BGS 21: 41-48.

Wüthrich, C., Nilsen, J. & Schaub, D. 1997b. Effects of increased cloudiness and water table changes on carbon balance in Arctic peatlands. Annales Geophysicae, Suppl. 15: 357.

Wüthrich, C., Möller, I. & Thannheiser, D. 1998. Soil carbon losses due to increased cloudiness in a high Arctic tundra watershed (Spitsbergen). pp. 1165-1172. In: Lewkowicz, A.G. & Allard, M. (eds), Proceedings, Seventh International Conference on Permafrost, Yellowknife, June 23-27. Nordicana, Université Laval.

Tasks for vegetation science

1. E.O. Box: *Macroclimate and plant forms.* An introduction to predictive modelling in phytogeography. 1981 ISBN 90-6193-941-0
2. D.N. Sen and K.S. Rajpurohit (eds.): *Contributions to the ecology of halophytes.* 1982 ISBN 90-6193-942-9
3. J. Ross: *The radiation regime and architecture of plant stands.* 1981 ISBN 90-6193-607-1
4. N.S. Margaris and H.A. Mooney (eds.): *Components of productivity of Mediterranean-climate regions.* Basic and applied aspects. 1981 ISBN 90-6193-944-5
5. M.J. Müller: *Selected climatic data for a global set of standard stations for vegetation science.* 1982 ISBN 90-6193-945-3
6. I. Roth: *Stratification of tropical forests as seen in leaf structure* [Part 1]. 1984 ISBN 90-6193-946-1
   For Part 2, see Volume 21
7. L. Steubing and H.-J. Jäger (eds.): *Monitoring of air pollutants by plants.* Methods and problems. 1982 ISBN 90-6193-947-X
8. H.J. Teas (ed.): *Biology and ecology of mangroves.* 1983 ISBN 90-6193-948-8
9. H.J. Teas (ed.): *Physiology and management of mangroves.* 1984 ISBN 90-6193-949-6
10. E. Feoli, M. Lagonegro and L. Orlci: *Information analysis of vegetation data.* 1984 ISBN 90-6193-950-X
11. Z. Sŷesták (ed.): *Photosynthesis during leaf development.* 1985 ISBN 90-6193-951-8
12. E. Medina, H.A. Mooney and C. Vzquez-Ynes (eds.): *Physiological ecology of plants of the wet tropics.* 1984 ISBN 90-6193-952-6
13. N.S. Margaris, M. Arianoustou-Faraggitaki and W.C. Oechel (eds.): *Being alive on land.* 1984 ISBN 90-6193-953-4
14. D.O. Hall, N. Myers and N.S. Margaris (eds.): *Economics of ecosystem management.* 1985 ISBN 90-6193-505-9
15. A. Estrada and Th.H. Fleming (eds.): *Frugivores and seed dispersal.* 1986 ISBN 90-6193-543-1
16. B. Dell, A.J.M. Hopkins and B.B. Lamont (eds.): *Resilience in Mediterranean-type ecosystems.* 1986 ISBN 90-6193-579-2
17. I. Roth: *Stratification of a tropical forest as seen in dispersal types.* 1987 ISBN 90-6193-613-6
18. H.-G. Dässler and S. Börtitz (eds.): *Air pollution and its influence on vegetation.* Causes, Effects, Prophy´laxis and Therapy. 1988 ISBN 90-6193-619-5

Tasks for vegetation science

19. R.L. Specht (ed.): *Mediterranean-type ecosystems.* A data source book. 1988 ISBN 90-6193-652-7
20. L.F. Huenneke and H.A. Mooney (eds.): *Grassland structure and function.* California annual grassland. 1989 ISBN 90-6193-659-4
21. B. Rollet, Ch. Hägermann and I. Roth: *Stratification of tropical forests as seen in leaf structure*, Part 2. 1990 ISBN 0-7923-0397-0
22. J. Rozema and J.A.C. Verkleij (eds.): *Ecological responses to environmental stresses.* 1991 ISBN 0-7923-0762-3
23. S.C. Pandeya and H. Lieth: *Ecology of Cenchrus grass complex.* Environmental conditions and population differences in Western India. 1993 ISBN 0-7923-0768-2
24. P.L. Nimis and T.J. Crovello (eds.): *Quantitative approaches to phytogeography.* 1991 ISBN 0-7923-0795-X
25. D.F. Whigham, R.E. Good and K. Kvet (eds.): *Wetland ecology and management.* Case studies. 1990 ISBN 0-7923-0893-X
26. K. Falinska: *Plant demography in vegetation succession.* 1991 ISBN 0-7923-1060-8
27. H. Lieth and A.A. Al Masoom (eds.): *Towards the rational use of high salinity tolerant plants*, Vol. 1: Deliberations about high salinity tolerant plants and ecosystems. 1993 ISBN 0-7923-1865-X
28. H. Lieth and A.A. Al Masoom (eds.): *Towards the rational use of high salinity tolerant plants*, Vol. 2: Agriculture and forestry under marginal soil water conditions. 1993 ISBN 0-7923-1866-8
29. J.G. Boonman: *East Africa's grasses and fodders.* Their ecology and husbandry. 1993 ISBN 0-7923-1867-6
30. H. Lieth and M. Lohmann (eds.): *Restoration of tropical forest ecosystems.* 1993 ISBN 0-7923-1945-1
31. M. Arianoutsou and R.H. Groves (eds.): *Plant-animal interactions in Mediterranean-type ecosystems.* 1994 ISBN 0-7923-2470-6
32. V.R. Squires and A.T. Ayoub (eds.): *Halophytes as a resource for livestock and for rehabilitation of degraded lands.* 1994 ISBN 0-7923-2664-4
33. T. Hirose and B.H. Walker (eds.): *Global change and terrestrial ecosystems in monsoon Asia.* 1995 ISBN 0-7923-0000-0
34. A. Kratochwil (ed.): *Biodiversity in Ecosystems: Principles and case studies of different complexity levels.* 1999 ISBN 0-7923-5717-5
35. C.A. Burga and A. Kratochwil (ed.): *Biomonitoring: General and applied aspects on regional and global scales.* 2001 ISBN 0-7923-6734-0

KLUWER ACADEMIC PUBLISHERS – DORDRECHT / BOSTON / LONDON

MIX
Papier aus verantwortungsvollen Quellen
Paper from responsible sources
FSC® C105338

If you have any concerns about our products,
you can contact us on
ProductSafety@springernature.com

In case Publisher is established outside the EU,
the EU authorized representative is:
Springer Nature Customer Service Center GmbH
Europaplatz 3, 69115 Heidelberg, Germany

Printed by Libri Plureos GmbH
in Hamburg, Germany